S0-ABU-977

Analysis for Drugs and Metabolites, Including Anti-infective Agents

METHODOLOGICAL SURVEYS IN BIOCHEMISTRY AND ANALYSIS

Series Editor: Eric Reid

Guildford Academic Associates
72 The Chase
Guildford GU2 5UL, United Kingdom

This series is divided into Subseries A: Analysis, and B: Biochemistry.
Enquiries concerning Volumes 1–18 should be sent to the above address.

Recent Titles

Volume12(A): Drug Metabolite Isolation and Determination
Edited by Eric Reid and J.P. Leppard
(includes a cumulative compound-type index)

Volume 13 (B): Investigation of Membrane-Located Receptors
Edited by Eric Reid, G.M.W. Cook, and D.J. Morre

Volume 14(A): Drug Determination in Therapeutic and Forensic Contexts
Edited by Eric Reid and Ian D. Wilson

Volume 15(B): Investigation and Exploitation of Antibody Combining Sites
Edited by Eric Reid, G.M.W. Cook, and D.J. Morre

Volume 16(A): Bioactive Analytes, Including CNS Drugs, Peptides, and Enantiomers
Edited by Eric Reid, Bryan Scales, and Ian D. Wilson

Volume 17(B): Cells, Membranes, and Disease, Including Renal
Edited by Eric Reid, G.M.W. Cook, and J.P. Luzio

Volume 18(A): Bioanalysis of Drugs and Metabolites, Especially Anti-Inflammatory and Cardiovascular
Edited by Eric Reid, J.D. Robinson, and Ian D. Wilson

Volume 19(B): Biochemical Approaches to Cellular Calcium
Edited by Eric Reid, G.M.W. Cook, and J.P. Luzio

Volume 20(A): Analysis for Drugs and Metabolites, Including Anti-infective Agents
Edited by Eric Reid and Ian D. Wilson

How to obtain future titles on publication

A standing order plan is available for this series. A standing order will bring delivery of each new volume immediately upon publication. For further information, please write to:

The Royal Society of Chemistry
Distribution Centre
Blackhorse Road
Letchworth
Herts. SG6 IHN

Telephone: Letchworth (0462) 672555

Methodological Surveys in Biochemistry and Analysis,
Volume 20

Analysis for Drugs and Metabolites,

Including Anti-infective Agents

Edited by

Eric Reid
Guildford Academic Associates, Guildford, United Kingdom

Ian D. Wilson
ICI Pharmaceutical Division, Macclesfield, United Kingdom

British Library Cataloguing in Publication Data
International Bioanalytical Forum, (8th; 1989; Guildford, England)
Analysis for drugs and metabolites, including anti-infective agents.
1. Drugs. Analysis
I. Title II. Reid, Eric III. Wilson, Ian D. IV. Series
615.1901

ISBN 0-85186-956-4

Based on proceedings of the Eighth International Bioanalytical Forum entitled Analysis for Drugs and Metabolites, Including Anti-infective Agents, held September 5–8, 1989 in Guildford, UK.

Published by the Royal Society of Chemistry, Thomas Graham House, Science Park, Cambridge CB4 4WF

Printed in Great Britain by Bookcraft (Bath) Ltd

Senior Editor's Preface

The session on assay validation (sect. #**A**) was a highlight of the 1989 Bioanalytical Forum which has given rise to this book. In respect of this and the other Forum material, the Editors have endeavoured, as in the past, to transform what could have been a mere 'Proceedings' volume, inevitably lacking smoothness and quality, into a 'reader-friendly' integrated book. Editorial trimming of fat off the meat, and rendering the content clearer or more informative in places, have been taken in good part by authors concerned.

Over the past decade it has become easier to gain publication texts as well as spoken presentations, especially from staff in the pharmaceutical industry. This is a healthy sign that sound bioanalysis is no longer a 'back-room' product which is taken for granted: the skills and judgement involved are gaining welcome recognition. Regrettably, however, some of the now numerous publications from company laboratories appear not in 'hard' journals but in a new breed of 'throw-aways'; such citations have mostly been excised in the present editing exercise.

The section (#**B**) on anti-infectives, including anti-parasitics and antivirals, is wide-ranging; but comprehensive-ness is obviously an unattainable aim (despite editorial 'top-up' at the end of the section), as was likewise the case for therapeutic classes that have featured in past volumes (anti-cancer, psychoactive, cardiovascular and anti-inflammatory drugs). However, articles that seemingly do not match a particular method-development need may nevertheless give guidance or ideas where there is some similarity in analyte or matrix properties to a reader's particular analytical problem. Such analogies in respect of the analyte may also be drawn from the feature-based Analyte Index, which has structural categories, especially relevant to solvent extraction, derivatization and GC detection, that date back to the 1978 volume but are still pertinent although HPLC now predominates as the end-step. The layout of the General Index also follows the pattern used in past volumes, to facilitate entry-searching.

Publishers' publicity for past volumes in the 'Analytical' subseries has not done justice to their usefulness as a reference source. Any reader who is only now becoming aware of this will find a list in a later 'Note' (#**ncC-3**) and may ask the Editor to amplify. Now we have a new publisher, assuring no inordinate delay in publication.

As a traditional book feature, there are 'Notes & Comments' ('**nc**') items which serve partly to give an impression of debate at the Forum and partly as bibliographic reinforcement. Many articles revolve on method-development strategy. This features at the start of the first bioanalytical volume (1976) in an article by R.G. Cooper that still holds good:- 'Development of Analytical Methods: General Philosophy'. One of his points, that a newly developed method needs try-out by those who will use it routinely, is echoed in the present volume, sometimes with the term "reproducibility" although sometimes this term is used as a questionable synonym for "precision". Whilst this Editor sees signs of computerized obsessiveness in present-day validation and quality-assurance policies and, moreover, is a heretic (not alone) as regards the vogue for having an internal standard, he is glad that one sin is now seldom perpetrated, namely the listing of near-nil values as '0' rather than as, say, '<0.1'.

Acknowledgements.- Valued support for the Forum came from U.K. pharmaceutical companies - Beecham, Glaxo and ICI. Many speakers made little or no call on Forum funds. Suggestions for Forum themes came from Honorary Advisers - U.A.Th. Brinkman, J. Chamberlain, H. de Bree, L.E. Martin, J.D. Robinson and R. Whelpton. Thanks are due to certain publishing bodies, as acknowledged where applicable, for permission to reproduce Figs. Vol. 10 (publ. Ellis Horwood) furnished the cover 'logo'.

Conventions and abbreviations.- For temperatures (°), °C is generally implied. Adherence to old-fashioned terms such as 'μg/ml' and 'M' (rather than 'μg.ml^{-1}', mol.L^{-1}) reflects editorial policy. Well-known terms such as GC, HPLC, S.D. and r (correlation coefficient) are used without definition. Other recurring abbreviations, usually listed in the articles concerned, include the following.-

Ab, antibody
EC, electrochemical (detection)
GC detector types: EC, electron-capture; FID, flame-ionization [cf. NMR usage!]; NPD, nitrogen-phosphorus
MS, mass spectrometry (modes include EI, electron impact)
HPLC modes: NP, normal phase; RP, reversed phase:, IE(C), ion-exchange (chromatography)
OPA, *o*-phthaldialdehyde
QC/QA, quality control/quality assurance [samples *implied?*]
RIA, radioimmunoassay
SPE, solid-phase extraction

A plea, with hindsight: shun the term reproducibility *(*'robustness' *or* precision *implied?).*

Guildford Academic Associates,
72 The Chase, Guildford,
Surrey GU2 5UL, U.K.

ERIC REID

25 March 1990

Contents

The 'NOTES & COMMENTS' ('**nc**' items) at the end of each Section include comments made at the Forum on which the book is based, along with some supplementary material.

List of Authors

Primary author

G.C. Bolton - pp. 353-354
Beechams Pharmls. Medicinal Res. Centre, Harlow, Essex

K. Borner - pp. 131-144
Klinikum Steglitz der Freien Univ., Berlin, F.R.G.

R. Calvert - pp. 3-7
General Infirmary, Leeds

G.S. Clarke - pp. 75-78
Bristol-Myers Squibb Internat. Dev. Lab., Moreton, Merseyside

L.A. Damani - pp. 341-342
King's Coll. (Chelsea), London

H. de Bree - pp. 221-225
Duphar, Weesp, The Netherlands

D. Dell - pp. 9-22 (& see Eggers)
Hoffmann-La Roche, Basle

M.V. Doig - pp. 157-162
Wellcome Res. Labs., Beckenham

R.J. Dolphin - pp. 247-256
Shandon Southern Products, Runcorn

M.C. Dumasia - pp. 305-308 (& see Houghton)
Horseracing Forensic Lab., Newmarket

H. Eggers - pp. 203-204
Hoffmann-La Roche, Grenzach-Wyhlen, F.R.G.

G.L. Evans - pp. 285-290
Glaxo Group Res. Greenford, Middx.

F.Y.K. Ghauri - pp. 321-324
Birkbeck Coll., London

S.S. Good - pp. 173-183
Wellcome Res. Labs., Res. Tr., NC, U.S.A.

Co-authors, with relevant name to be consulted in left column

C. Advenier - Varoquaux
J. Alexander - Hill
G.D. Allen - Bolton
J. Ayrton - Evans
J. Bleck - Sewing
P.A. Bombardt - Peng
E. Borner - Borner
W.M. Bothwell - Peng
U.A.Th. Brinkman - Lingeman
R.D. Brownsill - Vose (ii)
A. Brugman - Lingeman (ii)
D.G. Buckley - Whelpton (ii)
J. Campestrini - Lecaillon
K.S. Cathcart - Peng
D. Chang - Town
U. Christians - Sewing
P. Cordonnier - Varoquaux
M. Courtney - Peng
P. Damm - Lehr
G. Dean - Vose (i)
E. de Clercq - Szinai (i)
D. Decolin - Nicolas
G. J. de Jong - Lingeman (ii)
P. de Miranda - Good
J.P. Dubois - Lecaillon
R.O. Epemolu - Damani
A. Farjam - Lingeman (ii)
J.W. Firth - Vose (ii)
S.E. Fowles - Pierce
R.W. Frei - Lingeman (ii)
E. Gacs-Baitz - Szinai (ii)
K. Ganzler - Szinai (i) & (ii)
W.A. Garland - Town
L. Gerald - Quinn
I. Geuens - Woestenborghs (ii)
A. Ginn - Houghton
C. Gooijer - Lingeman (i)
D.B. Gower - Houghton
P. Grossi - Evans

Primary author

P. Hajdukiewicz - pp. 67-68
Jenmark Consultants, Bexley, Kent

H.M. Hill - pp. 23-36
Hazleton UK, Harrogate, N. Yorks

R. Horton - pp. 93-101
Beecham Pharmls. Chemotherapeutic Res. Centre, Betchworth, Surrey

E. Houghton - pp. 291-301 (& see Dumasia)
Horseracing Forensic Lab., Newmarket

C.A. James - pp. 117-122
Upjohn, Crawley, W. Sussex

G.S. Land - pp. 49-56
Wellcome Res. Labs., Beckenham

J.B. Lecaillon - pp. 265-270
Ciba-Geigy, Rueil-Malmaison, France

K-H. Lehr -pp. 205-206
Hoechst, Frankfurt, F.R.G.

H. Lingeman - pp. (i) 355-363, (ii) 365-370
Free Univ., Amsterdam

P.V. Macrae - pp. 145-152
Pfizer, Sandwich, Kent

A. Nicolas - pp. 271-278
Centre de Médicaments, Nancy, France

C.J. Olliff - pp. 333-336
Brighton Polytechnic, W. Sussex

G.W. Peng - pp. 197-200
Upjohn, Kalamazoo, MI, U.S.A.

D.M. Pierce - pp. 163-171
as for Bolton

R.P. Quinn - pp. 185-194
as for Good

E. Reid - pp. 313-316
Guildford

D.W. Roberts - pp. 257-263 (& see Ruane)
ICI Pharmls., Alderley Park, Ches.

R.J. Ruane - pp. 343-345
as for Roberts

Co-authors, with relevant name to be consulted in left column

G.W. Hanlon - Olliff
A.J. Harker - Evans
H. Hartwig - Borner
J. Hegedus-Vajda - Szinai (i) & (ii)
P. Heizmann - Eggers
J. Heykants - Van Rompaey & Woestenborghs (i) & (ii)
R.C. Hider - Damani
R.A.V. Hodge - Pierce
S. Holly - Szinai (ii)
M.W. Horner - Seymour
P.R. Hurst - Whelpton (i)
A.J. Hutt - Olliff
R.Hyland - Olliff
C. Janssen - Woestenborghs (ii)
A.E. Jones - Doig
H. Ko - Peng
H. Lenoir - Woestenborghs (ii)
P. Leroy - Nicolas
H. Lode - Borner
N.Lopez - Vose (ii)

J. McBride - Olliff
R.D. McDowall - Land
J. Maltas - Evans
D. Marshall - Houghton
A. Mehta - Calvert
M. Nash - Bolton
J.K. Nicholson - Ghauri, Wade

N. Oldfield - Town
B.S. Orban - Quinn
M. Pays - Varoquaux
M. Pellegatti - Evans
L.J. Phillips - Hill
P.J. Phillips - Wilson (ii)
H.E. Proud - Bolton

D.J. Reynolds - Good
M.C. Rouan - Lecaillon

Primary author	*Co-authors, with relevant name to be consulted in left column*
M.A. Seymour - pp. 309-312 as for Dumasia	G. Siest - Nicolas
K-Fr. Sewing - pp. 201-202 Med. Höchschule, Hannover, F.R.G.	A. Soldaat - Lingeman (ii) T. Spurway - Wilson (ii)
R.J. Simmonds - pp. (i) 69-74, (ii) 215-218 (& see James, Wood) Upjohn, Crawley, W. Sussex	P.M. Stevens - Vose (ii) J. Strohmeyer - Sewing
I. Szinai - pp. (i) 219-220, (ii) 317-320 Chem. Res. Inst., Acad. of Sci., Budapest	
R.J.N. Tanner - pp. 57-64 Glaxo Group Res., Ware, Herts.	S. Tadepalli - Quinn P. Teale - Dumasia, Houghton, Seymour
C. Town - pp. 109-116 Hoffmann-La Roche, Nutley, NJ, U.S.A.	P. Timmerman - Lingeman (ii)
F. Van Rompaey - pp. 37-47 Janssen Res. Foundation, Beerse, The Netherlands	M.P. van Berkel - de Bree N. van de Merbel - Lingeman (ii) R.J. van de Nesse - Lingeman (i)
O. Varoquaux - pp. 123-130 Centre Hospitalier de Versailles, Le Chesnay, France	N.H. Velthorst - Lingeman (i) Zs. Veres - Szinai (i)
C.W. Vose - pp. (i) 207-210, (ii) 211-214 Hoechst Pharmls., Walton Milton Keynes, Bucks.	
K.E. Wade - pp. 325-331 Birkbeck Coll., London	
R. Whelpton - pp. (i) 279-284, (ii) 337-340 London Hosp. Med. Coll., London	
I.D. Wilson - pp. (i) 79-82, (ii) 347-352 (& see Ghauri, Roberts, Ruane, Wade) ICI Pharmls., Alderley Park, Ches.	
R. Woestenborghs - pp. (i) 153-156, (ii) 241-246 (& see Van Rompaey) as for Van Rompaey	
S.A. Wood - pp. 103-108 (& see James, Simmonds (i)) as for Simmonds	C.M. Walls - Vose (ii) A. Warrander - Wilson (ii) C.F. Winton - Pierce

Section #A

PRODUCING VALID AND ACCEPTABLE ANALYTICAL RESULTS

#A-1

PROBLEMS AND PITFALLS IN ANALYTICAL REQUIREMENTS FOR PHARMACOKINETIC STUDIES BY THE MEDICINES COMMISSION

R. Calvert and A. Mehta

Department of Pharmacy,
General Infirmary at Leeds,
Great George Street, Leeds LS1 3EX, U.K.

Documentation of analytical methods used for biological studies forming part of a product licence application is often limited to a small paragraph referring to the precision of the method or to a published reference. This contrasts sharply with the reported detailed work-up of analytical methods used for stability studies; this difference is increasing sharply with the need to ensure "essential similarity" of drug substances when applying for product licences for generic products.

Possible reasons for this contrast in approach are discussed, and exemplified by applications which illustrate the point. Details of the information required by the licensing authority are reviewed with emphasis on the pre-analytical process and confirmation of specificity of the method. The increasing requirements for information on the kinetics of different isomers of racemic products are reviewed. The nature of the information is indicated, and proposals for future submissions outlined.

Most new drug applications, whether for a new chemical entity (NCE) or for an alternative formulation of an existing product, contain some pharmacokinetic data. This usually takes the form of data for plasma concentration *vs.* time. Other types of data, such as urine concentrations, are occasionally presented; but this is very infrequent.

When looking at the problems and pitfalls associated with the analytical aspects of such data, it is helpful to keep in mind the use to which the information is put. The Medicines Commission uses it to give reassurance that NCE's are absorbed efficiently, that the proposed dose regimens are appropriate, that the active agent is known and, for generics, that the formulation is as effective as those already on the market. This type of information can be obtained from pharmacokinetic studies by calculation of key parameters. Amongst these are, for i.v. administration, clearance, volume of distribution, fraction excreted unchanged and main routes of elimination; for oral administration, maximum plasma concentration and time to attain it, area under the plasma concentration/time curve, elimination half-life and absolute or relative bioavailability.

The Committee is not particularly concerned with the analytical methods used to obtain this information. It is very concerned with the reliability of the results presented, because key decisions as to the award of the product licence will be based on this information. The very nature of pharmacokinetic data gives cause for concern when making such decisions. Pharmacokinetic data when presented as the average results for a set of individuals often look respectable, as when the oral bioavailability of two products is being compared. The S.E.M. bars are often omitted because, as some applicants say, this helps to avoid presenting a complex picture. Examination of the individual data in detail can show that there is little inter-subject variability, and indeed the plotted mean values fairly represent the data, or as is more often the case we find a wide range of individual values for C_{max} and T_{max}. It is not uncommon to find that the S.D. is 40-50% of the mean value.

It is the latter type of data that causes problems for the Committee and highlights the role of the analyst since the Committee needs reassurance from the submission that the data are real and not an artefact of the analytical process. Applicants must provide sufficient information in the application to resolve any doubts in this area.

Unfortunately, this is an area which is often given less than adequate coverage in the report; it is unclear whether the quality assurance (QA) department or the product licence department is carrying responsibility for this problem. Many applicants give the validation studies for methods used in stability studies and in product-release specifications in some detail. In contrast, bioanalytical methods often receive a 10-line description with very little validation data, or at worst merely a reference to a published method again without validation data, implying that the literature method worked perfectly in their laboratory. Even in the better applications, which describe the analytical method in detail and give good validation data for the method, it is rare to find, for the whole analytical process, an appreciation which, with detailed consideration, does it justice from the sampling process right through to the presentation of results.

The pharmacokineticist is, then, looking for reassurance that the data presented can be relied on to distinguish true variation from that introduced during the analytical process. This requires the company to adopt a QA approach to the whole of the analytical process, not just to the final analysis, and to involve people with a QA background in the design and monitoring of pharmacokinetic studies.

Major pitfalls are present at many of the steps from blood sampling to data analysis. The problem for the Medicines Commission is that most submissions seem to ignore the existence of these pitfalls and omit any reference as to how the company has checked each area of the process and satisfied itself that it will be robust under field conditions and will contribute to giving results which are a true reflection of the situation and not an artefact of the process. Some specific areas of concern are now considered.

THE SAMPLE

Sample collection timing.- A brief outline of the sampling-procedure protocol is required, giving details of default procedures and action taken. **The actual sampling** needs to be specified: particularly how the sample was collected, and whether precautions were taken to avoid dilution with anticoagulants or cross-contamination from previous samples. **Choice of equipment** is also crucial. The type of collection container and syringe can affect the assay result [1, 2] by a variety of mechanisms such as adsorption, chemical interaction and release of contaminating substances from vessels - the tube cap or wall, or the syringe used in sample collection. These interactions are infrequent but unpredictable and must therefore be looked for and evaluated.

Stability on storage.- What happens to the sample after collection is a key factor that is rarely referred to. There will be *data* in abundance on the stability of the compound in its various formulations and packaging, but rarely a mention of its stability in the body fluid in which it is collected nor the conditions of storage, e.g. how long before freezing and how fast thawed. What is needed is reassurance that these steps in the process are not introducing uncertainty. An extreme example is 5-fluorouracil (5-FU). For many years it had a reported half-life of 10-20 min; it was only in 1987 that stability studies with plasma showed rapid degradation of 5-FU and led to a revision of its pharmacokinetics [3].

Other examples are decomposition of nitrazepam, clonazepam and cocaine when stored at 4° in blood. Rational approaches to investigating a drug's stability in biological samples have been presented [4, 5].

Pre-analysis sample care.- The details of how the frozen sample was brought to room temperature, and validation of the process, are required. This is particularly important with urine when large volumes are involved. Again our concern is for the elimination of uncertainty rather than looking for problems.

THE ANALYSIS, AND OVERALL VALIDITY

The analytical method used is frequently well described, however little attention is paid to the method of preparation and stability of standards used for calibration and check measurements: were these stock solutions stored for several days/weeks or freshly prepared?

The final stages of the overall process are usually well documented. These, however, are just the tip of the iceberg. What is needed is consistent and painstaking application of QA throughout the whole process from protocol planning to presentation of the results. Thereby we can be confident that differences or similarities in sets of data are valid and not a consequence of artefacts introduced by the methods used. The U.K. authorities will take an increasing interest in this area with firmer application of the CPMP† guidelines. This can make life difficult for some companies since much of the data may have been collected several years ago in different centres working to different standards. Probably the view will be taken that it is today's standards which must be met for today's product licence.

ASPECTS OF GROWING REGULATORY INTEREST

There are several new and challenging areas developing for the analyst involved in pharmacokinetic studies. Three main areas of concern to the Medicines Commission are sensitivity, the necessity for reporting free drug concentrations and the need for data on individual isomers.

If the sensitivity of the analytical process can be increased, then better data can be presented for several half-lives after single-dose administration. This can reduce the need for multiple-dose studies and give reassurance that the proposed dosing regimen is appropriate.

Our understanding of the role of protein binding and the significance of free drug concentrations has increased in recent years. This area is significant for drugs where binding characteristics may change in disease states. Information on binding makes possible the prediction of potential problems.

†Committee for Proprietary Medicinal Products

The need for data for individual isomers is already here. Many registration authorities are questioning the wisdom of giving racemates when only one isomer has the desired pharmacological activity. Why give half the dose as an ineffective, potentially harmful compound? For such an approach the authorities will ask for justification which will require pharmacokinetic data for each isomer and evidence of lack of interaction between the isomers. Generalizing such information will present a real challenge to analysts. If the questions are not asked within the company whilst data are in preparation, then there is a possibility of rejection or of requests for such information when the application is submitted.

CONCLUDING COMMENTS

In summary, there is a major role for the application of QA concepts developed in the laboratory to the whole process of generating pharmacokinetic data. Such an approach ensures that the data presented on pharmacokinetic studies represent the true position on drug handling and, on the basis of the data presented, the Medicines Commission can make decisions with confidence.

References

1. Wood, M., Shand, D.G. & Wood, A.J. (1979) *Clin. Pharmacol. Ther. 25*, 103-107.
2. Shang-Qiang, J. & Everson, M.A. (1983) *Clin. Chem. 29*, 456-461.
3. Murphy, R.F., Balls, F.M. & Poplack, D.G. (1987) *Clin. Chem. 33*, 2299-2300.
4. Adam, H.K. (1981) in *Trace-Organic Sample Handling* [Vol. 10, this series] (Reid, E. ed.), Horwood, Chichester, pp. 291-297.
5. Timm, U., Wall, M. & Dell, D. (1985) *J. Pharm. Sci. 74*, 972-977.

#A-2

VIEWS ON METHOD VALIDATION

D. Dell

F. Hoffmann-La Roche & Co.,
CH-4002 Basel, Switzerland

This survey of validation criteria complements accompanying presentations (notably #A-3, hereafter referred to as 'H.M. Hill'*) and, moreover, alludes to pertinent articles in Vol. 10 of this series [1]*. Consideration is given only to particularly controversial aspects: calibration, precision and quantification limit (QL®), and to a peripheral topic - drug stability - which bears on method validation.*

An Appendix summarizes the diverse current practices, e.g. concerning QL's, as established by Questionnaire at the Bioanalytical Forum which led to this book.

There is no universal consensus amongst analytical chemists concerning which characteristics have to be evaluated during method validation. The following criteria, however, appear most often in such considerations: precision, accuracy, QL, linearity, recovery from matrix, specificity, stability of analyte in matrix, and robustness of the method.

All these topics feature in accompanying Forum-based articles; the present focus is on the most controversial topics, including stability. Although it may be argued that drug stability has nothing to do with method validation, it is considered relevant to discuss it here, because the precision of a method influences the conclusions that may be drawn concerning stability. As sources of information for precision calculations, analysts use QC samples, replicate analyses of real samples, and calibration data. The use of the latter is controversial, and will be discussed.

**EDITOR'S NOTE.- Ensuring Reliable Results, Especially for Drugs in Blood* is the relevant section in [1], based on the 1979 Bioanalytical Forum; as this vol. is unfortunately out of print, pertinent articles (some might warrant seeking a photocopy) besides those now cited are listed at the end of the present article. Also relevant is art. #A-6 (this vol.) by R.J.N. Tanner as the leading discussant at the Forum.

®*Abbreviations include:* DL, detection limit (equivalent to Hill's term LOD); QL, quantification limit (= LOQ); QC, quality control; S.D. (*or* s), standard deviation, preceded by R. if relative.

Some official definitions of QL are dealt with, all of which are based on a statistical evaluation of the blank signal. Forum discussions indicated that the majority of analysts do not use these definitions, basing their definition on the lowest concentration giving an acceptable precision and accuracy.

STABILITY

An oft-encountered statement in publications is: "substance X was found to be stable in plasma at -20°C for six months". Rarely is the experimental design or the statistical treatment of the data mentioned. Commonly the same samples are analyzed at intervals, attributing any concentration decrease to degradation; this approach is perhaps acceptable in cases where 'very serious' degradation is encountered (~20%, or more). However, lesser degradation is often important, especially in pharmacokinetics; if, under these circumstances, the precision and accuracy of the determinations is not taken into account, combined with a sound statistical treatment, misleading conclusions could be drawn concerning the stability of the substance.

Adam [2] was one of the first to report a statistical method for handling stability data. Calibration lines, obtained by analyzing stored spiked plasma samples, were compared with similar lines from the analysis of freshly prepared samples. Using a t-test, the author calculated the significance of the differences between the slopes, and the 95% confidence interval for these differences; thus, he was able to attribute a significantly lower slope for stored samples to degradation (Fig. 1).

Timm, subsequently, developed a procedure [3] which is considered superior to the t-test approach for the following reasons.- A decrease of concentration on storage may be statistically significant, yet irrelevant from a pharmacokinetic point of view (Fig. 2, b). On the other hand, an insignificant change of concentration may reflect the imprecision of the method without being able to exclude, with high probability, a relevant instability (Fig. 2, e). The experimental design is slightly different from that of Adam [2]: responses of freshly prepared samples are compared with those of the stored samples at the same concentrations (at least 5 replicates each), and confidence intervals for the true % response differences are calculated (see [3] for details of statistical calculations).

The procedure is characterized by a consideration of two types of error: **error A:** the substance is stable, but the analytical error leads to the decision 'not stable'; and **error B:** degradation to a relevant extent, but the analytical

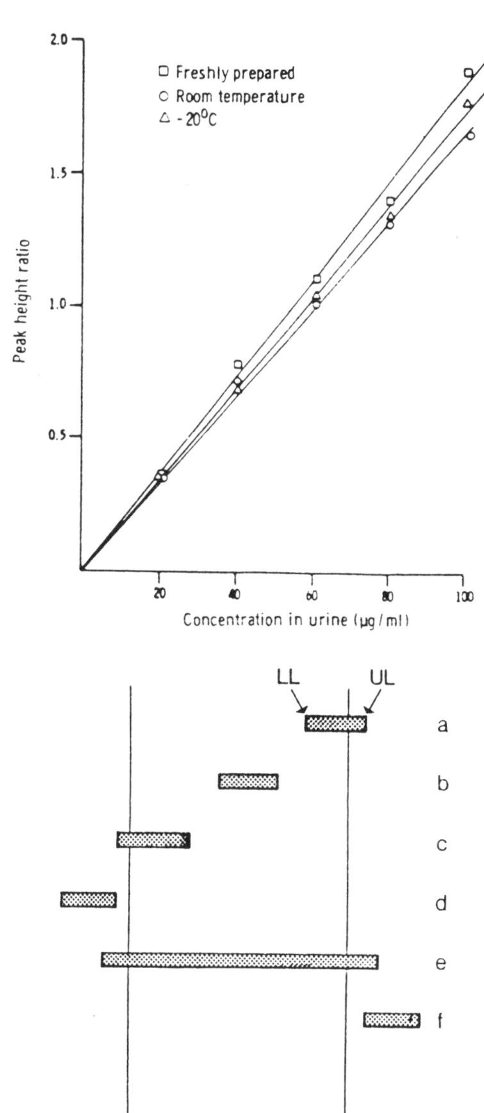

Fig. 1. Stability of a drug analyte (ICI 100,795) in urine stored for 2 weeks.
From Adam [2], courtesy of Ellis Horwood as publisher (Vol. 10, this series).

Fig. 2. Relationship between significant and relevant change of response (resp.; decr. = decrease) on storage. Bars characterize the ranges of the 90% confidence intervals for Δ, the true % resp. differences between stored and freshly prepared samples (LL= lower, UL = upper limit of the confidence interval).
a, change of resp., not significant and not relevant;
b, decr. of resp., significant but not relevant;
c, decr. of resp., significant and possibly relevant;
d, decr. of resp., significant and relevant;
e, decr. of resp., not significant but possibly relevant;
f, increase of resp., significant.
From [3], with permission from the American Pharmaceutical Association.

error leads to the decision 'stable'. The probabilities of these two types of error may be calculated: that of error **B** is kept constant at 5%, and that for error **A** is kept at a tolerable level by analyzing an appropriate number of replicates. In procedures based on the t-test, error **A** is fixed, but **B** is uncontrolled; thus, a substance may be classified as stable, with an unknown probability, even if a relevant degradation occurs.

It is important to be able to make a statement that any relevant degradation can be excluded with a certain high probability, and this is what Timm's treatment allows.

DETECTION/QUANTIFICATION LIMIT (DL/QL)

To attempt to review the many publications on this topic would be well beyond the scope of this article; the discussion here is confined to the most well-known definitions, as also surveyed in this book by H.M. Hill (see also R.J.M. Tanner, art. #A-6). As amplified by an excerpt on p. 31, McAinsh [4] discussed DL's in Vol. 10: he used the calibration line as a basis for calculating DL. The confidence limits for the regression line are calculated, and if the confidence interval for a measured concentration includes zero, then this concentration is defined as being below the DL (Fig. 3). Another way of expressing this is to say that the DL is the lowest measured concentration which is significantly greater than zero, or greater than the mean apparent blank sample result [5].

Uhlein [6] extended this approach by examining the relationship between S.D. (s) and concentration (x). In his experience of many methods he found this relationship to be linear:

$$s = mx + c$$

such that, as x approaches zero, s approaches c, the background value. A threshold concentration is eventually reached where s is so great that statistically significant differentiation from zero is not possible; DL is then defined as follows:

$$DL = t_{(n, 95\%)} \cdot s_{(x \to 0)}$$

(t is the one-tailed critical value of the t distribution for n determinations at 95% confidence). In our experience, the use of this treatment led to DL values whose R.S.D.'s were usually higher than was acceptable (30% or more).

The experience of most analysts, however, is that, although the relationship between R.S.D. and concentration (x) is fairly constant over a quite wide range, below certain low concentrations R.S.D. starts to increase markedly (below b, Fig. 4); using the same approach, one is working in this critical low concentration region, and the point of departure from constancy depends very much on how well the analytical system is functioning at the time of analysis. The day-to-day variation of QL, calculated in this way, can therefore be rather large. This could nevertheless be the basis for a definition of QL; the relationship between R.S.D. and x could be ascertained over a long period (several

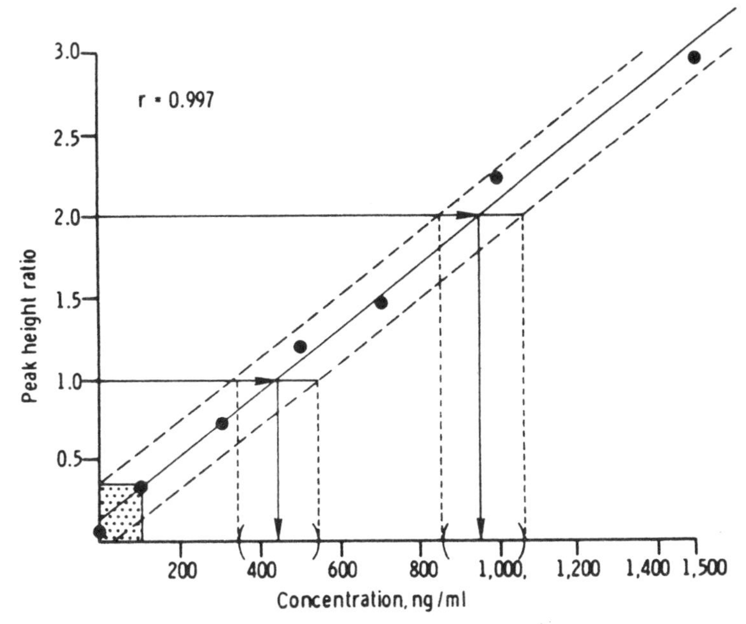

Fig. 3. The 'confidence-limit' approach to assessing detectability, based on a calibration curve (a GC assay) that intersected the y-axis above zero. The principle is to establish a 'non-detectable box' (speckled rectangle) housing small 'positive' concentrations arising from peak-height ratios larger than the blank. Taking into account that the top of the box (from which a perpendicular is dropped to the concentration axis) is set by the upper of the concentration confidence limits (interrupted lines; a 90% 2-sided limit), the top represents a 1-sided limit and it is 95% probable that a new blank value would fall below it. The 90% confidence limits for two notional test results are shown by vertical interrupted lines. *From McAinsh et al. [4] in Vol. 10, courtesy of Ellis Horwood.*

months), and this would allow a range for the QL (e.g. a→b in Fig. 4) to be defined, according to a maximum acceptable R.S.D. (or C.V.), defined by the analyst. In the example [7] shown in Fig. 4, the C.V. starts to increase markedly below ~100 pg/ml; accepting a C.V. not above ~20%, we may define a QL of 50-120 pg/ml for this assay.

Definitions from official bodies are also based on a blank signal (i.e. baseline noise plus any interfering peak) and its variation:

$$\text{DL} = \text{mean blank} + 3.\sigma_{\text{blank}} \quad \text{(IUPAC, [8])}.$$

Probably the best known of the official definitions is that of the American Chemical Society's Committee on Environmental Improvement ([9], Fig. 5). The main features are: (a) the

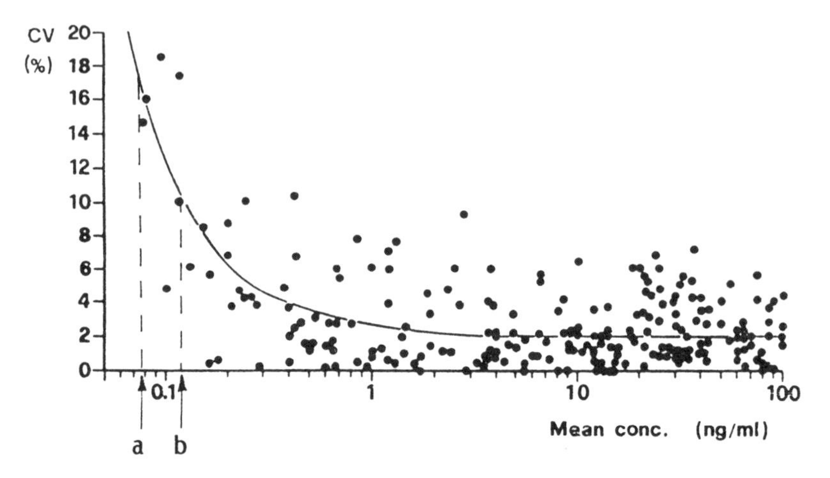

Fig. 4. Concentration dependence of the C.V. obtained from 219 duplicate determinations of a drug in clinical samples.

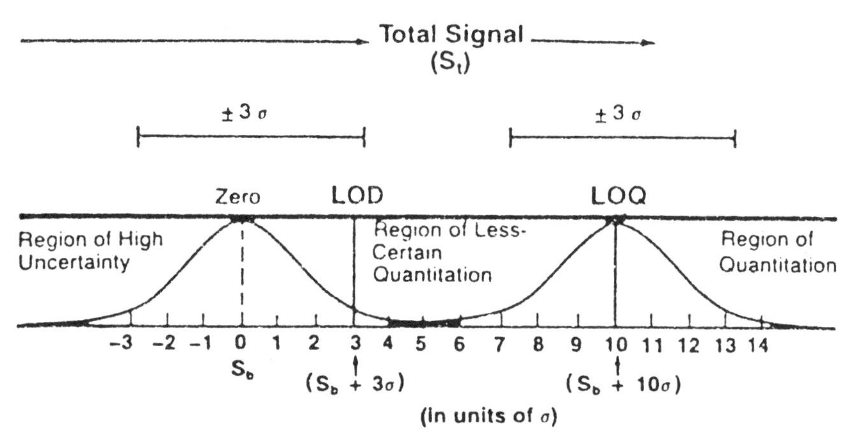

Fig. 5. Relationship of LOD (DL) and LOQ (QL) to signal strength : LOD located 3σ above the measured average difference between total (S_t) and blank (S_b) signals; LOQ 10σ above S_b. *From [9], by permission of the American Chemical Society.*

distinction between DL and QL, and (b) the separation into areas of 'high uncertainty', 'less certain quantification' and 'quantification'. It is unclear to the author to what extent these definitions, based on the variance of the blank (see also H.M. Hill's article), have found general acceptance among the analytical fraternity; it seems more usual to base the DL/QL on a multiple of the blank signal itself, ignoring the variance of the blank.

For pharmacokinetics, during method development, the chromatographic system is usually modified until separation of the analyte signal from any interfering endogenous-component signal is obtained. Under these circumstances, the blank signal consists only of the background noise; many analysts prefer not to work at the maximum detector sensitivity, so that this noise is very low and difficult to measure. Thus, working under these conditions, there is no blank to take account of in the calculation of QL, and the analyst defines his QL as the lowest concentration that can be measured with acceptable precision and accuracy over a long period. Each analyst defines what is acceptable for his laboratory; a Questionnaire (see Appendix) indicated that the majority of those responding would accept 15-20% for the R.S.D. and inaccuracy of the QL concentration.

During routine analysis, when one is confronted with samples from different patients, interfering peaks are sometimes encountered; if so, one of the official definitions could be used to calculate the QL on the basis of the interfering signal.

CALIBRATION

As H.M. Hill and co-authors deal comprehensively with this subject under the heading 'Linearity', only a little elaboration on weighting need be given here. McAinsh [4], in 1979, recommended that for calibration purposes a regression analysis should be undertaken wherever possible, because a 'best-of-eye' fit is in itself qualitative. This is an interesting remark, because it indicates that, at least up to ~10 years ago, many analysts were still constructing their calibration line manually. It is highly unlikely that more than a handful of analysts are working 'manually' these days; the increasing use of modern data acquisition and processing systems allows automatic generation of regression lines, with concomitant enhancement of the quality of quantification.

What is still not widely accepted, as the results of the Questionnaire showed, is the improvement in quality brought about by weighting the regression. Leppard, in Vol. 10 [10], was one of the first to point out the potential benefits of weighting when he stated that the overall precision for an assay will often be proportional to concentration. Under these circumstances, standard regression analysis often leads to an unreliable calibration line (poor fit at low concentrations). It is important, during method development, to investigate the relationship between precision and concentration since this will determine which weighting factor is most appropriate for the method (see Fig. 2 in the Hill article). Our consideration centres on Table 1.

No weighting factor:-

CALIBRATION RUNS UNITS : MCG/ML PLASMA

PEAK NAME	GIVEN RESP.	GIVEN CONC.	CALC. CONC.	% DEV.
Ro 13-5057	4.73	0.10	0.58	+484.1
Ro 13-5057	6.18	0.50	0.86	+ 72.7
Ro 13-5057	9.45	1.00	1.50	+ 49.5
Ro 13-5057	27.55	5.00	4.99	− 0.2
Ro 13-5057	56.70	10.00	10.61	+ 6.1
Ro 13-5057	241.30	50.00	46.24	− 7.5
Ro 13-5057	529.20	100.00	101.91	+ 1.8

THE AVERAGE ABS (% DEV) IS : 88.9

EQUATION IS : Y = 5.18 × CONC. + 1.71
COEFFICIENT OF CORRELATION : 0.999

Weighting factor 1/y:-

CALIBRATION RUNS UNITS : MCG/ML PLASMA

PEAK NAME	GIVEN RESP.	GIVEN CONC.	CALC. CONC.	% DEV.
Ro 13-5057	4.73	0.10	0.16	+ 57.9
Ro 13-5057	6.18	0.50	0.44	− 11.2
Ro 13-5057	9.45	1.00	1.09	+ 8.7
Ro 13-5057	27.55	5.00	4.65	− 6.9
Ro 13-5057	56.70	10.00	10.40	+ 3.9
Ro 13-5057	241.30	50.00	46.76	− 6.4
Ro 13-5057	529.20	100.00	103.48	+ 3.4

THE AVERAGE ABS (% DEV) IS : 14.1

EQUATION IS : Y = 5.08 × CONC. + 3.93
COEFFICIENT OF CORRELATION : 0.999

Weighting factor $1/\sqrt{y}$:-

CALIBRATION RUNS — UNITS : MCG/ML PLASMA

PEAK NAME	GIVEN RESP.	GIVEN CONC.	CALC. CONC.	% DEV.
Ro 13-5057	4.73	0.10	0.24	+139.9
Ro 13-5057	6.18	0.50	0.52	+ 4.5
Ro 13-5057	9.45	1.00	1.16	+ 16.1
Ro 13-5057	27.55	5.00	4.69	- 6.1
Ro 13-5057	56.70	10.00	10.38	+ 3.8
Ro 13-5057	241.30	50.00	46.41	- 7.1
Ro 13-5057	529.20	100.00	102.59	+ 2.5

THE AVERAGE ABS (% DEV) IS : 25.8

EQUATION IS : Y = 5.12 × CONC. + 3.50

COEFFICIENT OF CORRELATION : 0.999

Weighting factor $1/y^2$:-

CALIBRATION RUNS — UNITS : MCG/ML PLASMA

PEAK NAME	GIVEN RESP.	GIVEN CONC.	CALC. CONC.	% DEV.
Ro 13-5057	4.73	0.10	0.13	+ 30.3
Ro 13-5057	6.18	0.50	0.42	- 15.6
Ro 13-5057	9.45	1.00	1.08	+ 7.9
Ro 13-5057	27.55	5.00	4.72	- 5.6
Ro 13-5057	56.70	10.00	10.58	+ 5.8
Ro 13-5057	241.30	50.00	47.71	- 4.5
Ro 13-5057	529.20	100.00	105.60	+ 5.6

THE AVERAGE ABS (% DEV) IS : 10.8

EQUATION IS : Y = 4.97 × CONC. + 4.08

COEFFICIENT OF CORRELATION : 0.998

Table 1. Fits obtained ('% DEV') with different weighting factors (J-M. Varoqui & G. Wendt, personal communication), as amplified in the text.

The potential benefits as far as improved quantification at low concentrations is concerned are best illustrated by an example from our own laboratory (Table 1). The same set of calibration data was subjected to different weighting factors and the fits ('% DEV') compared. Evidently the fit improves, with the lowest analyte concentration, on going from a weighting factor of 1 successively through $1/\sqrt{y}$, $1/y$ and $1/y^2$. This, however, does not occur without cost to the fits at higher concentrations, although these deteriorations are, in this case, still acceptable.

This is an extreme example to illustrate the point; most analysts would not calibrate over such a wide range (1000-fold), and this probably explains why the fit, even with the factor $1/y^2$, was not satisfactory (30% deviation) at the lowest concentration. An alternative procedure, of course, would be to carry out two regressions over two concentration ranges (0.1-5.0 and 5.0-100, in this example). This, however, can be time-consuming; in addition, in pharmacokinetic work (especially toxicological) one is frequently confronted with very wide, and unknown, concentration ranges. Under such circumstances it is more efficient to carry out one wide-range calibration with weighting.

It is emphasized that the use of the appropriate weighting factor will always lead to improved quantification (admittedly sometimes only a small improvement) for the critical low concentrations.

PRECISION

This topic is also dealt with in detail by H.M. Hill. How precision is achieved with respect to what kind of samples are used to calculate this parameter is a very important consideration. Leppard [10] states that the vertical scatter about the calibration line gives an indication of assay precision. Many analysts including ourselves will argue that the imprecision of the regression line does not always reflect the imprecision of the measurement of unknown samples against such a calibration line. The main reason for not obtaining precision data directly from the calibration is that it is a circular argument: namely, since the responses making up the calibration line have an influence, of course, on the regression and fit, they themselves should not be used to judge precision.

Thus, most analysts use both spiked (QC) samples and replicates of real samples to assess precision. The question of whether to analyze in singlicate, or some number of replicates, has been adequately dealt with in successive

Forum-based books (H.M. Hill; S.H. Curry [11]; J. McAinsh [4]). On the whole, a second determination is worth carrying out from the point of view of an ameliorated precision, but a third determination usually confers little further improvement. It is argued that a very good reason for carrying out replicates, even for very precise methods, is that it can allow the recognition of methodological errors. If the analyst is unlucky and a single determination is wrong because of such a non-random error, then he needs a second determination to draw his attention to the first, wrong result. (For 'outliers' see G.S. Clarke, #ncA-3, and p. 87.)

We have compared, for particular sets of assay runs, the 'precision' obtained from calibration samples with that obtained from QC samples analyzed at the same time. The data in Table 2 represents just one of many examples all showing the same result: namely, that the precision from calibration samples is nearly always better than that obtained from QC samples. This serves to reinforce the above argument that calibration samples are not representative as far as overall method precision is concerned. The latter and other foregoing points on validation have been considered too in arts. #ncA-2 (R.G. Simmonds) and #A-6 (R.J.N. Tanner*).

Table 2. Precision data obtained from QC and calibration samples during the same analytical run (P. Heizmann and H. Eggers, personal communication; assays on plasma).

QC: µg/ml	n	R.S.D., %	*Calibration:* µg/ml	n	R.S.D., %
0.54	8	8.2	0.5	17	1.4
1.0	12	4.3	1.0	17	2.6
2.5	15	3.2	2.0	17	2.7
5.0	13	4.7	5.0	17	1.8
10.0	17	2.8	10.0	17	2.7
20.0	14	3.3	20.0	17	2.2
40.0	11	2.7	40.0	17	3.0

CONCLUDING COMMENT

Evidently, from accompanying articles and the Questionnaire, so far there is no complete consensus within the bioanalytical fraternity as to the 'correct' approach to any of the topics discussed above. Yet it is clear that a healthy debate is in progress and that we have moved on in the decade following Vol. 10.

* See p. 64 for some comments by D. Dell on Tanner's article.

References

1. Reid, E., ed. (1981) *Trace-Organic Sample Handling* (Vol. 10, this series), Ellis Horwood, Chichester: pp. 265-366 (Sect. F).
2. Adam, H.K. (1981) as for 1., pp. 291-297.
3. Timm, U., Wall, M. & Dell, D. (1985) *J. Pharm. Sci. 74*, 972-977.
4. McAinsh, J., Ferguson, R.A. & Holmes, B.F. (1981), as for 1., pp. 311-319.
5. Büttner, J. (1976) *Clin Chem. 22*, 1922-1932.
6. Uhlein, M. (1979) *Chromatographia 12*, 408-411.
7. Zell, M. & Timm, U. (1986) *J. Chromatog. 382*, 175-188.
8. IUPAC (1978) *Nomenclature, Symbols, Units and their Usage in Spectrochemical Analysis - II*, in *Spectrochim. Acta 33*, 241-245.
9. Keith, H.K., Crummett, W., Deegan, J., Libby, R.A., Taylor, J.K. & Wentler, G. (1983) *Anal. Chem. 55*, 2210-2218.
10. Leppard, J.P. (1981) as for 1., pp. 320-335. *Corrigenda, especially for equations:* Vol. 12*, *or contact Editor (Guildford).*
11. Curry, S.H. & Whelpton, R. (1978) *Blood Drugs and Other Analytical Challenges* (Vol. 7, this series), Ellis Horwood, Chichester, pp. 29-41.

*(1983) *Drug Metabolite Isolation and Determination* (Reid, E. & Leppard, J.P., eds.), Plenum, New York.

Editor's listing of pertinent Vol. 10 arts. besides those cited

(see footnote at start of present art.)

#*Analyses entailing sample processing: an overview* - E. Reid, 15-31.
#*Influence of blood-sampling procedures on drug levels in plasma* - O. Borgå, I. Petters & R. Dahlqvist, 265-269.
#*Some matrix-effects on the extraction of hydrophobic amines...... from plasma* - D. Westerlund & L.B. Nilsson, 270-273.
#*Fatty acids and plasticizers as potential interferences in the bioanalysis of drugs* - R.A. de Zeeuw & J.E. Greving, 274-283.
#*Modifications in the work-up procedure for drugs due to the biological matrix* - J. Vessman, 284-290.
#*GLP in bioanalytical method validation* - J.A.F. de Silva, 298-310.
#*Policies for internal standards* - K.H.Dudley, 336-340.
#*The usefulness of internal standards in quantitative bioanalytical methods* - J. Vessman, 341-346.
#*Critical factors in drug analysis* - S.H. Curry & R. Whelpton, 347-352.
#*Problems in getting valid results: an overview* - B. Scales, 353-362 (a notably useful art.).

APPENDIX: CURRENT METHOD-VALIDATION PRACTICES

Summary by D. Dell of Questionnaire replies

At the Forum (Sept. 1989) the Questionnaire was issued, with an assurance of individual confidentiality, to all participants (~100; a minority were from non-company laboratories, e.g. academic, hospital, dope-screening). There were **27 replies**, which was gratifying since some covered 2 or more participants from a particular laboratory: **23** of the 27 were from the **pharmaceutical industry**, U.K. and overseas. The objective was to ascertain whether there was a consensus concerning the three most difficult areas of method validation, namely calibration, precision and quantification limit (QL). The questions and replies were as follows.-

1. Are your policies on 2., 3. & 4. below -

   a) personal? .... "Yes", **2**

   b) standard (in your lab./dept.) .... "Yes", **15**
      - do rules (SOP's) exist? .... "Yes", **22**; "No", **2**

2. CALIBRATION
   a) Do you weight your regressions? .. "Yes", **9**
      "No", **14** (i.e. weight = 1)
      "Sometimes", **3**

   If "Yes" to 2(a), which weighting factor generally gives the best fit? ..... $1/x^2$, **7**; $1/x$, **1**; $1/y^2$, **2**; $1/y$, **1**

3. PRECISION: Which raw material(s) do you use to calculate it?

   a) from the calibration line data .... **15** [alone: **1**; + QC's: **14**]
   b) from the QC data **24**
   c) from replicates of real samples .. **12**

3. QUANTIFICATION LIMIT: Is your definition -

   a) based on the 'blank' (background) signal? **6**
   b) based on an official definition? **2**
   c) your own? **26**

Comments on the replies

1. The vast majority are working with SOP's.

2. Only 30% are weighting their calibrations with a factor other than 1; of the 30%, most find that weighting with the reciprocal of the square of the response or concentration

gives the best fit, indicating that for those methods, the S.D. is directly proportional to the concentration.

3. Most analysts are using QC data to calculate precision, although more than half use the calibration data additionally.

4. It was very interesting to note that very few analysts are routinely using one of the 'official' definitions for the QL. Of the 26 who are using their own definition, 19 define the QL as the lowest calibration point giving an R.S.D. of not worse than 20%, and 7 say that it is the lowest concentration giving an R.S.D. which is not worse than 15%. Although this was not fully discussed at the Forum, one may assume that the lowest concentrations giving these R.S.D.'s are determined by analysis of QC samples during method development.

The general outcome of this Questionnaire is comforting, in that it reveals that, as far as precision and the QL are concerned, most analysts see eye-to-eye. Only for calibration-line weighting is there no consensus. The situation is difficult to evaluate, however, because whether one needs to weight with a factor other than unity depends, for instance: (a) on the acceptability of the size of the deviation (see Table 1 in the foregoing article) for the lowest calibration point (i.e. 'acceptability' is often subjective); (b) on the concentration range of any particular calibration, which will then largely determine the dependency of precision on concentration and, therefore, which weighting factor is most appropriate. Thus, a much more detailed survey of 'calibration habits' would have to be undertaken before any hard conclusions could be drawn.

The above comments have been supplemented by R.J.N. Tanner - p. 88.

#A-3

QUANTITATIVE CHARACTERIZATION OF ANALYTICAL METHODS

L.J. Phillips, J. Alexander and †H.M. Hill

Hazleton U.K., Otley Road,
Harrogate, N. Yorks. HG3 1PY, U.K.

The relationship between what constitutes a validated method and obtaining valid data using that method is considered. Each parameter normally required by the regulatory authorities is mathematically evaluated, and the limitations discussed. Assay precision (reproducibility) is considered in the context of running singlicate rather than replicate assays. Factors relevant to accuracy and selectivity (specificity) include the purity of the standard and the presence of interfering substances. Also relevant to the acceptability of data are sensitivity, recovery, linearity and analyte stability.

Once the method has been shown to be valid in respect of the foregoing characteristics, it is essential in relation to acceptance criteria to show that these parameters remain 'under control' throughout the lifetime of the assay. Emphasis is placed on simple statistical and mathematical calculation usable routinely at the bench level.

Quantifying the characteristics of an analytical method during validation provides the regulatory reviewer with an objective evaluation of its adequacy and the analyst with reinforced confidence in its performance. In addition it provides a quantitative means for monitoring the assay during its use. The parameters normally required to be evaluated and quantified by the authorities include linearity, accuracy, precision (reproducibility/repeatability), sensitivity, specificity, recovery and analyte stability. Perhaps the pivotal component is linearity i.e. analyte concentration *vs.* response: this relationship must be correctly defined to ensure accuracy and can also affect both apparent precision and sensitivity.

Sensitivity, likewise an important parameter, should be defined so that it is independent of any mathematical model. The variety of literature definitions for sensitivity and recovery is one of the major difficulties in comparing and evaluating different assays. Specificity (or selectivity) of an assay is probably the most difficult parameter to evaluate and usually can only be inferred by a process

†*main author; new address at end of art.*

of elimination. Finally analyte stability, although not a function of the analytical method, can significantly influence its development and conduct.

The objective of any validation procedure is to ensure that the most appropriate calibration function is chosen and that the method's characteristics are within pre-defined limits. During the lifetime of the assay it is essential to show that these characteristics remain 'in control'. Accordingly, assays may be accepted or rejected on a batch basis on the grounds of the acceptability of the calibration curve and quality control (QC) samples. It is important to define acceptance criteria in such a way that simple mathematical and statistical calculations can be applied routinely at the bench level without presenting conflicting data.

The procedures described here for validating and evaluating analytical methods relate to chromatographic assays on samples from pharmacokinetic, bioavailability and bioequivalence studies. There is applicability to immunoassays also, albeit complicated by differences in calibration functions.

DEFINITION OF PARAMETERS

Accuracy

An often used measure of how close the observed value is to the true or nominal value is bias: e.g. -5% if the true and observed values are respectively 100 and 95 ng/ml. An alternative measure is the observed value expressed as a % of true. Sometimes this is termed recovery; but this term could be confused with extraction recovery of the drug from the matrix.

Although accuracy is conceptually straightforward, it largely depends on the quality of the reference standards and equipment used to prepare all solutions and samples. Equally important, as discussed in connection with linearity, is the definition of the calibration curve: the choice of model as well as the number and concentration levels of calibration points used. Calibration standards of known amounts of the analyte(s) must be dispersed in the appropriate biological fluid or solid matrix. These standards and the procedures used in preparing them must be of impeccable quality, which the results will reflect.

The area of standard and reference materials is beset by confusing nomenclature. IUPAC [1] identifies 5 grades of purity, 3 directly applicable to bioanalysis:
- **C.** Primary standard: 'commercially available substance of purity 100 ±0.02%';

- **D.** Working standard: 'commercially available substance of purity 100 ±0.05%';
- **E.** Secondary standard: 'a substance of lower purity than Grade **D** but can be standardized against Grade **C** material'.

Rarely do standards used in bioanalysis attain grades **C** and **D**: frequently our 'primary' standards match only **E**, but provided that the actual concentration is known then their use is acceptable. This terminology fits more closely the IFCC [2] definition, viz. 'a substance of known purity which is used as a basis for comparative measurement of the substance in a specimen (biological matrix)'. It seems best to allude to IUPAC standards in respect of grades but to IFCC definitions for primary and secondary standards. What matters is that the analyst should know the purity of the standard material, procured with analytical certification from the commercial or other source. Such documentation should define purity, whether it is a salt, free base or acid, the amount of water of crystallization, etc. It is also important that these standards be stored under ideal and well-documented conditions.

Primary standards (in the IFCC sense) should ideally be prepared in aqueous solution at such concentrations as produce only small volume changes when added to the biological matrix, thus minimizing any effect on extractability or accuracy. Volume changes should be <5% - preferably <2% where methanol or ethanol is the primary solvent. Organic solvents may cause protein precipitation and occlusion of the drug in the coagulate, resulting in changes in extractability. To maximize accuracy it is advisable to have as large a batch of the biological material as is feasible, using larger, more accurate measuring devices.

Precision (reproducibility[†]/repeatability)

Precision is a measure of the random distribution of individual measurements (x) around the calculated mean. This parameter is usually evaluated by taking a number of samples (spiked or real) containing the same amount of analyte through the entire procedure: minimally n = 4, but n = 6 or more is customary. Precision is generally expressed as relative standard deviation or coefficient of variation (R.S.D., C.V.), viz. S.D. [denominator: n - 1] as % of the mean value. As the C.V. is likely to vary with concentration*, it should be determined over the concentration range expected in the actual samples. This range should include samples

* *Editor's note*.- This 'grey area' was touched on in Vol. 10 by J.P. Leppard [5] and was considered in Vol. 7, with other assay aspects pertinent to the present art., by S.H. Curry & R. Whelpton in Vol. 7 (as for [4]; pp. 29-41): at low concentrations the C.V. is sometimes higher. †*See comment in Preface*

at the LLOQ*, a value near the LLOQ, a mid-range value, and a value ~80% of the highest likely amount. These data may be further presented in terms of the 95% confidence limits (mean ± t S.D.'s, where t is the appropriate quantile of Student's t-distribution).

While some indication of the precision of a single measurement is thereby obtainable, increasing the number of replicates on that sample improves the precision (assuming variability is purely random, which is usually true): e.g. with 95% confidence limits = ±10% for a singlicate, the confidence limits become ±7.0%, ±5.8% and ±5.0% for 2, 3 and 4 replicates respectively. Fig. 1 shows precision data for a xanthine derivative in urine assayed in singlicate, duplicate and triplicate illustrating this trend.

Improving precision by increasing the number of replicates may be a cost-effective alternative to changing the methodology, in spite of doubling or trebling the number of assays. Replication serves, moreover, as a way to accept or reject data. It also enables precision with real samples (through replicating 10-20% of these) to be compared with that for the spiked QC's, using an equation [3] discussed by Chamberlain [4]:

$$S = \sqrt{\frac{\Sigma d^2}{2n}}$$

where d = value for a duplicate expressed as % of the mean, and n = no. of samples assayed in duplicate; S represents precision of duplicates over the concentration range matching that of the QC samples used in the study.

Linearity

Linearity is widely used to describe the concentration-response relationship. However, this is often not truly linear, and diverse curves have been adopted for calculating sample values. It is essential to consider the variables and assumptions which underly this parameter, since other parameters such as accuracy, apparent precision and sensitivity depend on the choice of calibration curve. Curves may be (in a preferred or most generally accepted order): linear, linear forced through the origin, power (log/log), log/lin, lin/log, split linear, quadratic or polynomial. They may be further modified by application of weighting factors. Initially all relationships should be evaluated using a simple linear format and adjusted accordingly.

For this linear format which Leppard [5] considered with respect to determining precision from calibration curves, $y = mx + b$, where y is the dependent variable (e.g. peak

---

* LLOQ = lower limit of quantitation

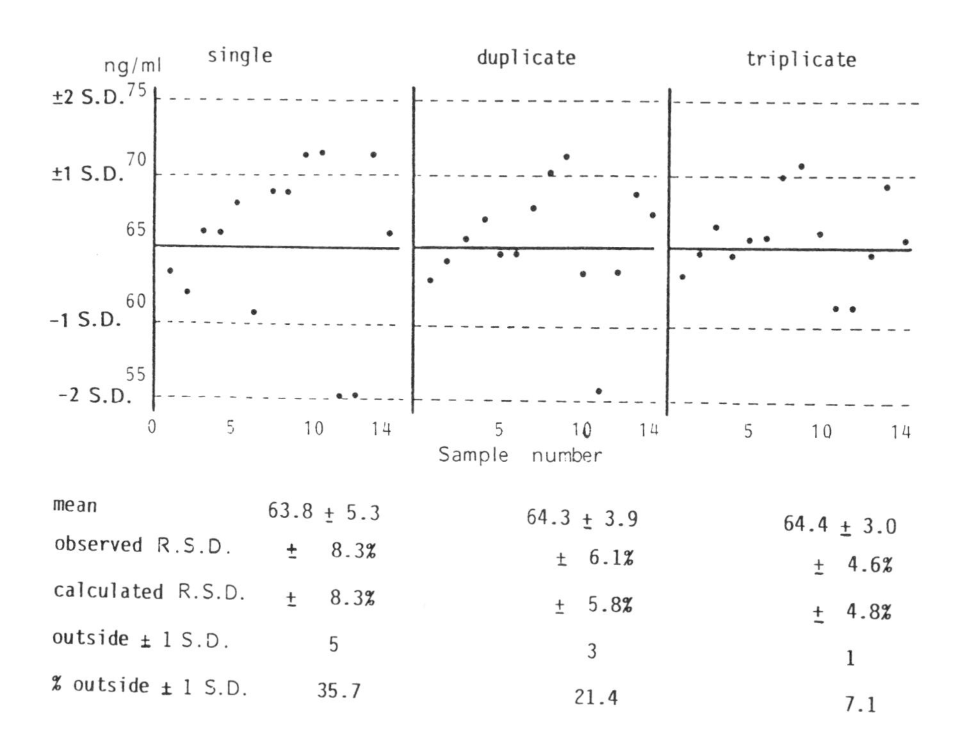

Fig. 1. Illustration of the improved precision (S.D.) by increasing replication (S.D. comparisons, actual *vs.* theoretical), the S.D. ranges having been calculated from the data in each graph.

height or area, absolute or *vs.* an internal standard), b is the intercept of the y-axis for x = 0, and m is the slope. The line which most closely fits the values obtained is calculated using the method of least squares, whereby the sum of the squares of all the vertical distances of the points from the line are minimized relative to the y value. This approach applied to calibration curves assumes that (1) x values are without error, all the error being associated with the response (y); (2) deviations are mutually independent; (3) deviations are normally distributed and have the same variance irrespective of x. Having calculated the parameters b and m, it is essential to evaluate how good the best fit is. Diverse indicators are available for evaluating this:

- (a) correlation coefficient $(r) = \dfrac{\Sigma xy}{\sqrt{(\Sigma x^2)(\Sigma y^2)}}$;
- (b) difference, **i**, between the observed and nominal response derived from the regression line, i.e. residuals;
- (c) $\Sigma e\mathbf{i}^2$, i.e. sum of residuals squared;
- (d) S.D. of the points from the calculated line;
- (e) as for (d) but using the residuals normalized to % of nominal [6].

Perhaps the most widely used indicator is r. It is at best an indirect measure of the quality of the line of best fit, merely confirming that the measured response is related to the concentration of the analyte. Such limitations are well documented by Mitchell & Garden [7]:
- r does not necessarily change when the curve shape does;
- r does not provide a quantitative comparison of two curves, e.g. an r value of 0.99 may be 20% more precise than r = 0.98, while with r = 0.95 the curve is hardly usable in most situations;
- r does not indicate whether the model fits the data.

In order to determine the line of best fit, examination of residuals from different models provides a means of evaluating the quality and applicability of the calibration curve. Fig. 2 shows typical plots from several calibration curves. The variation of the data is contained within the envelopes. By plotting residuals an improvement in the choice of calibration model may be tested. The closer residuals correspond to those in Fig. 2.1, the better the fit.

Using r and an evaluation of residuals, it is possible to determine whether the simple linear regression model is correct and whether any values are outliers that should be discarded (re-assay?). If the model is shown to be incorrect then a more appropriate model should be considered.

In defining a calibration curve it is important to consider the range of expected concentrations, this being dictated by the dose, individual variability and sampling times. Since it is unwise to extrapolate, it is essential to define as broad a range as possible. However, the mean value for the curve should be best in the region where most of the results will fall, since this is the region of highest precision. While more calibration points improve precision, the number of real samples must often be reduced to compensate. Hence most data are generated using 5-7 calibration points per curve.

To avoid problems with 'low end' calculations, some analysts use a calibration factor derived from a curve forced through zero. The penalty for this expediency may be reduced apparent precision, systematic inaccuracy over the calibration range, and a distorted LOQ.

The best criteria for choosing amongst types of calibration model are those that provide the minimim variance and the best estimate of the nominal data. With the widespread use of microcomputers and the abundance of literature on linear and non-linear calibration curves, it should be possible for all laboratories to fully investigate the nature of the concentration-response curve and choose the appropriate model.

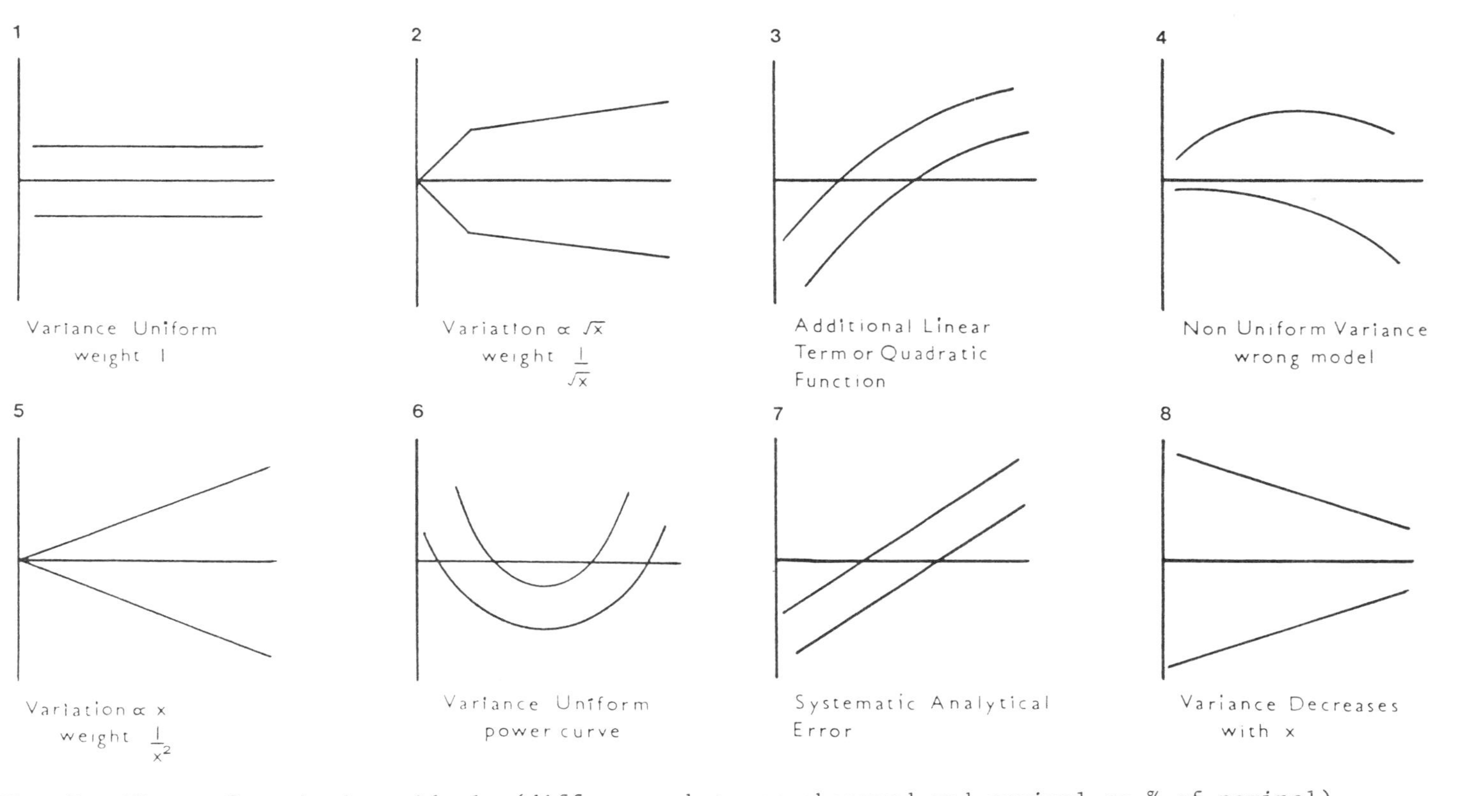

Fig. 2. Plots of typical residuals (difference between observed and nominal as % of nominal), enclosed in envelopes. In reality variances can be a combination of the idealized forms presented here. However, a review of these plots may give some indication as to the appropriate model to use.

Sensitivity

This term is frequently used without any definition. However, the variety of names is matched only by the number of definitions, these including:
- minimum measurable quantity (MMQ);
- minimum detectable quantity (MDQ);
- lowest limit of reliable assay measurement (LLORAM);
- limit of reliable determination (LRD);
- limit of detection (LOD);
- lower limit of quantitation (LLOQ);
- minimum reportable concentration (MRC);
- detection limit (DL).

Definitions by IUPAC [8] and Keith & co-authors [9] based largely on the work of Kaiser [10] differentiate between the detection limit (LOD) and the quantitation limit (or, as used here, the LLOQ). These definitions are based largely on the variability of the blank. Consequently the LOD is defined as the mean of the blank signal plus 3 times the S.D. of the blank, i.e. x + 3 S.D. Using similar criteria, LLOQ = x + 10 S.D. Whilst LOD is useful for evaluating the presence of residues of contaminants, it is of little value in pharmacokinetic or biopharmaceutic studies where the data are used to derive kinetic parameters.

As summarized opposite, the sensitivity issue has been well reviewed by McAinsh & co-authors [11] in Vol. 10 of this series. They reviewed a number of criteria and concluded that the use of confidence limits provided a useful criterion but was still problematical. Since the calculated LLOQ based on confidence limits will vary depending upon the spacing and concentrations defining the calibration curve, it is not an independent determination. Similarly, use of the regression intercept as a reference point for calculating sensitivity is subject to definition of the calibration model. Determination of sensitivity using half bottom standard also produces equivocal data depending upon whether the concentration or the ratio intercept on the y-axis is used.

Use of the bottom standard concentration minimizes much of the equivocation produced by the other definitions since it does not involve extrapolations. The suggested definition is as follows.

Criteria for defining the LLOQ:

#lowest calibrator with between-run C.V. of 10-20%;

#bias ±10-20%;

#signal should be at least 5 times interference (blank).

This definition is largely empirical and independent of the calibration model. In addition to fulfilling these criteria as a check, the determined value should agree with the Americal Chemical Society's Environmental Subcommittee definition [9], x + 10 S.D. Such agreement can be confirmed at the end of the study by evaluating the baseline/interference at the retention time of the study blanks which usually exceed 20 in number.

Recovery

This is a measure of the comparison between the amount of drug present in the final extract and that present in the original biological matrix. Usually it is expressed as a percentage. In addition to determining the recovery of the drug it is advisable to confirm that recovery of the i.s. parallels that of the drug.

Recovery should be established on at least 6 processed samples at each of the QC concentrations by a comparison of reference and processed samples. This provides an indication of the % recovery. However, it is preferable to determine the 'absolute' (true) recovery using volume-corrected data. The difference between observed and absolute can be significant, causing unnecessary alarm to the reviewer. Thus for a two-step extraction procedure with 80% of the analyte phase being returned at each stage and with an extractability of 90% for each partition, the absolute recovery would be >80%, whilst the observed recovery would be ~50%.

Interpolation by Senior Editor *concerning* **Sensitivity:** *EXCERPTED POINTS from McAinsh et al.* [11], *Vol. 10 being out-of-print*

From the authors' précis.- "Many approaches have been used, e.g. half the bottom standard or twice background, but all are qualitative and do not reflect the inherent scatter in any calibration line. A confidence-limit approach based on a statistical analysis of a calibration plot is proposed which allows an estimation of the experimental error in any one observation. Where the error is such that the confidence interval for the estimated concentration includes zero, the measured concentration is considered non-detectable. Yet this approach is not without its difficulties, and it is felt that there is still some way to go before the limit of detection problem is finally resolved." Their example of this approach is a GC calibration curve with its y-axis intercept higher than the blank; the 'non-detectable box' is reproduced in the present vol. (Dell, #A-2). The 'twice-background' approach "would in fact produce a negative concentration for the limit of detection"!

Analyte stability

Several aspects have to be considered.
- (1) **Stock solution.-** Stability must be checked at least at room temperature for 24 h and weekly at 4°. Where supplies of the drug, internal standard or metabolites are limited, it may be necessary to store them at -15° to prolong their survival.
- (2) **Post-extraction/injection stability.-** In the context of autoinjector operation, at room temperature or sometimes at 4°, stability should be evaluated at the applicable temperature for 24 h, or for longer if there may be continued storage.
- (3) **Stability in the biofluid** may be governed by enzymes, pH, light, oxygen, and notably by temperature - should be checked from room temperature down to -80°, taking account of the compound's inherent stability, of requisite storage times, and of circumstances surrounding sample collection.
- (4) **The freeze-thaw cycle** generally has to be evaluated, taking account of day-to-day variations, e.g. in the temperature and thawing time employed.

Since time and stability permutations are limitless, the conditions investigated should match those prevailing in the analysis of samples. A typical stability matrix is shown in Table 1, but it is difficult to provide detailed guidelines since these will depend on factors such as the assay and the type and number of samples. Each sample should be analyzed in replicate, usually 6, and quantitated against a freshly prepared calibration curve. Analysis of the data derived from such studies may be performed as set out by Timm, Wall & Dell [12] and Adams [13].

Specificity (Selectivity)

While the extraction and chromatographic procedures diminish the chances of encountering interfering (maybe co-chromatographing) peaks, additional steps taken to ensure the absence of interferences increase the analyst's confidence in the purity of the peak. More selective detectors, e.g. mass-selective (MSD) and diode-array (DAD), improve the specificity, but the degree of certainty increases with concentration, and they are inapplicable to some analytes. The analyst must therefore examine all possible sources of interferences - lest pharmacokinetic profiles be jeopardized - and satisfy himself as to the purity of the chromatographic peak.

Aspects needing investigation include:
- screening of the matrix from a wide variety of individuals to confirm matrix 'cleanliness', or at least identification of sources of interference so that they may be avoided;

Table 1. Evaluation of analyte stability: illustrative protocol.

Temp. (RT = room temp.):- TIME	Stock solution		Injection solution			Biological matrix					
						Blood		Plasma/Urine			
	RT	+4°	RT	+4°	-20°	RT	+4°	RT	+4°	-20°	-80°
0	*	*	*	*	*	*	*	*	*	*	*
1 h	*	*				*	*	*	*		
2 h						*	*	*	*		
12 h	*	*				*	*	*	*		
24 h			*	*	*						
48 h			*	*	*				*		
1 wk	*	*					*				
2 wk											
1 mth									*		
2 mth											*
6 mth									*		
12 mth											*

- processing of samples from individuals dosed likely co-prescribed drugs (preferable to evaluating pure standards, since metabolites are embraced);
- evaluating any interferences from co-administered drugs 'forgotten' by the volunteer, e.g. aspirin, paracetamol (acetaminophen) or caffeine;
- investigating potential sources of interference from metabolites of the drug itself, which is best administered prior to the definitive study. Metabolite interference may be merely an inference from peak shape and may need cross-validation under different chromatographic conditions.

VALIDATION PROCEDURES

In order to evaluate the characteristics discussed above, they should be measured under the same or similar conditions as apply to a 'real life' study. The constraints in respect of batch size are: (1) chromatographic run time (limit 24 h); (2) extraction technique/complexity; (3) column and detector stability; (4) sample stability.

Evaluation is carried out using a minimum of 4 batches to permit calculation of meaningful statistics for between-run precision and accuracy. The first batch is designed to provide initial intra-batch acceptability of these characteristics as well as stability data for the sample in the auto-injector. Further batches are designed to evaluate between-run precision and accuracy as well as to determine recovery, stability and confirmation of the sensitivity and specificity.

Confirmation of the chosen concentration-response relationship is made by comparison of the residuals from each of the calibration curves.

The first batch comprises test concentrations, each with at least 6 samples, namely: (a) the decided LLOQ, (b) low QC, (c) mid-range QC, (d) high QC, and (e) a range of blank matrices from different individuals; it also incorporates 3 calibration curves and a range of reagent blanks. The range of the calibration curve should be chosen to cover the maximum number of observed sample concentrations compatible with the detection system. The distribution of calibrators should be concentrated in the range where most concentrations are expected, usually in the first half of the curve. At least 5-7 non-zero calibration points should be used depending upon the range. Having chosen the calibration range, cardinal to setting up QC's, (b) is set at 3-5 times the LLOQ, and (c) and (d) at, respectively, 40-60% and 70-90% of the highest calibration value.

Within-batch sample order is: reagent blank(s), standard solution(s), calibration curve in sequence from low to high, matrix blanks, samples (a), (b), (c) and (d) arranged in random order together with a further calibration curve randomly distributed through the QC samples, calibration curve in sequence from high to low, and a standard solution.

The reagent blank confirms that the system is clean, while the standard solutions confirm both the response of the system and the retention times of the analytes and internal standards. Comparison of the first, second and third calibration curves can provide information on the stability of samples in the autoinjector, the possibility of system drift, and carry-over within the injection system.

Within-run accuracy and reproducibility may be determined from the QC's. One can evaluate whether the variability at the decided LLOQ has the required precision and accuracy. Precision and accuracy data from a single batch confirm the validity of the method on a single occasion; between-batch data provide a more realistic evaluation of the assay in routine use. Hence for statistical evaluation sufficient data must be provided - at least 4 batches, not necessarily exact repeats of the first batch. Further batches need contain only one calibration curve, 4 each of the QC concentrations and matrix blanks together with reagent blanks. In addition, batch size should be made up by inclusion of samples designed to evaluate stability, selectivity and recovery. The exact nature of these will depend upon specific assay requirements.

ANALYSES ON 'REAL' SAMPLES

The validated characteristics being demonstrably within acceptable limits, real samples can now be analyzed. The samples for a 'production' batch may be put in the following sequence: reagent blank, matrix blank, standard solution(s), 'LLOQ test', and then calibration and 6 QC samples randomly interspersed amongst the test samples. The 'LLOQ test' sample is added early in the batch to confirm the system's capability to achieve the LLOQ, especially important where this is close to the limits of the system.

Application of acceptance criteria

Usually the concentration values from individual batches are subjected to pharmacokinetic and/or statistical analysis. Hence it is essential to confirm the acceptability of the generated data, achievable by applying criteria for acceptability of (1) the calibration curve and (2) the QC samples. Decisions on which criteria to use may appear subjective, but their application must be as objective as possible.

Calibration curve acceptability calls for (a) $r = 0.9900$ or better, (b) matrix blank with signal <20% of LLOQ, and (c) other measures of goodness of fit, which may indicate significant differences between nominal and back-calculated calibration standards. Acceptable criteria for excluding outliers should be defined as part of the study protocol. If calibrated samples fall outside these limits they may be removed from the regression analysis. However the minimum number of calibration points required to define the curve should be decided before the study starts. Criteria for defining fit, (c), have been discussed in connection with linearity.

QC sample acceptability.- Assuming the calibration curve is acceptable, the acceptability of the QC samples can then be evaluated. Setting limits of ±1 R.S.D., the limits of the low, medium and high QC's are ±20%, ±15% and ±10% of nominal respectively. On this basis, for normally distributed data ~66% of samples should fall within 1 R.S.D. Thus for 6 QC's the expected number of acceptable QC's would be 4. However, so as to minimize bias to the higher concentrations, at least one QC of each concentration should be acceptable. The fourth acceptable QC may be of any concentration.

Replicate acceptability.- Where each sample is assayed in duplicate, acceptability may be judged by calculating, as % of the mean, the difference between the duplicates or (half as large) the difference between the higher duplicate and the mean. An acceptability range, say ±7-15%, may then be set empirically.

CONCLUDING COMMENTS

Criteria based on calibration and QC samples as well as replicates allow particular batches to be objectively accepted or rejected and between-run precision and accuracy to be determined. The criteria, although subjective, should be set simply and logically and applied objectively so as to obviate conflicting data. The foregoing method validation program and batch acceptance criteria come from subjective thoughts. Flexibility must be maintained, yet it is difficult and confusing to compare analytical methods and sample concentrations of diverse origin. Internationally agreed standard basic criteria would therefore be advantageous in defining and validating methods as well as accepting results.

Acknowledgement.- Thanks are due to colleagues past and present as well as clients who have contributed directly or indirectly to the validation program and acceptance criteria herein.

References

1. Analytical Methods Committee (1965) *Analyst 90*, 251-255.
2. IFCC - Provisional Recommendation on Quality Control in Clinical Chemistry (1975) *Clin. Chim. Acta 61*, F25-F38.
3. Snedecor, G.W. (1952) *Biometrics 8*, 85-87.
4. Chamberlain, J. (1978) in *Blood Drugs and Other Analytical Challenges* [Vol. 7, this series] (Reid, E., ed.), Ellis Horwood, Chichester, pp. 55-59.
5. Leppard, J.P. (1981) in *Trace-Organic Sample Handling** [Vol. 10, this series] (Reid, E., ed.), as for 4., pp. 320-335.
6. Knecht, J. Stork, G. (1974) *Fresenius Z. Anal. Chem. 270*, 97-99.
7. Mitchell, D.G. & Garden, J.S. (1982) *Talanta 29*, 921-929.
8. Long, G.L. & Winefordner, J.D. (1983) *Anal. Chem. 55*, 712A-724A.
9. Keith. L.H., Crummett, W., Deegan, J., Libby, R.A., Taylor, J.K. & Wentler, G. (1983) *Anal. Chem. 55*, 2210-2216.
10. Kaiser, H. [translated by Menzies, A.C.] (1968) Two papers on *The Limit of Detection of a Complete Analytical System*, Adam Hilger, London.
11. McAinsh, J., Ferguson, R.A. & Holmes, B.F. (1981), *as for* 5., pp. 311-319.
12. Timm, U., Wall, M. & Dell, D. (1985) *J. Pharm. Sci. 74*, 972-977.
13. Adams, H.K. (1981), *as for* 5., pp. 291-297.

**Editor's note.*- This book is now out-of-print. For Leppard's article, corrections are listed at end of Vol. 12 (1983), Plenum, New York.- See list at start of the present vol.

Dr H.M. Hill is now at: Phoenix International Life Sciences, 2330 Cohen St., St. Laurent (Montreal), Quebec, H4R 9Z7 Canada.

#A-4

DATA HANDLING AND QUALITY CONTROL IN THE BIOANALYTICAL ENVIRONMENT

F. Van Rompaey, R. Woestenborghs and J. Heykants

Department of Drug Metabolism and Pharmacokinetics,
Janssen Research Foundation,
B-2340 Beerse, Belgium

Although computerization, requisite to satisfy today's high sample load, improves laboratory efficiency and productivity, a risk is involved that increased throughput and less personal supervision impairs quality. Hence a thorough QC must be performed on the analytical data. In our laboratory chromatographic data are now processed by a CDS operating in a network of microcomputers (facilitating optimal utilization of the CDS's) served by a microVAX minicomputer. Similarly RIA data are processed by appropriate computer programs. The final concentrations, obtained by chromatography or RIA, are transferred on-line to a Macintosh microcomputer for additional calculations and reporting; thereby clerical errors are obviated, but the quality of the final results is not always improved. The requisite QC entails visual inspection of chromatograms and critical evaluation of calibration data and of results for replicates. After a method has been validated and approved, all QC-relevant data are put into a LIMS tailored to QC; its capabilities include graphical presentation of accuracy and precision data by means of Shewhart QC charts.*

The continuously increasing number of samples together with the demand for faster and better analyses has definitely altered the image of analytical laboratories. Since automation and computerization is the ideal way to improve a laboratory's efficiency and productivity, analytical instruments are being equipped with autosampling devices, simple integrators are being replaced by sophisticated CDS's, and large laboratories now make use of LIMS's.

**Abbreviations*.- QC, quality control; CDS, Chromatography Data System; LIMS, Laboratory Information Management System; QCDMS, Quality Control Data Management System; SAS®, Statistical Analysis System.

OPERATING PROCEDURES AT THE JANSSEN BIOANALYTICAL LABORATORY

Scheme 1 shows how the laboratory operates, with general information on the handling of samples, sample information and analytical data from sample reception to final report. At the reception desk, relevant sample information is written down in an ordinary notebook. At the end of every week, this information is added to a database by means of a user-developed LIMS-program (BALSAM: Bio-Analytical Laboratory Sample Analysis Management) written in SAS and operating on an IBM mainframe computer. It allows preparation of sample lists, work-schedules and suchlike. A detailed description is beyond the scope of this article. After the laboratory manager has defined priorities and divided the work among the analysts, samples are coded, extracted and analyzed by chromatography or RIA. The resulting raw data are processed by means of commercially available software packages and some user-developed programs running on IBM microcomputers. The analytical data obtained are then subjected and added to a QCDMS, another LIMS-program written in SAS. Finally, if the analyses meet the pre-defined QC criteria the results are reported.

HARDWARE AND SOFTWARE CONSIDERATIONS

In order to process analytical data from analytical instrument to final report without having to write out any of the intermediate results, a well-structured hardware and software configuration is cardinal. The hardware set up in our laboratory is shown in Scheme 2. Analogue data acquired from analytical instruments are digitized and captured by IBM microcomputers (IBM PC® for RIA data and IBM AT® for the more complex chromatographic data) for processing. Chromatographic data are then processed using a CDS (Nelson Analytical 2600® Chromatography software) instead of simple integrators, resulting in numerous advantages:
- ability to process the same chromatogram with various integration parameters, allowing optimum integration conditions to be selected;
- reprocessing of poorly integrated chromatograms, either individually or in batches, thus improving the quality of the analyses;
- plotting of several chromatograms overlaid or stacked, facilitating comparison of chromatograms;
- subtraction of chromatograms to perform blank corrections in cases of overlapping peaks;
- calculation of calibration curves within the CDS environment, resulting in very fast and efficient processing of data from chromatogram to final concentration;
- the facility to link user-developed programs to CDS-software, making it possible to adapt the final computer output to the desire of the user;

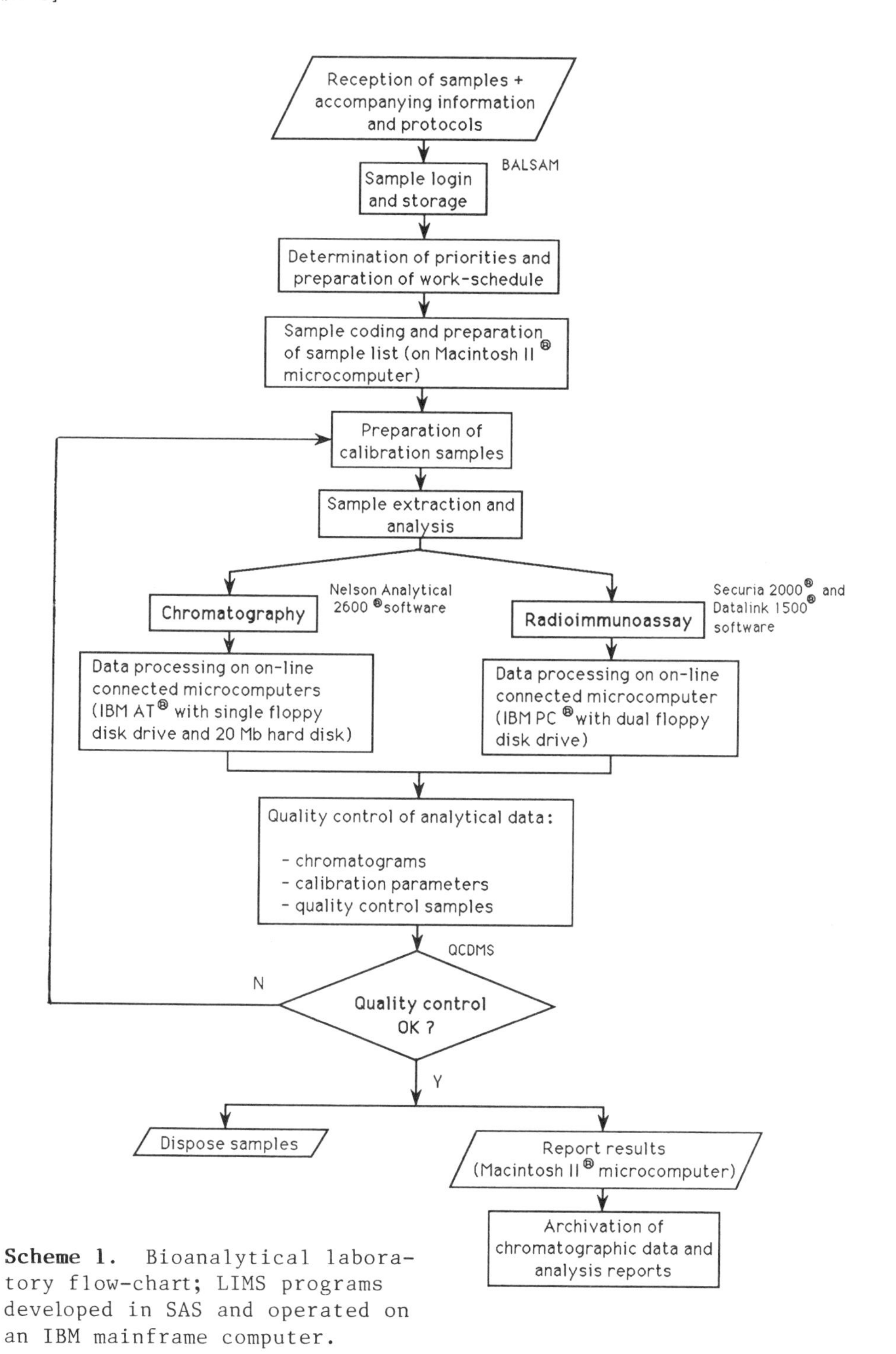

Scheme 1. Bioanalytical laboratory flow-chart; LIMS programs developed in SAS and operated on an IBM mainframe computer.

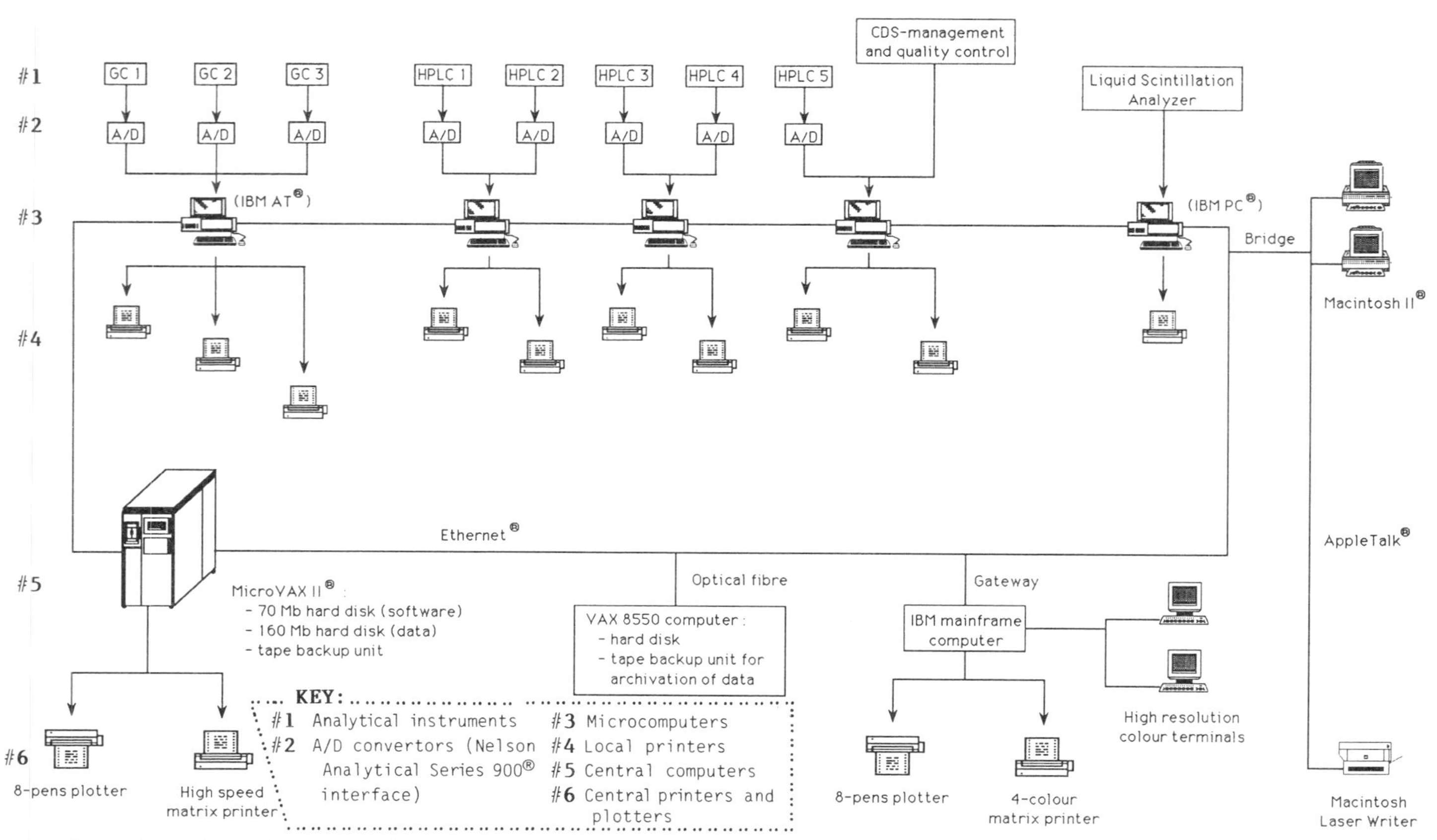

Scheme 2. Bioanalytical laboratory hardware configuration.

- the ability to import CDS data files into the commercially available software, allowing further data processing without having to enter any of the results manually.

Each individually operating microcomputer is linked to an Ethernet® computer network, served by a MicroVAX II® minicomputer which is used to store all data generated during chromatographic analyses. This approach gives great flexibility since all data can be accessed on any of the chromatography work-stations. For the comfort of the analysts, chromatograms and integration reports are printed on local ink-jet printers. After the final results are reported, all data are archived on a central VAX mainframe computer.

In addition, Ethernet® is connected to AppleTalk®, a computer network for Macintosh microcomputers, and to an IBM mainframe computer. The ability to transfer data from one computer environment to another allows the most suitable system for each job to be selected. Thus, in our laboratory all final reports are generated on the very user-friendly Macintosh microcomputers. Besides, computer networking has the advantage that most expensive devices, such as high-speed printers, laser writers and plotters, can be shared amongst all network users.

COMPUTERIZED DATA PROCESSING

Scheme 3 outlines how the data are actually processed and transferred on-line from analytical instrument to final report. After creation of a sample list and corresponding sample file (containing sample description and sequence no.) on a Macintosh II® microcomputer, samples are analyzed. The combination of commercially available and user-developed software yields a results file, containing sample sequence no. and measured concentration. Combining these two files through their unique link, the sample sequence no., results in a final report, and a report file which may be accessed by the study directors for pharmacokinetic analyses. Since during the whole procedure no data have to be written out, clerical errors and rounding errors belong to the past.

All QC data generated during sample analysis unfortunately still have to be transferred to our QCDMS-database. In the near future, however, when the required software is developed, this step will be performed on-line as well.

INTERNAL QUALITY CONTROL

With the advent of automated laboratory instrumentation and computerized data processing, resulting in an increased sample throughput, extensive QC procedures have become obligatory. A thorough QC of the analytical procedures,

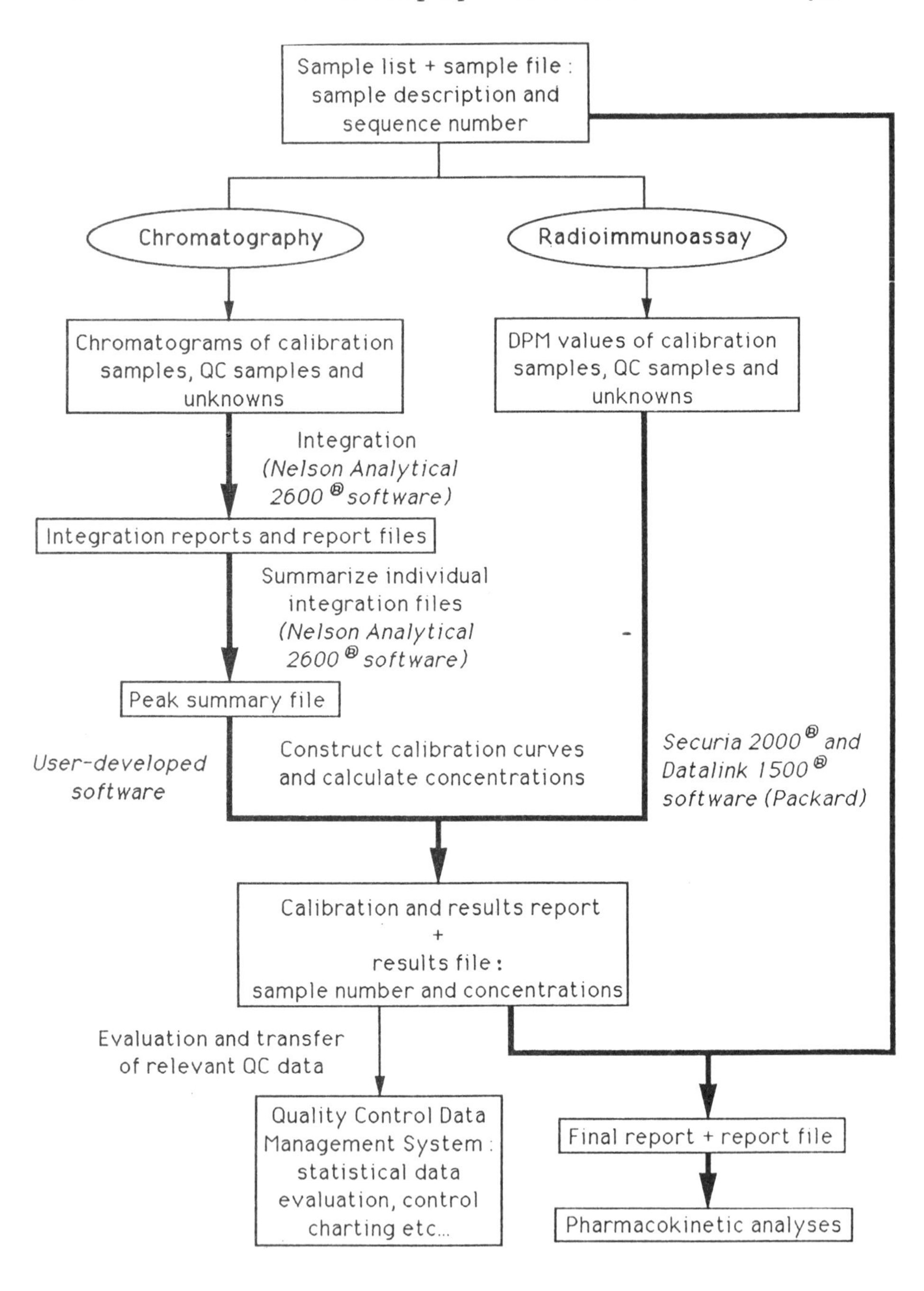

Scheme 3. Bioanalytical laboratory data processing.

however, does not always imply that the results correspond with the investigator's expectations. Since analytical results can never be better than the quality of the sample, the importance of sample collection, treatment, shipment and storage should never be under-estimated. Only if this is taken into account is the analytical QC really meaningful.

In bioanalytical laboratories where samples are analyzed by chromatography or immunoassays in a concentration range of several orders of magnitude and with detection limits often <1 ng/ml, QC encompasses the visual inspection of chromatograms, evaluation of calibration data and, as the most widespread method for monitoring analytical quality, the analysis of spiked QC samples.

Quality-control samples

QC samples are prepared by spiking blank material, composed of a matrix as close as possible to the real samples, with an accurately known amount of the analyte. The spiked pools are subsequently divided into small aliquots and kept frozen at -20° until just before use. The reference material should of course be of the highest possible purity. The number of different QC samples that are prepared depends on the application.

During method validation a series of at least 10 samples, covering a range from blank to the highest anticipated concentration in clinical and pharmacokinetic samples, are analyzed blind. At least 2 duplicate pairs, one at a low and one at a high concentration, are included. Results are evaluated by calculating the accuracy (difference between measured and spiked concentration, expressed as % relative error) and precision (reproducibility of measured concentrations of duplicate samples, expressed as S.D., range or, most frequently, as % C.V.). Student's t-test is performed on the accuracy data so as to check whether the mean relative error differs significantly from zero. Additionally, a least-squares regression analysis of measured *vs.* spiked concentrations is carried out and regression parameters are evaluated. The various situations that can occur are summarized in Fig. 1.

The presence of random errors in the analytical method leads to a scatter of the points around the regression line. Under ideal conditions this results in an intercept and slope exactly equal to zero and unity, respectively. In practice, however, this situation will never occur. A constant or translational error yields an intercept different from zero. A proportional or rotational error on the other hand leads to a slope different from unity. Both types

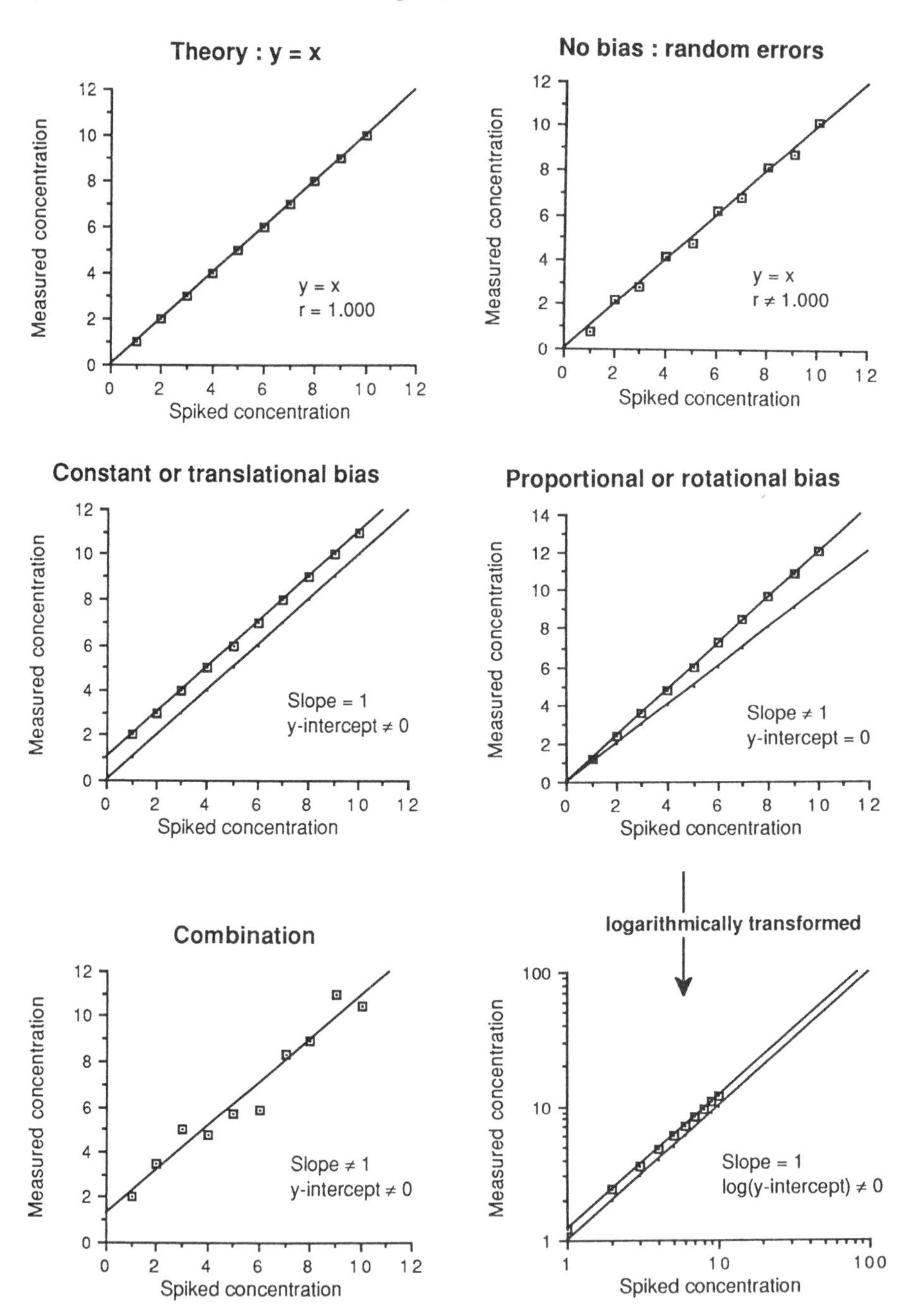

Fig. 1. Error characterization by means of regression analysis.

of error may also appear together. A Student's t-test on intercept and slope allows one to define whether or not the translational and proportional errors are statistically significant.

Control charting

As soon as an analytical procedure has been validated and approved for routine use, the quality is continuously monitored by including 2 pairs of duplicate spiked QC samples in each series of unknowns, one at a low (sub-therapeutic) level and one at a high (therapeutic) level. Results are evaluated by calculating the accuracy (as % relative error) and precision (as % error range), and by the maintenance of Shewhart QC charts [1, 2] for relative error, as exemplified in Fig. 2.

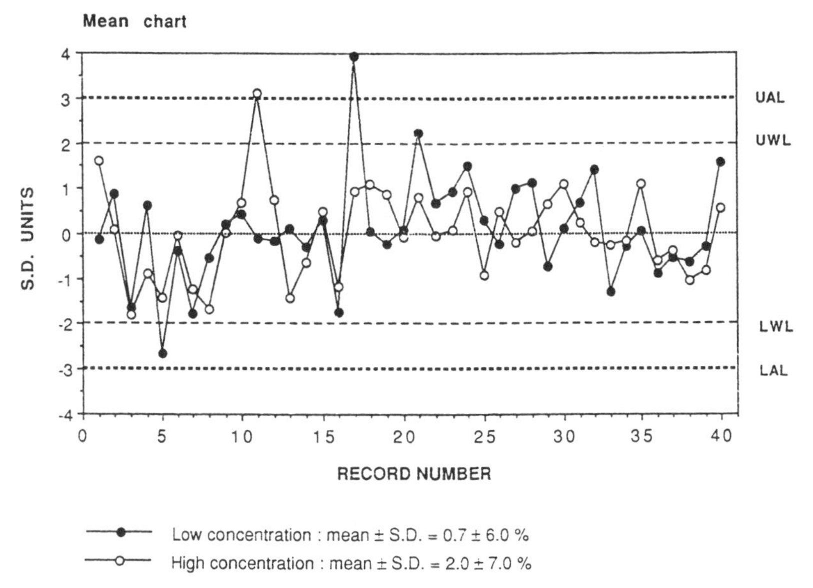

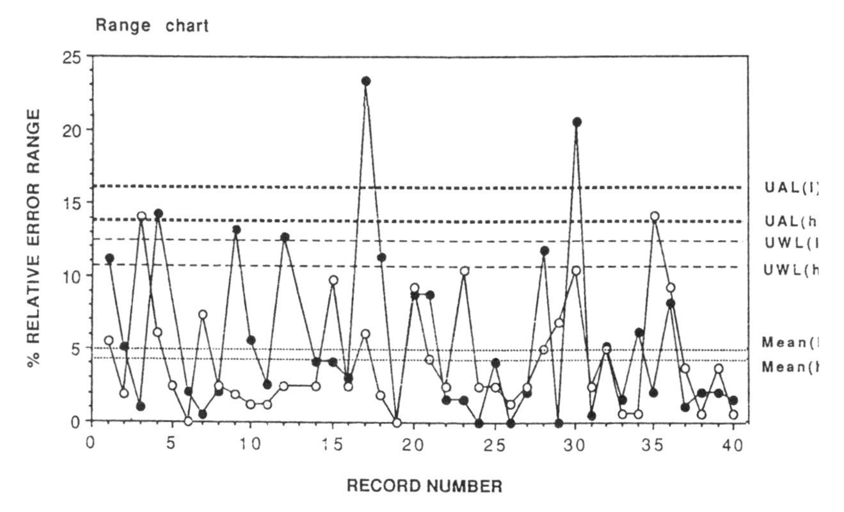

Fig. 2. Example of a Shewhart QC chart for relative error. As amplified in the text, 'Warning' and 'Action' Limits are indicated; l and h refer to the concentrations, low and high.

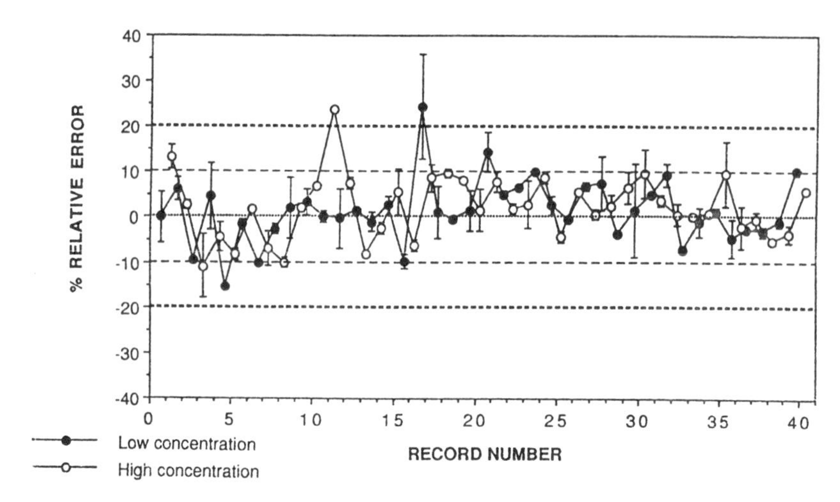

Fig. 3. Example of an error QC chart.

The mean chart is a graphical presentation of the mean relative error obtained for duplicate analyses as a function of time. Warning and action limits (WL and AL) are set 2 and 3 S.D.'s from the mean, yielding 2 upper and 2 lower limits (UAL, UWL; LWL, LAL). From normal distribution theory, the probability of observing a value outside these limits is very small: only 1 in 20 for WL's and 2.7 in 1000 for AL's. To track changes in variability, a range chart is constructed in a similar fashion but using specific factors available in standard references. To interpret the charts and to define when preventive action should be taken, certain rules are applied. Examples of these rules indicating that the analytical procedure is out of statistical control are: one or more successive points outside the AL's; 2 or more outside the WL's in the same direction; 3 or more outside the WL's in either direction; 7 or more above or below the target value.

A drawback of the use of Shewhart QC charts is that for procedures with a very high precision, points falling outside the acceptable limits can still be very accurate. Therefore error charts are maintained in addition to the Shewhart charts, as depicted in Fig. 3. The mean relative error of duplicate analyses is plotted *vs.* time and the error range is indicated by vertical bars. The following rule-of-thumb criteria are applied.- As long as QC samples yield relative errors <10%, results are accepted; if repeatedly 10-20%, investigation is needed, and >20% entails rejection. In some situations, as at very low concentrations, these criteria are impracticable, and rules are adapted to feasible limits which can be derived from Shewhart QC charts.

Although QC samples are prepared independently of calibration samples and by a different person, they are always prepared in the same environment. This involves the risk that some shortcomings and systematic errors will never be discovered. Hence participation in inter-laboratory schemes should be encouraged.

Because spiked QC samples are free of interfering compounds and metabolites, it is a good practice to analyze aliquots of a pool of real samples as well. This applies especially to immunoassays where metabolites may show cross-reactivity with the parent compound. A major problem, however, is that very often not enough samples are available to prepare sample pools; hence control charts are often restricted to spiked samples.

Besides QC samples, control charts of calibration parameters are maintained as well. For chromatographic assays this relates to response factor data, while for RIA it encompasses parameters such as total dose, non-specific binding, zero-binding and regression parameters. In situations where QC samples yield results outside acceptable limits, this can be very helpful in attempting to pinpoint where the problems may have occurred.

CONCLUSIONS

Although there is no doubt that the introduction of computerized data handling improves the efficiency and productivity of a laboratory, it has little effect on the quality of the final results. The analysis of QC samples and the maintenance of QC charts is a generally applied method for alerting to analytical errors. Nevertheless, one should always bear in mind that QC samples are often ideal samples, free of unknown interferences, and hence might behave differently from real samples. Furthermore, QC samples prepared in the same environment as calibration samples may mask shortcomings and systematic errors. The importance of participation in inter-laboratory comparisons cannot be overemphasized. Both organizing and participating laboratories may enjoy its benefits.

References

1. Duncan, A.J. (1974) *Quality Control and Industrial Statistics*, 4th edn., Richard D. Irwin, Inc., Homewood, IL.
2. Grant, E.L. & Leavenworth, R.S. (1972) *Statistical Quality Control*, 4th edn., McGraw Hill, New York.

#A-5

VALIDATION REQUIREMENTS OF BIOANALYTICAL METHODS IN RELATION TO POSSIBLE INTERFERING AGENTS

G.S. Land and R.D. McDowall

Department of Bioanalytical Sciences,
Wellcome Research Laboratories,
Langley Court, Beckenham, Kent BR3 3BS, U.K.

To ensure that results from a new bioanalytical method are correct, validation tests have to be performed. Two major sources of interference may affect assay specificity: endogenous constituents of the matrix, and exogenous substances introduced after initial sampling or during analysis. This article gives examples of some of the interferences that can occur, together with a standard approach to validation.

Although interferences represent a relatively small area of method validation, they can strongly affect all other aspects of validation, particularly precision and accuracy at the limits of analytical technology. They can arise from one or both of two sources: endogenous material in the sample, and material introduced during sampling, storage, extraction or analysis. They can jeopardize confidence that only the analyte(s) intended are being measured. Here we consider only the specificity (selectivity) of bioanalytical methods used in drug discovery and development. Other authors have dealt with regulatory aspects of method validation [1] (see also pp. 67 & 89], and with validation in general [2, 3], as considered throughout this section. Table 1 lists four phases into which drug discovery and development can conveniently be divided, and the changing validation requirements in respect of selectivity.

VALIDATION REQUIREMENTS

Selectivity requirements depend on factors such as phase of development, animal species, biological matrix and analytical technique (Fig. 1).

Influence of the analyte's pharmacology

The pharmacological profile of a new chemical entity (NCE) will determine the dose and hence the likely range of concentrations to be measured. This determines the most suitable analytical technique(s). For some drugs the plasma concentrations may be of the order of µg/ml, whilst others (such as receptor-based drugs) may be active at pg/ml levels in plasma such that methods offering very high sensitivity (e.g.

Table 1. Selectivity requirements in method validation.

Development stage	Assay objective	Validation requirements
#*1*: discovery	Comparison of some related compounds	Similar selectivity towards all compounds examined
#*2*: pre-clinical	Toxicology/metabolism support to regulatory standard	Reliably selective; apt for blood/plasma/other animal matrices
#*3*: clinical - healthy volunteers	Definitive clinical studies; high throughput; usable in other labs.	Excellent selectivity in human biofluids - plasma, urine, saliva, CSF; perhaps faecal homogenates
#*4*: clinical - patients	World-wide routine monitoring of patients varying in medication & dietary background	Reliable selectivity in the presence of numerous substances probably not hitherto encountered

GC-ECD and immunoassay methods in general) may be the only candidates. The larger the dose the easier it is to fulfil the analytical goal and the less demanding are the selectivity demands of an assay.

Changing analytical objectives during drug development

Assay objectives and requirements differ in each phase of drug development and hence require different criteria for selectivity validation, as now considered for three main functional areas.

The research programme.- This aims to evaluate a series of candidate compounds with respect to comparative metabolism and biological activity. Information obtained leads to synthesis of further candidate compounds; promising compounds may undergo more detailed metabolic, pharmacokinetic, bioavailability and toxicological studies. An assay method must have rapid sample throughput and fast turn-around time, so that the results of one experiment are available to influence drug design without delay. Various animal species will be used during this work, e.g. rabbit, dog, mouse and rat. One may observe species differences in drug metabolism or in sample-matrix composition, calling for matrix revalidation to ensure no change - affecting selectivity - due to a new endogenous interference or metabolite.

To ensure selectivity, mass spectrometry (MS), direct or GC- or HLPC-interfaced, can clinch identity. However, when time is limited the analysis of blanks with the same

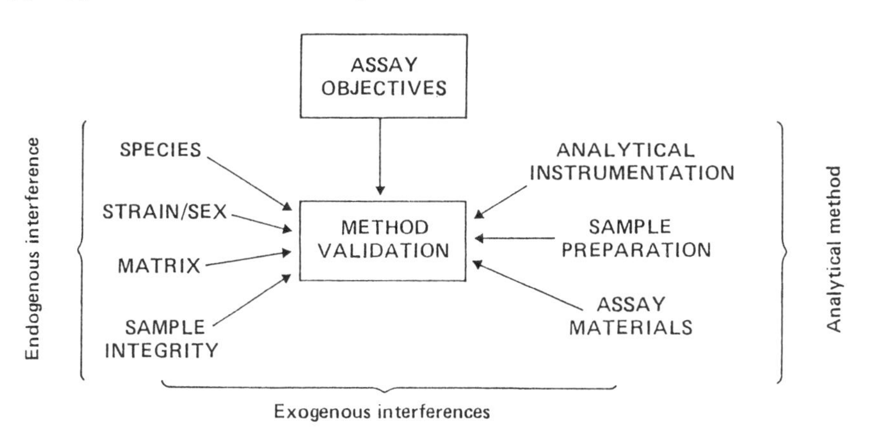

Fig. 1. Factors affecting validation in relation to selectivity.

matrix, pre-dose or from similarly fed animals, is the easiest way of identifying interferences with the possible exception of metabolites. Chromatographic peak shape may disclose any closely eluting interferences. HPLC peak-purity assessment by diode-array detection or 'on-the-fly' spectral scanning can help identify the presence of metabolites. Spectral scanning can also help assess peak purity in (HP)TLC.

If an internal standard (i.s.) is used, pre-dose and a few 'typical' drug-containing samples must be run through the method with and without it, to find whether it is liable to any interferences. Ideally each analytical run should contain a blank without i.s., to confirm that no exogenous interferences are present. Ideally the matrix should also be able to measure a number of similar compounds as possible alternative analogues. Whilst similarity in physicochemical properties within a class of candidate compounds reduces the risk that selectivity for a particular member will be compromised, there should be an assessment similar to that for an i.s.

Pre-clinical development, entailing study of safety, absorption, distribution, metabolism and elimination in two or more species, sharply increases the need to validate the assay, since 'GLP' regulations usually apply (OECD, Japan and USA). The precision and accuracy of an analyte ideally should be within ±15% but may, for pharmacokinetics, be poorer at lower concentrations if measurement variability is known. Selectivity is vital, as the data will be used to assess drug absorption and exposure in animals and to extrapolate to man, for dosage also with due regard to any toxicity in chronic dosing. Method revalidation may be needed if sensitivity has to be improved for pharmacokinetic measurements.

Exemplifying a species-dependent artifact, the i.s. (3,4-dihydroxybenzylamine) used in a catecholamine assay [4] was bound to protein in plasma from the Bedoin Goat, although not in human plasma for which the method had originally been developed. A simpler problem often observed is the wide variation in endogenous components from one animal species to another, sometimes necessitating changes to a method that had previously been very reliable.

To give the highest degree of confidence in chromatographic peak purity and, in general, assay selectivity, it is important to apply as powerful a means of identification as possible, such as mass spectrometry (MS) or UV spectral scanning. Capillary zone electrophoretic (CZE) chromatographic separation is likely to be increasingly used, and high-field NMR [cf. #ncC-5 & -6, this vol.], although limited in sensitivity, can be very informative about the absence of interferences and can help prove the identity of any metabolites. If applied to samples from metabolism studies employing a radio-label, such techniques can further affirm that only one analyte is being measured, and the comparison of results with total circulating ^{14}C can reinforce evidence for selectivity. Another way to assess selectivity is to compare the chosen method, say immunoassay, with an alternative such as HPLC, HPTLC or GC.

Clinical development.- Normally safety, metabolism, pharmacokinetics and pharmacodynamics in humans are assessed first, then therapeutic activity and safety in patients. The demand for assay sensitivity increases as the drug dose is reduced from that used in toxicology to that required for therapy. In dosage trials and later pharmacokinetic studies (up to 3-4 half-lives), concentrations at least 10-fold lower than therapeutic levels will have to be measured. Any relevant metabolites present may have to be selectively determined at levels lower than for the parent drug.

With a lowered dose there is more likely to be interference by matrix components; while the sample can be larger than in the animal work, interferants may limit this benefit. Important pharmacokinetic parameters, often hinging on assay values near the detection limit, may be determined from the worst bioanalytical data! Near this limit, therefore, soundness of assay is more important than in the pre-clinical stage and, moreover, effects of endogenous and exogenous interferences can be exaggerated, with manifestations ranging from mere false-positive results, or false-negatives due to binding onto surfaces (glass/plastic), to endogenous interferences (proteins/phosphates) making an otherwise linear assay response non-linear.

Consideration also has to be given to use of the method to analyze samples from patients from many different countries. The assay will have to be tested as much as possible against different dietary/racial backgrounds or conjecturally co-administered drugs whose metabolism will have to be taken into account too. If, moreover, the method is a bioassay, one must check for possible interference in both the assay and the treatment by pharmacologically active metabolites [5] or endogenous/exogenous interferences. Ideally one should include samples from control (e.g. placebo) patients, or pre-dose samples from the actual patients. Besides, if other laboratories are measuring the analyte(s) an inter-laboratory quality assurance scheme will need to be established, to demonstrate that the methods used are producing results of acceptable accuracy and precision (±10%).

Influence of assay method and matrix type on selectivity

The type and degree of interference depends on the technique used for detection and separation, e.g. TLC and densitometry *vs.* immunoassay or HPLC using various detection methods. The value of cross-validating different assay methods in establishing selectivity is illustrated by the following example. Acrivastine, a novel H_1-antihistamine, was measured in human plasma by RIA (chosen for high sensitivity, 5-50 ng/ml). The method appeared highly selective, but during further development an HPLC method was used to measure acrivastine in urine, and a metabolite was discovered which resembled the parent drug in RIA reactivity. The metabolic site was a part of the molecule which the RIA could not differentiate. With difficulty, an assay was developed for acrivastine and its metabolite (BW270C) in plasma: BW270C was found at significant concentrations (5-15 ng/ml), and the RIA was in fact measuring both.

Derivatization can increase sensitivity and reduce selecvity when endogenous or exogenous components with a similar reactive group (endogenous analogues or phthalate esters) give a product that elutes close to the analyte(s) of interest. Interferences can arise from pre-analytical sample processing. Complexing of an analyte with an exogenous or endogenous component can give anomalous results whose cause is difficult to identify.

Matrix influences.- Different matrices (such as blood, serum, plasma, saliva, CSF, urine and bile) have different compositions which may affect the assay in different ways. Thus, whereas plasma is protein-rich and near-constant in pH, urine is salt-rich, variable in pH and has many polar low-mol. wt. constituents. Moreover, if collection periods are short, individual samples of urine will vary widely in interferences, including exogenous ones reflecting influences such as diet, disease state and amount or type of liquid intake. Additionally, urine and faeces vary in amount; if low, such that there might be a high level of interferant, the assay has to be continually checked in this respect. If such samples are collected at the start of a weekend, they may be at room temperature for 72 h; or they may suffer exposure to sunlight. Hence method validation must include analyte-stability checking. There is also the possibility of interferences arising from degradation of bio-constituents or, by leaching, from the vessels.

A SYSTEMATIC APPROACH TO VALIDATING SELECTIVITY

Initial and later interference-checking with 'blank' matrices

Endogenous interferences.- Final validation should involve processing a number of animal or clinical analyte-free samples. If from patients, especially treatment-resistant or with a life-threatening disease, there may be various co-medicants, maybe long-term, which method validation must take into account[†]. Checking co-medicants or their metabolites must be done with representative samples from patients as well as with normal plasma appropriately spiked. An example is now given.

Lamotrigine assay, as an illustration.- Before assessment for monotherapy is allowed, a novel anti-epileptic such as lamotrigine must be shown by 'add-on' therapy to benefit patients resistant to other forms of anti-epileptic treatment. During the clinical programme the assay method that had been developed and validated for plasma was installed at a number of outside laboratories. One changed the method from the NP-HPLC system to a RP-HPLC system; there was good accord when the same samples were analyzed in a validation exercise, but 'lamotrigine' was later found in plasma samples from certain patients not dosed with the compound, using the modified method, but not with the original method when re-applied. The source of this interference, seen only in a few patients, could not be identified and the mobile phase used for RP-HPLC had to be modified to obviate the co-elution. This illustrates that even where an assay is both validated internally and cross-validated with another method, many samples not containing the drug have to be assessed before one can be certain that no interferences will arise; any method modification necessitates re-validation for that matrix.

Exogenous interferences.- As these can arise from any of the materials, vessels and pipetting devices used in an assay, 'background' must be monitored using reagent blanks routinely. To help identify the source of a novel interference, one must know details of materials used (batch no., grade,

[†]especially if drug-clearing organs are dysfunctional

source), including vessels and sample containers for which any time-dependent leaching must be assessed. Additives such as citrate, EDTA and heparin must also be checked. An easily overlooked problem in extracting analytes from the biological matrix is the addition of stabilizers to solvents during manufacture. In diethyl ether possible additives to prevent peroxidation include 2% ethanol (Merck), 2-3 ppm pyrogallol (Rhone-Poulenc) and <0.1 ppm butylated hydroxy-toluene (Aldrich). Such additives, which may interfere in the assay, may not be mentioned in the analytical specification for the solvent or reagent. In capillary GC with FID the traces after injecting solvents show that few are truly pure. (As indicated in #ncC-3, this topic-area has repeatedly featured in the Forum-based books - *Ed.*)

An HPLC assay for fenoldopam in human plasma [6] was transferred from a U.S. laboratory to a U.K. laboratory. The method entailed ether extraction, evaporation and reconstitution in mobile phase followed by HPLC with electrochemical detection. As the ether used in the original assay (Malinkrodt) was unobtainable in the U.K., analytical-grade ether from Rhone-Poulenc was substituted. The establishment of the method was impeded due to the presence of large interfering peaks that were not seen in the original laboratory. This was traced to the ether and a different batch was assessed. The problem did not disappear, and it was ascertained from the manufacturer that pyrogallol was an additive, later confirmed to be the interferant.

Middleditch & Zlatkis [7] have detailed commercial stabilizers in common use, and Middleditch [8] described numerous additives or contaminants that can interfere with analytical procedures. It should be remembered that the stabilizers are there to prevent the solvent from decomposing. However, the solvent itself can react with the analyte being extracted. During chloroform extraction of plasma, certain tricyclic antidepressants reacted to form carbamate derivatives, giving anomalous results [9]. (This carbamate formation [10], and other interactions, are mentioned by E. Reid in Vol. 12, p. 264.)

Instrumentation-derived interference

How reliable is the system? Can false results originate from malfunctioning instrumentation? Full assurance hardly comes from obligatory technical assessment, routine instrument calibration and the use of test mixtures whose behaviour is well known. Possible trouble-areas include column switching, GC-MS and HPLC-MS interfaces, and incorrectly set data acquisition systems. Surprisingly, the accuracy of sample loops can be out by as much as 20%, well beyond manufacturers' tolerances [11].

Interference associated with instrumentation but external to the laboratory is exemplified by an assay where solvent extracts, dried under N_2 (more likely than air to minimize any oxidation) with warming, were taken up in ethanol and analyzed by GC. After smooth method performance, small interfering peaks that prevented quantification at low levels abruptly appeared, eventually tracked down to a non-volatile residue from the site N_2 supply. Presumably, in plumbing work, oil or other semi-volatile substance had entered the system. An activated carbon filter put in the line removed the problem. Similarly, a 'poisoned' ECD can result from impure GC carrier gas.

Reliable analyte recovery

Analyte binding to (e.g.) phthalates, as found [9] with tris(2-butoxyethyl)phosphate present in purified distilled water, is well known to impair recovery in some assays, and can be hard to discover. Impaired recovery of analyte and i.s. due to the matrix was encountered in studying lamotrigine penetration into the CNS using surgeon-derived brain and plasma samples. In the pre-HPLC extraction into ethyl acetate at alkaline pH, recovery from plasma is >90%, but with spiked rat-brain homogenates the recovery was 84% for lamotrigine and only 70% for the i.s.; this explained an observed difference of ~14% in the ratio-derived slope between plasma and brain. Besides the matrix, plastic or rubber is a well-known source of losses, by adsorption (cf. art. #B-9).

Concluding comments. - Interferences can be endogenous or exogenous. Whilst initial validation inspires confidence that an assay is specific for the analytes of interest, constant checking is required to preserve confidence during routine operation at all drug-development stages leading to documented method-validation data that meet 'GLP' standards.

References

1. Brooks, M.A. & Weinfield, R.E. (1985) *Drug Develop. & Indust. Pharm. 11*, 1703-1728.
2. Taylor, J.K. (1983) *Anal. Chem. 55*, 600A-608A.
3. Inman, E.L., Frischmann, J.C., Jiminez, P.J., Winkel, G.D., Persinger, M.L. & Rutherford, B.S. (1987) *J. Chromatog. Sci. 25*, 252-256.
4. Garty, M., Steinmetz, Y., Rosenfeld, J.B. & Goldstein, D.S. (1988) *J. Chromatog. 430*, 123-127.
5. Garattini, S. (1985) *Clin. Pharmacokin. 10*, 216-227.
6. Boppana, V.K., Heinemann, R.K. *et al.* (1984) *J. Chromatog. 317*, 463-474.
7. Middleditch, B.S. & Zlatkis, A. (1987) *J. Chromatog. Sci. 25*, 547-551.
8. Middleditch, B.S. (1989) *Analytical Artifacts GC, MS, HPLC, TLC and PC*, J. Chrom. Library, Vol. 44, Elsevier, Amsterdam.
9. Williams, D.T., Lebel, G.L. & Benoil, F.M. (1981) *J. Assoc. Off. Anal. Chem. 64*, 635-640.
10. Webster, R., Noonan, P., Markos, C., Bible, R., Jr., Aksamit, W. & Hriber, J. (1981) *J. Chromatog. 209*, 463-466.
11. Coburn, S.P. (1987) *Clin. Chem. 33*, 1297.

#A-6

THE VALIDATION OF BIOANALYTICAL ASSAYS

R.J.N. Tanner

Glaxo Group Research, Ware, Herts. SG12 ODJ, U.K.

Editor's Note.- *The author was the leading discussant following the 'validation' presentations at the Sept. 1989 Forum, and now alludes to remarks by other discussants (see also later, p. 88). The article opens with key questions that govern validation procedures such as those presented in preceding articles (see also the Questionnaire that concludes D. Dell's article, #A-2). The procedure used in the author's laboratory is then described.*

Bioanalytical assays are used to determine the concentration of analytes, particularly drugs, in biological fluids. The results of these analyses are frequently presented to governmental regulatory authorities in support of applications to market new drugs or new formulations. It is essential that analysts know how the assays perform in routine operation. The reliability of the analytical data must be established, because they are used to calculate a range of pharmacokinetic parameters that are critical to the review of the performance of a drug or a formulation. The scientific importance of validating bioanalytical assays has been recognized by the regulatory authorities, and some guidelines have been published [1] and others drafted [2].

Having characterized the performance of an assay, an analyst must then demonstrate that the performance is maintained during the analysis of samples. This article discusses the validation procedure and QC* control process that has been in operation in the author's laboratories, and its sub-contracted laboratories, since 1986. Attention is also paid to discussion points made at the Forum that led to this book.

VALIDATION

The FDA guidelines on assay validation are very brief: ".....documentation should be provided of the sensitivity, linearity, specificity, and reproducibility of the analytical method, including sample chromatographs, recovery studies etc." [1]. Fortunately, interpretation of these guidelines, and

---

**Abbreviations*.- i.s., internal standard; r, regression coefficient; FDA, Food & Drugs Administration (U.S. Public Health Service); QC, quality control; QL, limit of quantification.

further explanation of the *etcetera* required, have been published by members of the FDA [3, 4]⊗. The assay parameters that have to be determined as part of a validation exercise are: accuracy, precision, sensitivity, specificity, linearity, recovery, repeatability and, if the assay is to be used in more than one laboratory, reproducibility [4]. The requirement to establish these parameters is not usually the subject of discussion, but the procedures used to calculate them are frequently debated by analysts, for example:

#How wide can the calibration range be?
#How do you regress the calibration data?
#Do you need an i.s.?
#How do you determine the QL?
#How do you report the data?

One set of answers to these questions is presented in the following sections.

A VALIDATION PROCEDURE

The following validation procedure is designed to provide the data necessary for the calculation of the assay parameters.

The calibration range

Firstly it is assumed that the assay has been developed and optimized and that an estimate of the QL has been made. The concentrations of the calibration standards are chosen with the proposed QL as the lowest standard and the proposed upper limit of the assay as the top standard. Forum discussants agreed that the lowest standard should be at the QL for the assay and that there should be no extrapolation below the calibration range when calculating analytical results. There was less unanimity about extrapolation above the top of the calibration range; some people were prepared to extrapolate over a considerable range, provided that linearity had been demonstrated over the extended range. Others extrapolated only a small amount, ~10% above the top standard, or not at all. Samples containing analyte at a concentration well above the calibration range are diluted with control matrix and re-assayed. It may be possible to predict which samples these will be, and therefore they may be diluted automatically prior to analysis.

The calibration standards are prepared in control biological material from stock solutions derived from a single weighed sample of the analyte. Successive dilutions of stock solutions are monitored spectrophotometrically or chromatographically. A second set of stock solutions derived from a second weighed sample of the analyte are used to

⊗See pp. 89-90

prepare QC control samples and samples used to test the assay during the validation procedure.

The number of calibration standards required will depend on the range of the calibration line. Routinely 5 calibration standards are prepared in duplicate for a 20- to 30-fold calibration range. A wider range would require more standards. The standards are not uniformly spaced throughout the range: usually there are more at the lower concentrations.

The method of calculating the equation of the calibration line was extensively debated at the Forum. Linear regression of the calibration data is the major technique used, and many analysts favour weighted regression, such as $1/x$ or $1/x^2$. Weighting is particularly appropriate for large calibration ranges where the residual at the top concentration, say 5% at 100 ng/ml, can outweigh the response at 1 ng/ml.

Internal standards

These are used to correct for analytical processes that may vary from sample to sample. However, for many analyses the assay-to-assay reproducibility is precise and accurate and an i.s. is not required[⊗]: indeed, it may even reduce the performance of the assay, because the errors involved in adding, extracting and measuring the response of the i.s. are added to the errors in determining the analyte [5, 6]. Thus the need for an i.s. should be assessed before time is spent in identifying one.

A suitable i.s., if required, must have physical properties sufficiently different from the analyte to enable it to be detected independently from it - and from any endogenous compounds - yet be sufficiently similar to behave like the analyte during processing and analysis. Many HPLC assays, including those with a liquid/liquid extraction step, do not need an i.s., provided that the steps in the procedure are controlled. The transfer of a measured volume of the liquid phase is usually more precise than attempting to transfer all the phase. Some analytical techniques, however, do require an i.s., e.g. quantitative LC-MS where the detector response can change rapidly. In this type of assay, a co-eluting isotopically labelled analogue of the analyte is the best i.s. [7].

Within-batch (intra-assay) validation

The within-batch validation of an assay is carried out by analyzing, as unknowns, 6-fold replicate samples prepared in control matrix at each of the calibration sample

[⊗]a view held too by I.D. Wilson: see #ncA-4

concentrations. These spiked samples are analyzed against a duplicate calibration line. The results provide the data used to calculate the within-batch accuracy, precision and linearity of the assay. These are usually better than the between-batch parameters. Accordingly, if the assay is subsequently used to analyze samples from a bioequivalence study, the variation in the data due to the assay can be minimized by assaying both sets of samples from one person in the same batch.

Between-batch (inter-assay) validation

The between-batch validation is carried out on 4 different days. At least 2 of these batches must be similar in size to the batches of real samples that will be analyzed when the assay is running routinely. The batches comprise the following.-

(a) A duplicate set of calibration standards. The data from these are used to construct the calibration line against which all the remaining samples are measured.

(b) A duplicate set of spiked samples at each of the calibration standard concentrations made up in control matrix. Comparison of the results for these samples over the 4 days gives the between-batch accuracy and precision data.

(c) Duplicates of stock solutions containing the analyte at concentrations similar to the spiked plasma samples. These stock solutions are injected direct and so must be made up in an appropriate solvent. Comparison of the analytical results for these stock solutions with those for the spiked samples gives the recovery of the assay procedure throughout the calibration range. Recovery does not have to be 100%, but it does have to be reproducible and consistent throughout the range. An inconsistent recovery would indicate a problem with the assay that may be resolved by further method development, or by the use of an i.s.

(d) Duplicate QC samples at 3 concentrations. The QC samples are made up in bulk and are used to monitor the performance of the assay when used routinely - prior to which, however, it is wise to check that they have been prepared accurately. The data from the QC samples will provide information that is used to establish the acceptance criteria for the routine assays.

(e) At least 6 different samples of control matrix. Discussion at the Forum confirmed the need to prove that matrix components, which may differ between individuals, do not interfere with an assay. In routine analysis this is confirmed by the analysis of a control, or t = 0, sample taken immediately before administration of the drug.

(f) Stock solutions of metabolites of the analyte, if available, and administered precursors such as pro-drugs. The chromatograms for these stock solutions will show whether the analyte is adequately resolved from other drug-related material. If not, a subsequent batch in the validation series should contain control matrix samples spiked with the interfering compound. If analysis of these samples shows that the compound is lost during the sample work-up, then the assay may still be reckoned specific; if not, the assay requires further validation. If a metabolite is only just resolved from the analyte it may be necessary to build resolution checks into subsequent assays to ensure that the assay remains specific. If the drug's metabolites are unavailable, but are known to be present in significant amounts in the matrix taken from dosed individuals, the specificity of the assay may have to be deduced from available information such as pKa, log P or absorption characteristics. An acidic metabolite is less likely than a basic metabolite to interfere in the assay of a basic drug.

(g) If the analyte is administered as part of a combination product, or is likely to be co-administered with other drugs, then it will also be necessary to demonstrate that these other compounds do not interfere with the assay.

(h) For at least 2 batches additional samples are added to make the batch size up to routine assay size. It is essential to demonstrate that the system can cope with a full-size batch, before any samples are committed. Potential problems such as insufficient disk space, build-up of back-pressure or drift of an electrochemical detector can be identified and overcome.

THE VALIDATION REPORT

Regulatory authorities require the results of a validation exercise to be documented, e.g. in a formal company report, which is then included in the analytical section of the Chemistry and Pharmacy part of a submission.

The analytical data are used to calculate the accuracy, precision and linearity of the assay within- and between-batch. Particularly important are the results for the lowest concentration standard - as this will be the QL, provided that the bias and imprecision are $<15\%$. The calibration line must have $r > 0.99$. The specificity of the method is deduced from the chromatograms obtained for the control samples, in respect of endogenous interference, and for spiked control samples, in respect of metabolites and co-administered compounds. Comparison of the results for the spiked samples with those for the corresponding stock solutions assayed directly determines the efficiency and precision of the sample preparation procedure. The time taken to carry out the assays and to process the data will indicate the rate of sample throughput to be expected for real analyses.

The validation report also contains the acceptance criteria that will be used to judge the acceptability of future analytical batches. It is important that these be clearly defined and easily applied so that as soon as a routine batch of analyses has been completed the analyst can quickly determine whether the assays were in control or not. If an assay goes out of control, further analyses should be suspended while the problem is resolved. Advance warning of a routine assay going out of control may be obtained from plots of the QC data [see Simmonds, #ncA-2]. Trends in the data, indicative of a problem, may show up before the assay actually goes out of control.

The validation exercise may have demonstrated that the assay is more precise at higher concentrations. Therefore the acceptance criteria established for the QC samples may be different for the low, intermediate and high concentrations. Our criteria permit a maximum bias of 15%, but some companies accept a bias of 20%.

QUALITY CONTROL (QC)

The quality, and therefore acceptability, of routine analytical results must be monitored to demonstrate that a batch of analytical data are valid. The first exercise is to examine each chromatogram to check that the peak integration was executed correctly. Occasionally changes in endogenous components in the sample or slight changes in the chromatography require that the data be re-integrated. If a chromatogram is unacceptable, the sample will have to be re-analyzed. The decision to re-integrate, or to re-analyze, must be recorded with the raw data. The next step is to calculate the equation of the calibration line which should have r at least 0.99. The line is then used to calculate the concentration of the QC samples. These results are then compared to the acceptance criteria. For a batch of analyses to be valid, 4 of the 6 QC samples must meet these criteria, with at least one at each concentration. Finally the analytical results for the samples are calculated. Results that appear markedly out of context with their neighbours are re-analyzed.

CONCLUSION

The requirement to validate and to control an assay is founded on a good scientific principle: an analyst must know how an assay has performed if he, or she, is going to make a scientific judgement based on the data produced.

This principle has been recognized by the drug regulatory authorities who require pharmacokinetic data to be supported by validation and QC data. A consensus view of what is required to demonstrate validation and control is beginning to emerge. Thus, discussion at the Forum identified the need to include all the routine calibration-line and QC data with suitable statistical summaries in the analytical report on a study. The Forum provided an ideal environment for further progress towards consensus.

References

1. U.S. Department of Health and Human Services, Public Health Service, Food and Drug Administration (1987) *Guideline for the Format and Content of the Human Pharmacokinetics and Bioavailability Section of an Application*, p. 8.
2. CPMP Working Party (August 1989) *Quality Control of Medicinal Products: Note for Guidance, Analytical Validation*, Commission of the European Communities, 111/844/87-EN.
3. Dighe, S.V. (1984) *Clin. Res. Practices & Drug Reg. Affairs 2*, 401-421.
4. Shah, V.P. (1987) *Clin. Res. Practices & Drug Reg. Affairs 5*, 51-60.
5. Curry, S.H. & Whelpton, R. (1978) in *Blood Drugs and Other Analytical Challenges* [Vol. 7, this series] (Reid, E., ed.), Ellis Horwood, Chichester, pp. 29-41.
6. Haefelfinger, P. (1981) *J. Chromatog. 218*, 73-81.
7. Oxford, J. & Lant, M. (1989) *J. Chromatog. 496*, 137-146.

Note by Senior Editor: *comments by D. Dell on points in the foregoing article appear* ***overleaf****, and the author's comments on a Questionnaire, supplementing comments by D. Dell, appear on* **p.** *88; see* **p.** *89 for some observations (cf. refs.* 3. *&* 4. *above) by FDA staff.*

COMMENTS by D. DELL on the foregoing art. (R.J.N. TANNER)
- *as sent by Senior Ed. to D. Dell* ***after*** *he had compiled his own art.*, #A-2

Calibration standards and quality controls

Many laboratories adopt an 'independent preparation' philosophy: one analyst makes up the calibration standards, and another analyst the QC samples, from the initial weighing right through to spiking and final dilutions. However, both analysts use the same batch of substance and control plasma. The purpose is to overcome any possible problems due to a 'personal bias' effect; i.e. the accuracy of the calibration or QC samples is assured. Such calibration and QC samples are analyzed together before routine analysis is started; we then use the values found for the QC samples from such determinations as the control values for subsequent routine analysis (i.e. we do not use the original spiked concentrations).

Between-batch (inter-assay) validation

(**c**) in Tanner's art.- Concerning recovery, many favour the use of 'matrix standards', instead of stock solutions made up in an appropriate solvent, as the 100% values. Matrix standards are prepared by adding a known amount of the substance in solution to an evaporated extract of control biological fluid. The purpose is to obviate any possible influence of the plasma components on the response of the substance, since such influence would lead to false results when 'simple standards', as described by Tanner, are used as the 100% values for calculation of the recovery. Many analysts have experienced the phenomenon of peak shape amelioration when a substance is chromatographed in the presence of biological-fluid components, compared to the injection of a pure solution. This may often be attributed to partial adsorption of the substance either on the walls of the vessel containing the sample solution or in the chromatography column itself. Such adsorption is often hindered by the presence of biological-fluid components, such as lipids and/or proteins.

(**e**).- We use 10 different samples of control matrix during validation, and provided that 8 of these samples show no 'serious' interference, we are satisfied that the method is specific enough, for that species, to allow further validation.

(**h**).- "The calibration line must have $r > 0.99$." Agreed, but r alone is a rather weak parameter for judging the quality of a calibration; more important, and revealing, are the deviations of the fitted values of the concentrations from the spiked amounts, especially at the low end of the calibration.

#ncA

NOTES and COMMENTS relating to

PRODUCING VALID AND ACCEPTABLE ANALYTICAL RESULTS

Forum comments relating to the preceding main arts. and to the 'Notes' that follow appear on pp. 83-86 and 88

Supplementary material furnished by Senior Editor:

- A calibration-line policy (E. Doyle): p. 86
- 'Outliers': p. 87
- FDA viewpoints: pp. 89-90
- Reinforcing citations: p. 90

NOTE that #ncB-9 (p. 221) is relevant to 'GLP'

#ncA-1

A Note on EXPERIENCES AT THE INTERFACE BETWEEN THE BIOANALYST AND THE REGULATORS

P. Hajdukiewicz

JENMARK Consultants, 26 Briar Road, Bexley, Kent DA5 2HN

Within a small non-innovative company, problems may occur at the bioanalyst-regulators interface, for several reasons. Generally, a small company that is not research-based may lack bioanalytical expertise, and so will make extensive use of either external consultants or one of the many specialist contract companies that provide these services. In addition, it is highly unlikely that significant knowledge on a particular drug molecule is available. The only recourse that such a company has is the information available in the public domain. Some is obtainable by literature-searching, but generally the overall picture obtained is sketchy. Another important concern, in relation to the company's size, is the cost of bioanalytical studies. In efforts to reduce this, an inadequate study design may be implemented, resulting in problems later raised by the regulators. However, such problems are commonly evident as potential problems at the design stage or on completion of a bioanalytical study. Examples of problems are now indicated in three areas.

1. Adequacy of power in bioequivalence studies.- In scrutinizing the design of such studies, regulators look for bioequivalence with the 'market leader' drug. In addition, within the statistical analysis they expect to see an assessment of the power of the study to detect differences between the two formulations. Some regulators ask for a power estimate for the AUC (area-under-curve) over the dosage interval, the sampling time or to infinity, whilst others ask for specific power calculations at specific sampling times. Assuming that bioequivalence exists, the main reason for inadequacy of power is that there were too few subjects, usually because of cost constraints. Thus, to reduce costs no pilot study may have been performed and hence no estimate of the number of subjects required could be made. In addition the arbitrary number chosen may be minimal, simply to reduce costs. The lack of power in one particular study was overcome by the fact that an analysis for the two active metabolites produced from each formulation was acceptable (the plasma time-curves being near-coincident).

2. Inadequate validation and the consequences.- For formulations that contain an active material of established efficacy, the regulators often accept a well designed bioanalytical study to support the formulation's efficacy, without the company undertaking formal clinical trials. This has commercial advantages for the company but is not without risk, since the efficacy must rely solely on the bioanalytical results. Companies may undertake clinical work, but very often this is only to support promotional activity and the studies are often not conclusive evidence of efficacy, either because of their design or because the disease itself shows a pronounced placebo response. Clinical studies are of course a heavy expense for a small company; so it is imperative that the bioanalytical study be fully validated. When it is contracted out, particularly close scrutiny must be applied to all aspects, with regular audits; the start of the study must await full validation of the methodology. Failure to adhere to this can result in significant investment by a company, and in hopes, based on questionable bioequivalence analyses, for a product which ultimately is doomed to failure due to lack of clinical efficacy.

3. Inadequate characterization of a raw-material source.- When a non-innovative company embarks on introducing a formulation of a known active substance the material is purchased from a different source than the original innovative company, and may well have been produced by a different synthetic route than the original. This will obviously result in a different impurity profile in the finished product. This must be considered and adequately investigated before embarking on bioanalytical studies. The regulators may not accept this new material source for various reasons, and an alternative source may be unacceptable until bioanalytical work has been done. A company may therefore have to repeat an extensive series of bioanalytical studies.

Fraudulent data centred on bioanalysis needs consideration, although not of direct concern to the bioanalyst. Several generic companies have been prosecuted in the U.S.A., and eventually this is bound to occur in Europe. Fraud may range from slight manipulation of data to wholesale invention. It is in this area that the experienced bioanalyst can be extremely helpful to a company, since detection is very difficult especially if unsuspected. Sometimes it is only by logical evaluation that it can be suggested that the 'findings' are not possible on theoretical grounds. In other cases suspicion may be generated by examination of the between- and within-subject variance for the drug *vs.* the comparator or literature values. Possibly the problem is soluble only by extension of the Medicines Surveillance Department's ambit to include bioanalytical evaluations of marketed products.

#ncA-2

A Note on

QUALITY CONTROL OF ROUTINE ASSAYS AT CONTRACT FACILITIES

R.J. Simmonds and S.A. Wood

Clinical Research and Development, Upjohn Ltd., Fleming Way, Crawley RH10 2NJ, U.K.

At Upjohn the main concern in respect of assays is with method development; all routine assays are contracted out. These assays are for novel drugs and their metabolites, exemplified for antibiotics in companion articles (#B-2 & #B-4). For such assays there do not exist established and tested QC[⊗] procedures, as would be the case with, for example, assays for generic drugs. QC procedures have to be devised for each new assay.

We have not, for such assays, produced data that have fully satisfied statisticians, and in this sense our QC/QA must be lacking. However, we have produced data that have satisfied colleagues such as clinicians and pharmacologists and enabled drug development projects to proceed.

It is the purpose of QC to assure us that the data are good enough for their intended use, and to convince the Regulatory Authorities that data are adequate for their purpose and have been produced in an appropriate manner. QC procedures will be different in detail for each assay and for each assay application. It follows that setting up appropriate QC procedures for data generation and review is not a task performable automatically. - Scientific judgement is required. It is not a secretarial task, and not something that can be devolved to an OCL's own personnel, certainly not to their QA department. They have to be told what is acceptable. Choice of an OCL is important, since a contract facility's expertise and experience in a particular type of assay will influence the QC/QA procedures that are necessary to support data. Chamberlain [1] has reviewed well the considerations in choosing an OCL. QC considerations need to be involved in the in-house development and validation of the method and formulation of analytical support as a protocol, and at the OCL in setting up the assay, running it routinely and reporting the data. Here I am taking a very broad view of what 'QC' means.

[⊗] *Abbreviations.*- GLP, Good Laboratory Practice; OCL, outside contract laboratory; QA/QC, quality assurance/control.

Method development is all-important, and the following considerations will ensure, as far as is possible, that routine QC can be kept to a minimum, or that - in terms that managers understand - subcontracting will be maximally cost-effective. Prescribing appropriate QC procedures hinges on a good in-house working knowledge of the already validated method. In the in-house design or modification of the method for OCL use, aims must include simplicity and, even at the expense of sensitivity, ruggedness; some speed may have to be sacrificed to assure performance. It is impossible to control a 'sloppy' method, and time involved in in-house validation may be amply repaid; no amount of QC can rectify a poor method, and overall assay performance may well be poorer at the OCL - an important point when setting standards for the routine assay.

Validation in-house must reckon that most OCL's are loathe to trouble-shoot; its purpose is to test the method, to comply with GLP, and especially to get to know the method so that there should be few surprises when the assay is put out to contract. A further consideration is that with a method fully validated in-house the validation at the OCL can be simplified.

The purpose of validation at the OCL is really to ensure that the method works there: it is not a full validation exercise as such, and the emphasis is on the practicalities of the method. Factors which need checking, *in turn*, include:
- (1) system performance, e.g. column capacity, efficiency, peak skew (A_S) (or column asymmetry) and detector response;
- (2) method performance including: extraction recovery, stability of samples and extracts, and interferences;

  and, using QC's and perhaps some blind samples provided by us and running 3 standard curves:
- (3a): overall assay performance, e.g. linearity, accuracy, repeatability, inter-day and intra-day variability, assay drift;
- finally (3b) results with at least one *real* sample.

Real samples are run only when (1)-(3a) have been satisfied in turn: they are ineffective for sorting out problems, and will not necessarily behave as well as the spiked samples already used.

A Bioanalytical Protocol, important to Regulatory Authorities, explicitly defines the work expected of the OCL:
- the scope and purpose of the Projects;
- who is responsible for what and where: e.g. (a) Project Leader with overall responsibility, normally the person who first developed the method; (b) the Liaison Scientist, usually the section leader at the OCL, responsible for the assay

and its routine operation at the OCL; (c) Project Deputy at the OCL; (d) number and nature of personnel involved in Project, reporting to the Liaison Scientist;
- number and nature of samples;
- experiments involved in setting up and validating the assay;
- running of routine assay: number of QC's and calibration samples, acceptable ranges for calibration samples; concordance between duplicates (the risk with singlets is inordinate, and random duplicates are statistically unsound);
- the reporting situation;
- default options: what to do when things go wrong.

This document not only fulfils GLP but also makes it as clear as possible what we expect the OCL to actually do. When dealing with OCL's there can be a problem of poor communication, and the Protocol needs to be commented on by the OCL in draft form. It then becomes a mutually agreed plan of what is necessary and feasible. Amendments can be made later, if necessary.

Routine-assay design obviously should aim at the best ratio that is feasible between actual samples and calibration and control samples. The design of a routine analytical run might optimally have the following guidelines.
- (1) The run size is defined by the number of samples that can be prepared in a day, typically 30-40 in duplicate (see foregoing comment).
- (2) The calibration curve entails 3-4 samples in triplicate to best 'pin down' the curve.
- (3) One QC is set up for each order of magnitude covered by the assay; the QC's are prepared in a batch and stored as aliquots with the samples.
- (4) A typical run comprises: (i) a newly prepared calibration curve, (ii) a reagent blank, (iii) QC's, and (iv) samples in duplicate, say 1 QC per 10 real samples whilst keeping constant the total numbers of QC's at each level (so that the averages of QC values from different runs can be meaningfully compared).

Criteria for particular parameters (those that have been tested and defined during setting-up the assay) indicate, from data obtained as above, whether the values we set for the routine assay are achievable in practice. Limits that we might set are exemplified:
- system performance: N = 10,000-12,000; A_s = 0.9-1.1; k' 2.5-3.0 (definable at beginning and end of run);
- method performance: recovery ≥80%, no interfering or long-running peaks, clean reagent blank;
- assay performance: (**i**) r (linearity) ≥0.99, slopes ±10% of expected, acceptable (small or zero) intercept; (**ii**) QC's:

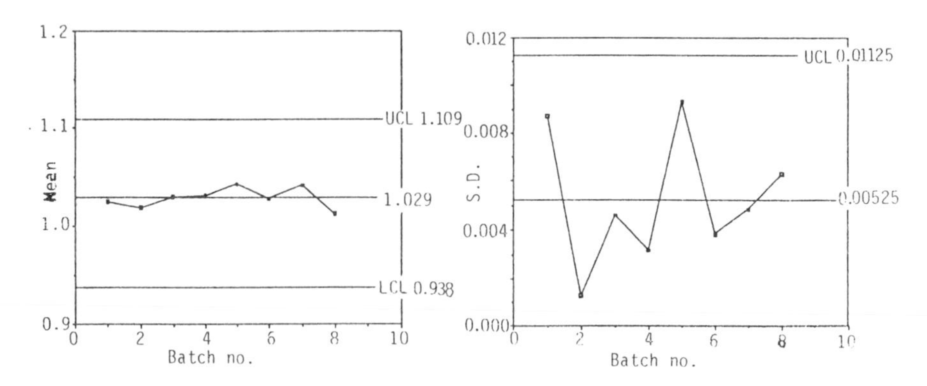

Fig. 1. Example of QC data: *left*, sample mean; *right*, S.D.'s. From trospectomycin studies (#B-2). UCL/LCL, upper/lower confidence limits.

#average concentration say ±5% of expected, and #no two successive QC values >2 C.V.'s from true values in same direction, nor a single QC >3 C.V.'s from true value (these assessments indicate whether a particular run needs repeating); #duplicates within 20% of each other.
QC limits should be based on values obtained during method development and in-house validation.

Retrieval of aberrant runs.- If a run discloses problems, its acceptability can be judged by a logical progression of checks:
- inject standards, check system parameters, correct faults if possible;
- re-inject QC's, check concentrations;
- plot sample position in run *vs.* concentration (any trends?);
- re-run some calibration samples, QC's and repeat samples in duplicate;
- re-run extracted samples;
- repeat complete assays;
- re-validate method: look for continuity of QC values.

In general, good robust analytical methods run without any disasters, but one must be on the alert for subtle and insidious long-term trends. Hence it is important to plot control graphs (example: Fig. 1) of QC data as they are generated during extended assay work. Ultimately we want to be able to say about the data, assuming that the method is sound at the outset, that the individual values do not depend upon *when* the assay was done and *where* in a routine assay run the sample was put. Relatively small random fluctuations are not considered problematic. We realize that such a problem could be circumvented by going to the trouble of randomizing samples. The possible juxtaposition of high- and low-dose samples that may then occur can obviously create problems, as OCL scientists are less able to judge aberrant results. Importantly, their morale may be jeopardized insofar as they are not being regarded as competent and experienced collaborators.

The outcome: reporting of data.- We are now in a position to define what we mean by 'valid data', reportable to the end users such as biopharmacists and clinicians. The procedures and criteria will have been explicitly stated in the Project Protocol, as follows.
- *For ongoing studies:* #results generated by routine assays that run within specifications comprise valid data, and can be certified by the OCL Liaison Scientist; #where problems are minor, data can be accredited by the Project Leader or Deputy, justified by scientific judgement; #only QC data are sent to the Project Leader, to assure that all is well and for liaison with clinicians, pharmacologists or other end-users of the data.
- *For finished studies:* #only in the Final Report do complete sets of certified data appear.

Even if there was reporting of data on a day-to-day basis, all projects need a Final Report, at least for Regulatory Requirements. This should feature the following, not necessarily in this order:
- Introduction: background to the project;
- formal description of method, suitably the Standard Operating Procedure (SOP);
- the Protocol and any amendments;
- the results of setting-up and 'validation' at the OCL;
- certified data;
- summarized QC data;
- QA compliance reports (from the Contractor's Department);
- a QA statement from the Contractor's own QA Department.

Concluding comments

The various points specified or advocated above are mostly quite obvious. Protocol amendments allow reasonable changes to be made in any respect, including GC and QA procedures. Flexibility and feedback are always necessary to set up methods well; otherwise QC can become a nightmare.

Getting good results from OCL's is, in our experience, best achieved by observing the following guidelines.
- Only subcontract assays that you have experience of.
- If necessary, simplify the method for subcontracting; sacrifice sensitivity for robustness.
- Treat OCL staff as collaborators of experience.
- Establish an agreed Project Protocol, helping ensure the requisite good communication.
- For routine assays set appropriate specifications, not too tight; leave room for scientific judgement.
- Define what is meant by 'valid data'.
- Check QC data as they are generated, and look for trends, whilst not being unduly concerned with random effects.
- Setting-up is cardinal and should not be skimped.
- Try to fulfil GLP by doing good science.
- The aim of QC is to back up scientific judgement; data need only meet *a* reasonable quality standard.

Reference

1. Chamberlain, J. (1985) *Analysis of Drugs in Biological Fluids*, CRC Press, Boca Raton, FL: see pp. 168-175.

#ncA-3

A Note on

QUALITY AND PRODUCTIVITY IN THE BIOANALYTICAL LABORATORY

Graham S. Clarke

Bristol-Myers Squibb, International Development Laboratory, Moreton, Merseyside L46 1QW, U.K.

With increasing stringency of product-registration requirements the workload in many bioanalytical laboratories has escalated, and it has become desirable to improve sample throughput without increasing the number of analysts. The efficiency of a laboratory can be determined by objective assessments of both the quantity and the quality of data produced: if both are to be improved, subjective assessments (e.g. "we are working very hard") must be superseded. Here we describe, with the outcome, how quality and productivity have been assessed in this laboratory.

Laboratory description.- At the beginning of this appraisal the section consisted of a graduate supervisor and two non-graduate analysts. Studies comprised 500-1000 biological samples (mainly serum or plasma) using methodology developed within the section. The laboratory had two HPLC systems, one automated with an autosampler and integrator and the other with manual injection and peak-height recording.

Initial evaluation.- Baseline data for quality and productivity were obtained late-1987 by retrospectively evaluating, for each study, the assay procedure and the parameters indicated by the column headings in Table 1. Pre-sample validation assays and pharmacokinetic analysis of results were excluded, but not calculation of results nor sample reassays as stipulated when a senior investigator considered that the original result was erroneous or failed to give a smooth pharmacokinetic profile. (For the problem of 'analytical outliers', see ***Appendix***.)

Table 1. Quality and productivity data: 1987, aggregated results for different analytes.

Assay procedure	Samples, total no.	Assay rate⊗	Repeat rate, %
Dilute; manual injection	300	13	0
Solvent extraction; manual injection	85	13	0
Solvent extraction; automated injection	500	17	17
SPE*, automated injection	1200	26	9
RIA (Radioimmunoassay)	600	35	0

*solid-phase extraction ⊗samples/analyst per day, excl. controls

Table 2. Quality and productivity data: 1988 ('Year 1').

Assay procedure (always automated injection and electronic data analysis)	Samples, total no.	Assay rate	Repeat rate, %
Dilute/inject (Analyte A)	43	6	0
Dilute/inject (Analyte B)	156	39	0
SPE/inject (Analyte C)	30	9	0
Solvent extraction/inject (Analyte D, Study 1)	530	15	15.7
Solvent extraction/inject (Analyte D, Study 2)	800	15	7.6
Solvent extraction/inject (Analyte D, Study 3)	530	16	8.0

Evidently (Table 1) for some methods where injection was manual there was no difference in the assay rate between solvent extraction and mere dilution. Thus the rate-limiting step was the chromatography, and a decision was made to develop methodology which had as short a run time as possible and so obviate this inefficiency. This was achievable by modifying the mobile phase or using a shorter HPLC column.

The finding with solvent extraction that using an auto-sampler and integrator increased the assay rate from 13 to 17 (+30%) was used in support of the purchase of a second automated HPLC system. SPE instead of solvent extraction further increased sample throughput (~+50%) and, moreover, reduced the repeat rate by almost 50%, providing an example of how improving quality also improves productivity. The highest sample throughput was achieved with RIA (Table 1), as expected: sample preparation was simpler than for HPLC, and there were not the time limitations of a chromatographic system.

Based on these observations, quality and productivity objectives were set for the group: a minimum of 13 samples/analyst per day, with a maximum repeat rate of 10%. So as to achieve this, the automated-equipment purchase proposition was implemented, and all methods developed were to be initially evaluated using SPE with short HPLC columns.

'Year 1' performance was evaluated at the end of 1988, as in the previous year, in relation to the objectives (Table 2). During 1988 the work pattern was dominated by 3 studies on the same drug substance, perforce with solvent extraction since attempts to develop a method using SPE were unsuccessful. As Table 2 shows, when the number of samples was low (<50) the assay rate was below the target: each analyst has to become familiar with the assay, and optimized sample throughput has to await this. Our efficiency data indicate that the number of samples required for familiarization is ~100. As anticipated, the analysis rate is sub-optimal early in a study, increasing with time to a maximum.

For each of the 3 major studies in 1988, the sample throughput objective was exceeded by at least 2 samples/day. This increase may seem insignificant, but this represents a saving of one person-week per study. In Study 1 the percentage of samples re-analyzed was greater than anticipated; but the cause was discovered to be an instrument problem: the autosampler was delivering inconstant sample volumes. Once this fault was corrected, acceptable repeat analysis rates were achieved in Studies 2 and 3. This highlights the importance of maintaining each component of the equipment in good condition and using system suitability procedures to monitor the performance of the analytical system. Any savings achieved by increased productivity can be lost through poor quality of the system.

'Year 2' performance is still being appraised, but several points are already emerging. The emphasis on developing SPE methodology has continued; 4 new methods were developed in 1989, all with SPE for sample preparation. One of these methods has served for a large study (>1000 samples) wherein the rate for samples/analyst per day was 23: this significant surpassing of the target rate was due not only to use of SPE but also to the short HPLC time (<6 min for both drug and metabolite). There was, however, a high repeat rate (15.3%); this emphasizes that getting high throughput needs care.

Concluding comments.- We have realized that by reviewing our approach to bioanalysis we can significantly improve the efficiency of producing data and their quality. Evidently method development can be directed objectively rather than rely on the subjective decisions of the analyst developing the method. We reckon that such quality and productivity studies are of value in all laboratories that aim to operate efficiently.

APPENDIX *relevant to data-quality consideration in foregoing art.*

USE OF THE NAIR CRITERION FOR DETERMINING OUTLIERS

G.S. Clarke (Bristol-Myers Squibb Internat. Dev. Lab., Moreton)

In biopharmaceutical analysis, as in other analytical areas, it may be difficult to decide whether to accept or reject individual values from a set of replicates. The bioanalytical literature lacks guidance, and practices in individual laboratories range from purely subjective to sophisticated statistical techniques. In our laboratory, analysis of a sample can yield a set of triplicate results of which two are close and the other is apparently an outlier - in which case it is rejected and the mean of the other two is used; otherwise the three are averaged. However, when two operators judge the same data, sometimes there will be disagreement. Among statistical tests examined to overcome this problem, the Nair criterion [1] was found to accord most often with the author's judgement. This criterion has been chosen because it allows extreme observations in only one direction to be rejected, a situation often addressed in biopharmaceutical analysis; its use in a basic computer program (i) enables easy use of the rejection criterion, and (ii) ensures between-operator consistency.

Application.- The requisite external estimate of the S.D. for the method is provided by controls that are co-assayed with the samples. At least 11 estimates are required for the control data to obtain a **t** value from the Extreme Studentized Deviate Table. A choice is made of the probability α that the analyst is willing to take of rejecting an observation that really belongs to the group. The Table furnishes $t\alpha$ (n, v) where n is the no. of observations for the sample and v is the no. of degrees of freedom for Sv, the S.D. for the control samples.

The algorithms used in the program are defined thus, for observations which are considered rejectable [2]:

- if too large, compute: $t = (X - \overline{X})/Sv$
- if too small, compute: $t = (\overline{X} - X)/Sv$

If **t** exceeds $t\alpha$ (n, v), the observation must be rejected and the mean of the remaining values quoted. If the observation cannot be rejected, all values must be included in the calculation of the mean.

References

1. Natrella, M.G. (1963; reissued 1966) *Experimental Statistics*, Natl. Bureau of Standards Handbook **91**, 17-5 [Washington, D.C.]: see p. 170.
2. Clarke, G.S. & Robinson, M.L. (1988) *J. Pharm. Biomed. Anal. 6*, 317-319.

#ncA-4

A Note on

OBSERVATIONS ON THE USEFULNESS OF INTERNAL STANDARDS IN THE ANALYSIS OF DRUGS IN BIOLOGICAL FLUIDS

I.D. Wilson

Drug Kinetics, ICI Pharmaceuticals, Mereside,
Alderley Park, Macclesfield, Cheshire SK10 4TG, U.K.

There is a widespread perception that an internal standard (i.s.) is an essential feature of bioanalytical methods based on chromatographic analysis. In our view the i.s. should be a close structural analogue of the analyte with, as near as possible, identical spectroscopic characteristics and very similar physicochemical properties (such as pKa and log p, i.e. partition ratio). Such provisos were coupled with cogent advocacy of i.s. use in an article by J. Vessman [1], and in another by K.H. Dudley [2] who warned against using an *N*-dealkylated or phenolic drug-related compound since it might turn out to be a metabolite of the drug - as was the case when the *m*-hydroxy isomer was tried as an i.s. for assaying a drug metabolite having a *p*-hydroxyphenyl moiety.

To perform its function of compensating for variability, the i.s. should be added to the sample before the first of the successive assay steps, which typically comprise aliquoting, diluting, extraction, chromatographic loading (manually or by auto-injector) and detection. Reminiscing about the early days of quantitative analysis by GC, sample introduction was then a major source of variability and the i.s. was necessary to enable this to be corrected for. However, with the improved accuracy and reproducibility of modern autosamplers for GC and HPLC much of this variability has been removed. Given that this is so it is difficult to support the still common practice of adding an i.s., often with little structural similarity to the analyte and with different spectroscopic properties, only at the end of sample preparation immediately prior to injection (the use of the unqualified term 'internal standard' then being arguably a misnomer [3]).

So deeply ingrained is the belief that the i.s. is both necessary and beneficial that only a few analysts (e.g. [1, 2, 4, 5] & Tanner, #A-6, this vol.) appear to have devoted much time to determining the effect of these compounds on the accuracy and precision of their methods. It should, however, be self-evident that for a method in which all the steps are accurately and precisely controlled the use

Table 1. Assay validation for a drug spiked into urine and, without *or with* i.s.-based correction, analyzed by HPLC (n = 6).

µg/ml	Mean ± S.D. (µg/ml)		Recovery, %		C.V., %	
0	0.5 ±0.3	*0.5 ±0.3*				
1	1.3 ±0.2	*1.3 ±0.2*	130	*130*	15.4	*15.4*
3	3.6 ±0.4	*3.5 ±0.4*	120	*116*	11.1	*11.4*
10	10.6 ±0.6	*10.3 ±0.5*	106	*103*	5.7	*4.9*
30	34.9 ±1.5	*33.9 ±1.8*	116	*113*	4.3	*5.3*
100	107.2 ±2.6	*104.1 ±2.9*	107	*104*	2.4	*2.8*
300	318.9 ±8.1	*316.1 ±9.4*	106	*105*	2.5	*3.0*
1000	1051 ±29.2	*1037 ±4.4*	105	*103*	2.8	*0.42*

of an i.s. will bring few if any benefits. Indeed it can be argued that for such methods additional errors may arise in a step such as preparing or adding the i.s. and will actually lead to a decrease in accuracy and precision.

Example #1.- This is illustrated by the results in Table 1 for samples spiked during validation of an assay for a carboxylic-acid drug. The method merely involved the aliquoting of samples into WISP (Waters Assoc.) autosampler vials, addition of alkali to hydrolyze ester glucuronides, and then direct injection of aliquots onto an RP-HPLC column. The i.s.-corrected values (Table 1, *italics*) are similar to values which ignore the i.s., perhaps not surprisingly in view of the simplicity of the method and the accuracy and precision of modern autosamplers. Provided that the quality of data provided by the method is carefully monitored by the judicious use of quality-control samples, the case for using an i.s. for this type of method hardly seems resounding.

Example #2.- The use of an i.s., even a close structural analogue, will not assuredly improve an inherently poor assay such as that for which Table 2 gives values without and *(italics)* with i.s.-based correction. The method involved liquid-liquid extraction of a weakly basic compound from plasma followed by GC-NPD. The assay clearly suffered from unacceptably poor precision. What is noteworthy is that the i.s., although carefully selected, hardly improved the results. These data were obtained in the course of method validation, and much effort was devoted to unsuccessful attempts to improve the situation. Before this could be achieved the development of the compound was terminated (a fortunate but unpredictable event!).

Table 2. Assay validation for an analyte spiked into plasma and, without ***or*** *with* i.s.-based correction, analyzed by GC-NPD (n = 6). ND, not 'detectable' (zero intercept).

µg/ml	Mean ± S.D. (µg/ml)		Recovery, %		C.V., %	
0	N.D.	*N.D.*				
4.5	6.6 ±2.3	*6.4 ±2.3*	147.0	*142.2*	34.8	*27.8*
15	16.1 ±2.3	*15.2 ±3.1*	107.3	*101.3*	14.3	*20.4*
45	44.9 ±10.6	*42.6 ±8.1*	99.8	*94.7*	23.6	*19.0*
150	141.4 ±17	*146.8 ±33.7*	94.3	*97.8*	12.0	*23.0*
450	466.6 ±69	*469.8 ±44.8*	104.0	*104.4*	14.8	*9.5*
1500	1464.8 ±150.5	*1490.2 ±78.1*	97.7	*99.3*	10.3	*5.2*

DISCUSSION

It is not our contention that i.s.'s are of little value and should therefore be abandoned, far from it. However, they will not invariably confer improvements in accuracy and precision and are certainly not a panacea. It should be part of any method validation to actively investigate the effect of the proposed i.s. on the overall performance of the method and to demonstrate its usefulness. On the basis of the outcome the analyst can then make an informed judgement on whether or not an i.s. is needed in that particular method.

It is our belief that, given the notably good reliability and performance of modern autoinjectors, there should rarely be a need for an i.s. in a simple 'aliquot and inject' assay and that even in much more complex assays i.s.'s may not be mandatory. Rather there is a danger of over-reliance on an i.s. to automatically correct for the failings of an assay, leading to complacency.

Of course there are additional benefits which arise from the use of an i.s. besides a potential improvement in quantification. For example, a spiked-in i.s. should appear in every chromatogram and its evident presence therefore provides some reassurance that (1) the method has been carried out according to the 'SOP', and (2) that the absence of a peak for the analyte whilst the expected i.s. peak is manifest means absence of the analyte from that sample rather than failure of the method. In addition, the peak shape found for the i.s. may be useful in demonstrating that the chromatography is still acceptable. Whether these comprise sufficient reasons to retain an i.s. in a method where it does not make a useful contribution to quantification is debatable.

Clearly there are circumstances where an i.s. is mandatory, particularly quantitative LC-MS where detector response can change rapidly (see Tanner's art., #A-6). Obviously an i.s. should always be used where its presence results in improved quantification.

References

1. Vessman, J. (1981) in *Trace-Organic Sample Handling* [Vol. 10, this series] (Reid, E., ed.), Ellis Horwood, Chichester, pp. 341-347. *(This vol. is now out-of-print.)*
2. Dudley, K.H. (1981) *as for* 1., pp. 336-340.
3. Reid, E. (1978) *as for* 4., pp. 61-76.
4. Curry, S.H. & Whelpton, R. (1978) in *Blood Drugs and Other Analytical Challenges* [Vol. 7, this series] (Reid, E., ed.), Ellis Horwood, Chichester, pp. 29-41.
5. Haefelfinger, P. (1981) *J. Chromatog. 218*, 73-81.

Comments on #**A-2**, D. Dell - METHOD VALIDATION *(see also p. 88)*

R. Whelpton remarked that sometimes we fit straight lines to what should be curves, and wondered how one can say whether a model requires weighted 'linear' calibration and a non-linear model, e.g. quadratic - to which **D. Dell replied** that this may be ascertained during method development, when the relationship between S.D. and concentration may be closely examined. **Added by U. Timm** (colleague of Dell): we use quadratic regression only when the curve for standards proves to be non-linear, as is often the case with GC-NPD; with HPLC-UV the curve is usually linear, and a quadratic regression curve will give too much weight to (e.g.) high-level points that are erroneous. **J.B. Lecaillon asked** whether log-log regression is similar (**Dell:** Yes!) to weighted $1/y^2$ regression, the latter being warranted by reproducibility results in a general review of his laboratory's methods; we disfavour a log-log approach with its hypothesis that the curve goes through the origin, although log-log regression could be advantageous if the curve is not a straight line. **Dell's response.**- We too prefer weighted linear regressions (we don't use log-log) and don't like to force a standards calibration through the origin; if it becomes clear that the relationship is non-linear we use a polynomial regression. Ref. [1] is helpful. **F. Van Rompaey was answered** on similar lines, in connection with compensating for heteroscedasticity. (Ref. [2] overleaf is pertinent. - *Ed.*)

P.R. Sallabank asked if in routine analysis the goal should be to run only singlicates and what checks would be needed. **Dell agreed** that singlicates save much time, but sometimes there is a 'wrong' result due to a methodological (i.e. non-random) error, or bias, as can come to attention through use of duplicates. If, after analysis of hundreds of samples, we feel that the method is robust (precise and accurate enough), we often take a decision to analyze only a portion of subsequent samples in duplicate (10% to 20%). **P. Logue remarked** that the preparation of Q.C.'s may be overdone: if a certain % of 'patient' samples are analyzed in duplicate, they will act as a secondary, independent and complementary source of Q.C. data. **Comment by J. Burrows:** Uhlein's approach to determine detection limits is still used, but less rigidly than before. The minimum S.D. is determined from the average of the 4 lowest calibration values. The detection limit is revised upwards, taking into account of accuracy limit (80-120% at the detection limit) and background (blanks). The assay's working range thus becomes higher.

Dell, replying to K. Borner: we do not accept a correction for losses during storage, believing that such a procedure is unreliable and can lead to inaccuracies. **Remarks by**

R.J. Simmonds.- The stability of spiked samples, whilst providing some information, may differ from that of 'real' samples (which contain metabolites etc.); so whilst statistics allows *firm* statements to be made about spiked samples, this only means that some *scientific judgement* can be made about the stability of 'real' samples - which needs checking. Statistics could give a false impression of certainty.

1. Agterdenbos, J. (1979) *Anal. Chim. Acta 108*, 315-323.
2. E. Doyle (1986) in *Development of Drugs and Modern Medicines* (Gorrod, J., Gibson, G.G. & Mitchard, M., eds.), VCH, Weinheim, & Ellis Horwood, Chichester, pp.534-540. *For Ed.'s amplification, see 2 pp. further on; likewise for further observations on D. Dell's presentation.*

Comments on **#A-3**, H.M. Hill - METHOD VALIDATION

H.M. Hill, answering G.L. Evans.- The assay validation exercise is performed by a single analyst, apart from the preparation of Q.C.'s which involves another analyst. The procedure does not have allowance for variability due to the analyst, but if this (or system variability) is expected to be significant, this aspect could be incorporated in the validation program. **Reply to V. McNally.**- One can afford to delete most calibrators from a set (of 7 or 8); in fact it is generally best to go for a larger number of single calibrators than for a smaller number of duplicates, which minimizes the need for acceptance/rejection criteria to deal with any pair that differ significantly. Duplication of 10-20% of test samples shows a satisfactory correlation with the precision of the assay. For pharmacokinetic samples (**answer to query by R.J. Simmonds**) one should not use the mean of the duplicates since there will be a difference in precision from the data generated from singlicate assays; it is preferable to designate *before* samples are assayed which in a pair will be the pharmacokinetic sample and which is to be the confirming duplicate. The latter, serving for QC purposes, should be reported distinctly from the pharmacokinetic data; in fact it is usual for the duplicate to be in a separate assay batch, analyzed for QC purposes (and the results accord satisfactorily with between-batch validation data obtained earlier).

R. Calvert remarked that it is extremely dangerous to discard outlier values without a full investigation of reasons for anomalous results; many drugs show profiles which do not fit 'models' when sample frequency is increased. **G.S. Land asked** how possible variations in the matrix of the species investigated are accounted for in the validation. **Hill's reply.**- In connection with assessment of the 20% LOQ during assay validation, the number of animal/human blanks looked at may amount to 20-50.

Comments on #**A4**, F. Van Rompaey - A LABORATORY SYSTEM
#**ncA-2**, R.J. Simmonds - CONTRACTING-OUT OF ASSAYS

Remark by C. Town to F. Van Rompaey, who had reckoned that their extensive computer system didn't need validating since much of it was commercial software.- This software should in fact be validated in-house. **F.V.R. answering B. Meyer.**- The sample list indeed includes calibration and Q.C. samples, added to the list by the analyst. **Answer to R. Whelpton,** who had asked whether background chromatograms could reasonably be subtracted.- At least this is reasonable for calibration standards, where one assumes constant background.

R.J. Simmonds, in reply to H. de Bree and P. Logue.- The net time gain through contracting out assays by methods validated in-house may, at the best, be 60%. Overall assay performance in the OCL improves with time. **Comment by P. Logue.**- Taking into account the burden of developing and validating an assay in-house, one would reckon that most people who turn to an OCL do so because of intractable methological problems or of lack of suitable equipment (**Simmonds:** N/A to *our* contracted-out work!); it seems strange to carry out all the hard work in-house and then turn this over to the OCL.

Comments on #**ncA-3**, G.S. Clarke - QUALITY & PRODUCTIVITY
also (Appendix) 'OUTLIERS'

Replies by G.S. Clarke: to Tanner (1): by "processing" of a sample we mean taking it to a ng/ml result, *not* through to input into (say) a kinetics package; **to Simmonds and Lehr:** (2) "efficiency" cannot be meaningfully judged for method development or even, on a per-analyst basis, for routine assays, especially as different analysts spend different amounts of time on these two aspects; (3) our per-day rate is 17-25 samples/person, aiming to be in the upper part of the 8-30 range which seemed, from 'audience participation', to be the norm in other laboratories. **Replies to S. Wood and D. Dell.**- An analyst should be able to cope with 100 samples/day (total, including calibrants etc.) in a routine assay entailing SPE and HPLC; automatic running continues overnight, and a half-day per week is earmarked for tasks such as writing-up, GLP, archiving, tidying-up and equipment maintenance.

Remark by J. Burrows.- In our labs. (Hoechst, UK) typically 1 week is allowed to set up and report a study; the analysis rate is set at 2 batches/week, each 40-60 samples, and most analysts easily exceed this target. **H. de Bree** reckoned that per-sample analysis time (including associated tasks, e.g. reporting and archiving) ranges from ½-h for a very simple assay (mere dilution and injection onto the column) to 1-1½ h for a complicated assay, exceptionally 2 h if, for example, high sensitivity is demanded. **Clarke, answering R. Calvert.-** Our pre-1989 data (art. #ncA-3) mostly do justify our claim that analytical quality as well as efficiency was improving. We have not considered using a 'quality code' as distinct from a 'quality standards/objectives' approach. **Reply to P. Arnoux.-** Despite ever-tightening requirements, quality assurance is still compatible with productivity. Poor quality would increase the number of samples that have to be re-assayed, and conversely (increasing one's confidence in the data). However, there must be a 'pay-off', and a balance struck between quality achievement and acceptability of the time-scale.

Concerning 'outliers' within pairs, **Logue asked Clarke** how, in a pharmacokinetic context, he deals with an outlier that is not ascribable to an analytical problem. His **reply**, that the result would have to be accepted for pharmacokinetic analysis, drew a comment from the audience that his approach could be very misleading.

A POLICY, *pertinent to validation views (cf.* #**A-2**, D. Dell), *published by a company bioanalyst (SK&F, U.K.)* -

#*Points from ref.* [2] *on p. 84**, E. Doyle - "A weighted least-squares method for fitting calibration lines to heteroscedastic data......" exemplified by data from a simple HPLC assay

The weighted as distinct from the ordinary least-squares (WLS *vs.* OLS) regression procedure copes with 'heteroscedastic' data as observed in establishing a drug-level assay: replicates were "more variable at high concentrations than at low concentrations......the accuracy of the unweighted line is unsatisfactory [at the lowest concentration] because of the poor estimation of the intercept" *[but Ed. puzzled: tabulated unweighted C.V.'s* ***decrease*** *with increasing concentration]*. Calculation of weights entails forcing the line through the origin (unnecessary if there are sufficient data) to ensure no negative S.D.'s, estimation of S.D.'s from the derived regression line, and then calculation of the variances and weights. In the model used to calculate the regression line, the usual equation, of (Y = a + bX) type, has two extra '+' items: "the error of the mean ratio at the jth concentration mean from the straight line (lack of fit)" and "the error of the ith response from the mean ratio at a concentration mean (pure error) respectively. Weighted calibration data are tested for fit to a straight line. A key outcome is obviation of an assumption in the OLS procedure that the data are 'homoscedastic', i.e. that variance for replicates is independent of concentration.

RELEVANT TO 'GLP', *in respect of lab. safety: art.* #ncB-9.

*Other arts. in [2] give regulatory guidance, not merely FDA.

'OUTLIERS': *Notes by Senior Editor*

A criterion for deciding whether to reject one value in a trio as a 'rogue' is advocated by G.S. Clarke (Appendix to #ncA-3) in whose laboratory the practice is to assay each sample in *triplicate*. This topic has received some consideration elsewhere, although typically absent from books on statistics. The option of never rejecting a result violates common-sense, and the option of rejecting a result which fails to meet a chosen probability criterion based on the variability for the particular sample [1,2] is off-track for <4 values/sample, as in biopharmaceutical analysis.[⊗] (A publication in the occupational hygiene field [3] may be more pertinent.)

The only clear consensus from debate at the 1979 Forum [4] was that no effort should be spared to track down a cause for any out-of-line result. J.A.F. de Silva's practice in pharmacokinetic studies was to assay in *singlicate* and, if there seemed to be a discontinuity in the fall-off curve, to repeat the assay. At least in pharmacokinetic studies with assay duplication the curve shape may justify rejecting a discordant value in a pair. Where a result obtained in *duplicate* is not part of a set, an inexplicably disagreeing *pair* may have to be rejected, as discussed at an RIA symposium [5, cited in 4].- Taking account of the S.D. of the response variable, "an essentially arbitrary choice must be made of a limit for the ratio of observed to 'expected' differences between duplicates; a sample with a ratio above this limit may properly be rejected". A *rough* measure of the S.D. is the square root of the difference between duplicates in the assay; merely on statistical grounds, ~11% of all pairs will differ by ~3 times the S.D., and the choice of limit should not be rigid [5].

1. Pantony, D.A. (1961) *Statistics, Theory of Error and Design of Experiment*, Lect. Series 1961, No. 2, Royal Institute of Chemistry, London, 38 pp. *[The Institute merged with the Royal Society of Chemistry, Cambridge.]*
2. Lark, P.D., Craven, B.R. & Bosworth, R.C.L. (1968) *The Handling of Chemical Data*, Pergamon, Oxford: see p. 122.
3. Taylor, D.G., Kupel, R.E. & Bryant, J.M. (1977) *Documentation of the NIOSH Validation Tests*. DHEW (NIOSH) Publication 77-185, Cincinnati.
4. Reid, E., ed. (1981) *Trace-Organic Sample Handling* [Vol. 10, this series; *now out-of-print*], Ellis Horwood, Chichester: see pp. 352 & 369.
5. Ekins, R.P. & Malan, P.G. (1978) *Ann. Clin. Biochem. 15*, 125-126.

The Editor would welcome responses from readers (literature, or practices adopted). [[⊗]but Hill (#A-3) favours 6 replicates

***Re* Dell's comments on Questionnaire (at end of art. #A-2):**
#SUPPLEMENTARY COMMENTS BY R.J.N. TANNER

After compiling his own article (#**A-6**) the author learnt the results of the questionnaire that was issued at the Forum by Dr. D. Dell. In the main it corroborates the views and opinions expressed in #**A-6**. However, some comments are warranted, supplementing those in Dr. Dell's Appendix.

(a) The use of 1/y or $1/y^2$ weighting for calibration line regression is not valid if calibration standards are in duplicate. The line would be unevenly weighted to that standard of a pair that had the lower response.

(b) It is essential that precision for the assay as a whole be derived from results for test samples that are totally independent of the calibration line. Thus assay precision data cannot be derived solely from calibration results. The test samples must be derived from a different weighed sample of analyte than that used to prepare the stock solutions for the calibration standards.

(c) If results will be quoted at concentrations down to the quantification limit, the performance of the assay at that limit must be established. Variability, or noise, in detector response to a control sample is not a major factor in determining the quantification limit for most bio-sample assays. However, variability in matrix, in sample preparation and in chromatography are major factors. Accordingly, test samples must be used to define the quantification limit at an acceptable precision and accuracy.

Re **#A-2**, D. Dell - VIEWS ON METHOD VALIDATION
#COMMENT MADE AT THE FORUM BY R.J.N. TANNER*, complementing the art. (#**A-6**) he compiled after seeing Dell's text*

Using calibration data to calculate the precision of a method tells one a lot about the precision of the calibrants but not much about the precision of unknowns analyzed against the calibration line. I prefer to analyze as unknowns, against the calibration line, replicate spiked samples based on an analyte weighing-out different from that done to prepare the stock solutions used for the calibrants. Within-batch precision is then carried out using 6-fold replicates at each calibration concentration. Between-batch (4-day interval) duplicates are analyzed at each concentration.

A benchwork form devised by E. Reid for sample preparation (liquid-liquid extraction emphasis, but adaptable to SPE): ***'SOP' Check-list****.* *Appeared on p. 28 of Vol. 10, now out-of-print; copy requestable.*

Some views on bioanalysis expressed by FDA staff
- Senior Editor's compilation, in lieu of a publication text from a Forum speaker, Dr C.T. Viswanathan (Divn. of Biopharmaceutics, Office of Res. Resources, Food & Drugs Admin.,Rockville, MD 20857); the following was his Forum Abstract, co-authored by J.P. Skelly:-

Bioanalytical Requirements in Drug Development

The Agency considers bioanalytical quantitative assay procedures in determining drug levels in biological fluids as of great importance, and examines such procedures and resulting data carefully. It is not uncommon to have studies rejected on the basis of inadequate analytical procedures. Standard approaches for the validation of any analytical method include generation of standard curves in the therapeutic concentration range resulting from the dosing regimen. Precision, accuracy and reproducibility[†] should be documented. The specificity of the moiety [parent drug/metabolite(s)] that is being quantitated should be established. Cross-reactivity should be checked for, and the possibility of endogenous interference should be studied. The linearity and sensitivity of detection as well as the information about recovery and reproducibility should be established. The stability of the assayed drug should be determined both on storage and throughout the assay work-up. The issue of stability is critical when derivatization procedures are used. Use of internal standard (i.s.) is encouraged, and the extraction ratio of the i.s. to that of the assay candidate will be necessary when structurally dissimilar compounds are used as i.s. Appropriate statistical methods should be applied for data analysis.

Points from two arts. *in a journal (publ. Marcel Dekker) besides those cited above and by R.J.N. Tanner (#A-6): see his refs. list (FDA staff authorship*)*

#S.V. Dighe *(1984; focus on bioavailability and bioequivalence)*

In each assay, ideally run interspersed controls at 3 concentrations, duplicated so as to give early warning of any within-batch imprecision. Such points should be stated in the Submission along with methodological and statistical validation.

#V.P. Shah *(1987; focus on bioanalysis)*

Besides minimum detectable quantity (response twice the background noise), the concentration needed for acceptable precision (MQL) should be stated; illustrative chromatograms are advantageous. For intra- and inter-laboratory precision the respective terms are repeatability and [not always used

---

**likewise:* De Camp, W.H. (1989) *Chirality 1*, 2-6 *(focus on stereoisomer development and bioanalysis)* [†]see p. 90, footnote

in this sense! - *Ed.**] reproducibility. A low recovery is allowable if the method is reproducible. For chromatography, derivatization is often by methylation, but other modes should be tried to verify specificity in view of possible metabolic methylation/demethylation. For RIA specificity, monoclonal antibodies are desirable. To verify stability during sample storage, frozen aliquots of standard should be assayed in triplicate at intervals; similarly for clinical samples, in relation to possible metabolite interference.

Note by Ed. *- Neither paper makes any recommendation on aliquot replication for a test sample in the assay.*

=============

Comments at the Forum on talk (see above) by C.T. Viswanathan

C.T. Viswanathan, replying to P. Logue.- On common-sense grounds, a higher R.S.D. may be acceptable close to the detection limit, especially if only a small minority of the actual samples fall in this region. **Reply to H. de Bree:** it is not the actual overall recovery of analyte in the work-up that matters, but rather its consistency, say a C.V. of ~15%. **Replies to J.B. Lecaillon and R. Woestenborghs.**- The method should be re-validated, over the whole calibration curve range, in each clinical study, and ideally should be validated for any active metabolite, especially if an important one. **Answer to V. McNally**, who had asked about the need to monitor drug degradation during the analysis: drug stability is a matter for the chemistry and pharmacy personnel. Concerning the desirability of an i.s. (discussants: **R.J.N. Tanner**, **R. Whelpton**), it must not be assumed that an i.s. will compensate for changes in, for example, detector specificity or the column; these should be tested for.

Citations contributed by Senior Editor

'Clarification of the limit of detection in chromatography'.- Foley, J.P. & Dorsey, J.G. (1984) *Chromatographia 18*, 503-511. Literature and models are considered, and a policy suggested.

'The misuse of correlation coefficients'.- Tiley, P.F. (1985) *Chem. Br. 21*, 162-163; also 351 (R. Hulme and R.C. Watkins – each reinforcing the expressed "very timely warnings"). It is an unjustified belief that the closer an **r** value is to unity, the better the fit of the results. Linearity is vital.

'Description of an analytical method in clinical biology'.- Editorial policy (1988) *Ann. Biol. Clin. 46*, 653-654. In the clinical biochemistry field, admirable criteria are set down (in English), with obvious applicability to drug-level studies (where an endogenous blank is less likely to be a problem).

**See Preface; can signify* precision *(p. 25) or* 'robustness'*! (Cf. p. 89)*

Section #B

ANTI-INFECTIVE DRUGS AND THEIR METABOLITES

#B-1

A COMPARISON OF HPLC AND BIOASSAY FOR β-LACTAM ANTIBIOTICS

R. Horton

Beecham Pharmaceuticals Research Division,
Brockham Park, Betchworth, Surrey RH3 7AJ, U.K.

HPLC and bioassay techniques for β-lactam antibiotics are compared, with especial reference to recently introduced penicillins and β-lactamase inhibitors. The following aspects are considered: detection limits, selectivity and interference, accuracy and precision, sample processing and preparation, assay time and throughput, and resources (human and equipment). The choice between the two techniques has to take account of the robustness of the bioassay on the one hand and the resolving power of HPLC on the other.

Agar diffusion bioassays have been used to quantify antibiotics in biological samples for many years. More than 100 years ago the Swiss microbiologist Garré [1] in his work on the antagonism between *Pseudomonas fluorescens* and other bacteria was the first to show that a zone of agar medium could be made inhibitory to the growth of one organism by the products of another. However, it was not till the 1940's that workers at Oxford used the technique to make quantitative measurements of penicillin [2].

More recently HPLC has been shown to have certain advantages over the bioassay. In this article comparison of HPLC and bioassay techniques for β-lactam antibiotics is made, with specific reference to recently introduced penicillins and β-lactam inhibitors.

LIMITS OF DETECTION

Cooper & Woodman [3] established the following formula for the size of circular inhibition zones around a cup cut in agar and filled with an inhibitory substance:

$$X^2 = 4DT\ 2.3\ (\log m_0 - \log m')$$

where X is the distance from the cup edge to the zone edge, D the diffusion coefficient of the substance, T the time taken for the zone to be fixed, m the concentration in the cup and m' the critical concentration of the substance

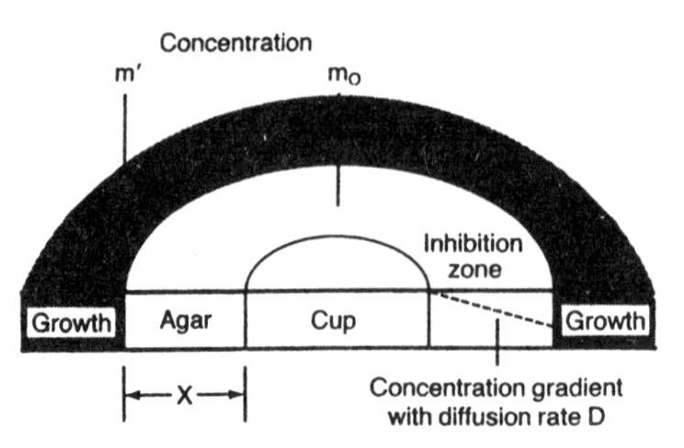

Fig. 1. Formation of a zone of inhibition by antibiotic diffusion into agar.

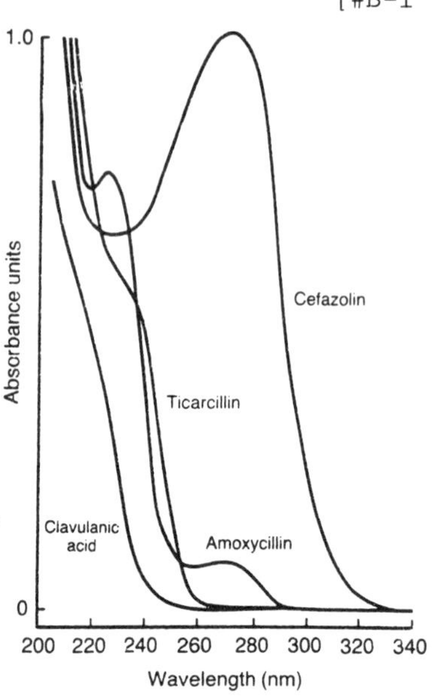

Fig. 2 *(right)*. Relative absorbance characteristics of amoxycillin, ticarcillin, clavulanic acid and cefazolin, each 8.0×10^{-5} M in pH 7.0 phosphate buffer.

which will inhibit the growth of the test organism under the conditions of diffusion (Fig. 1). With small cups (<8 mm diam.) X may be substituted for X^2 in the equation. Clearly zones of inhibition can be formed only if m_0 exceeds m'. Further, m' is stated to be ~2-4 times the minimum inhibitory concentration (MIC) of the substance for the test organism. Thus the detection limit in the bioassay is closely related to the antibacterial activity of the compound being assayed. As novel antibiotics are generally developed for their activity against a particular spectrum of microorganisms it follows that there will be a test organism that will allow detection of the compound at concentrations only just above the MIC. In practice, test organisms are highly susceptible laboratory strains with particularly low MIC's which allow therapeutic levels to be estimated at or below the MIC of the common pathogens. Table 1 shows the relationship between m' and MIC values for the test organisms used in the assay of some recently developed β-lactams. The β-lactamase inhibitor, clavulanic acid possesses little direct antibacterial activity and can be estimated biologically only in the presence of a penicillin.

Penicillins and clavulanic acid, unlike many cephalosporins, do not in general possess a specific UV chromophore [4] and are consequently less easy to detect than the latter group of compounds when using HPLC with UV absorbance monitoring (Fig. 2). Moreover, there is no relationship between absorbance and antibacterial activity; hence detection limits may not be low enough to allow measurement of therapeutic levels,

Table 1. Relationship between antibacterial activity and critical concentration in the bioassay for some penicillins and clavulanic acid. Organisms: S = *Staphylococcus saprophyticus* MCTC 8340; P = *Pseudomonas aeruginosa* NCTC 10701; K = *Klebsiella pneumoniae* NCTC 11228. MIC = minimum inhibitory concentration in agar (mg/L; inoculum, one drop of an undiluted 18 h broth culture); m' = critical concentration in the bioassay (mg/L).

Compound	Organism	MIC	m'
Amoxycillin	S	0.002	0.008
Ticarcillin	P	0.5	0.73
	S	0.25	-
BRL 3024	P	>512	-
	S	0.002	-
Temocillin	P	1.0	1.45
Clavulanic acid	K	0.002⊗	0.032⊗

⊗in the presence of 60 mg benzyl penicillin/L agar

entailing resort to the use of derivatization, other detection methods such as fluorescence, or preparation of samples in such a way as to reduce endogenous interference.

The approaches that have been adopted in HPLC and in bioassay to the quantitation of the β-lactam inhibitor clavulanic acid, co-dosed with amoxycillin and ticarcillin*, are perhaps instructive. This compound, besides having a low UV absorbance, has a relatively short HPLC retention time even in low-pH eluents, and co-chromatographs with poorly retained endogenous serum peaks [5]. Further, its antibacterial activity [6] is poor, the MIC's being in general above concentrations achieved in blood after normal oral dosing. It is, therefore, not possible to use a direct bioassay to detect the quantities commonly present in human samples.

Adaption of HPLC and bioassay approaches

One HPLC stratagem has been to use pre- or post-column derivatization to generate a chromophore. Foulstone [7] reported the use of ultrafiltration to remove plasma proteins followed by pre-column derivatization with imidazole and subsequent detection of a chromophore at 311 nm (Fig. 3). This reaction has the advantage that penicillins such as amoxycillin or ticarcillin will not react with imidazole in the absence of mercuric chloride, and that there is negligible endogenous interference at this wavelength. Watson [8] has reported the detection limit in this assay to be 0.1 mg/L.

*AUGMENTIN and TIMENTIN respectively (Trade Marks of Beecham Group plc)

Haginaka & co-authors [9] have reported a post-column derivatization procedure in which alkali is used to produce a chromophore at 257 nm and in which the retention of clavulanic acid is extended by ion-pairing with tetrabutylammonium bromide. This method was also stated to have a limit of reliable detection as low as 0.1 mg/L in plasma. Haginaka [10] also reported the reaction of benzaldehyde in acid buffer at 100° with clavulanic acid to give a fluorescent product (386 nm excitation, 460 nm emission). Thereby as little as 0.01 mg/L plasma could be detected.

In the bioassay an indirect method has been developed [11] in which a β-lactamase-producing organism (*Klebsiella pneumoniae*) is exposed to clavulanic acid with 5 mg/L benzyl penicillin present in the assay medium. Here m' is a concentration just sufficient to inhibit the β-lactamase present and hence render the test organism sensitive to the penicillin. In practice this concentration is ~0.04 mg/L.

Thus adaptions of both HPLC and bioassay techniques have enabled low detection limits to be achieved for this compound.

SELECTIVITY AND INTERFERENCE

There are numerous examples of the selectivity of HPLC and its ability to separate parent compounds from metabolites or impurities, or to quantitate the components of epimeric mixtures. Similarly, there is a large body of literature on the removal or avoidance of interference from endogenous sample components.

The bioassay, however, may also be used selectively. Thus the assay for the α-carboxy penicillin, ticarcillin, is designed to measure this compound in the presence of its decarboxylated analogue, BRL 3024 (Fig. 4). This analogue has significant activity against gram-positive assay organisms such as *Staphylococcus saprophyticus* but is inactive against the strain of *Pseudomonas aeruginosa* (NCTC 10701) selected for the assay of ticarcillin [12, 13]. Moreover, this strain is insensitive to clavulanic acid, enabling the selective assay of ticarcillin in the presence of this acid.

The work in our company [14] on the side-chain epimers of temocillin provides an interesting comparison of the selectivity of HPLC and the bioassay. The HPLC assay could determine total temocillin in human plasma as well as the levels of the *R* and *S* diastereoisomers (Fig. 4) which result from the unresolved asymmetric centre at the carbon bearing the carboxyl group in the side chain. Although dosed at an *R*:*S* ratio of ~65:35 the epimers have different pharmaco-

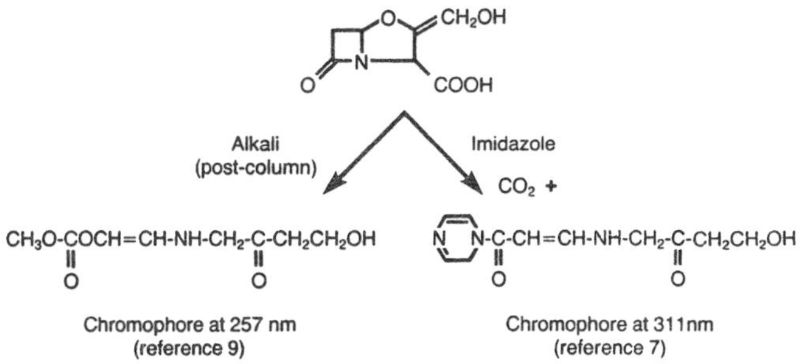

Alkali (post-column)

Imidazole

CO_2 +

CH_3O-COCH=CH-NH-CH_2-C-CH_2CH_2OH

N-C-CH=CH-NH-CH_2-C-CH_2CH_2OH

Chromophore at 257 nm (reference 9)

Chromophore at 311nm (reference 7)

Fig. 3. Chromophore generation from clavulanic acid.

Fig. 4 *(right)*. Penicillin structures.

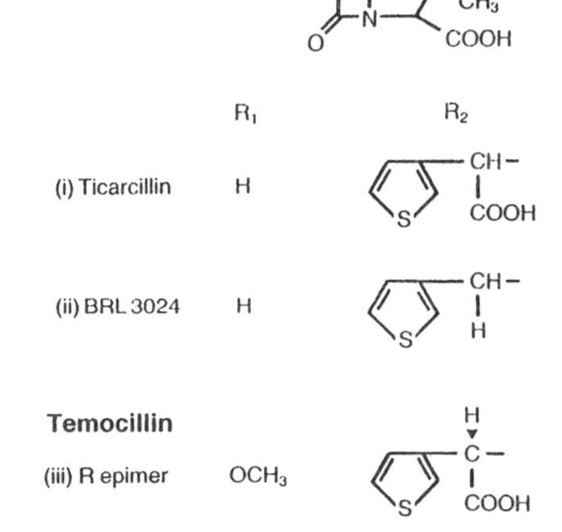

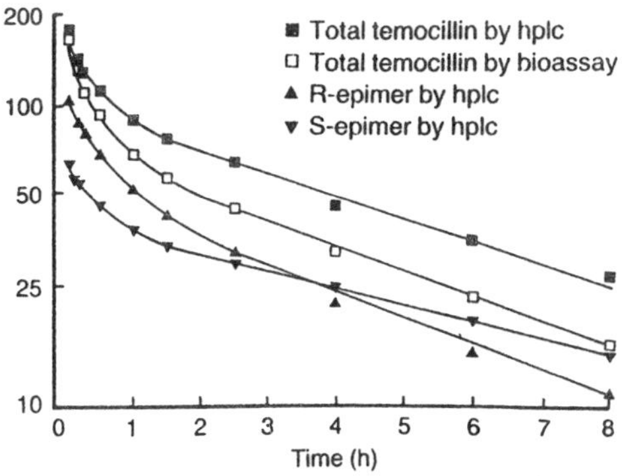

■ Total temocillin by hplc
□ Total temocillin by bioassay
▲ R-epimer by hplc
▼ S-epimer by hplc

Time (h)

Fig. 5 *(above)*. Plasma profiles of temocillin and its epimers after a 1 g i.v. dose of temocillin.

kinetic properties (Fig. 5). Further, much of the antibacterial activity appears to reside in the *R* epimer. The total temocillin concentration values estimated by bioassay reflect most closely those of the *R*-epimer. Correction of these values for the composition of standard material used in the bioassay (65% *R*) brings them into close agreement with the HPLC assay of this epimer. Thus the bioassay can selectively estimate the active isomer whilst HPLC can quantify both epimers as well as total temocillin.

The bioassay is clearly unable to detect biologically inactive metabolites such as the penicilloic acid of amoxycillin. Haginaka [15], however, has described an HPLC method involving post-column derivatization with hypochlorite which allows the simultaneous quantitation of amoxycillin and the epimers of its penicilloic acid, and has shown that in urine these metabolites account for >16% of the administered dose.

Because of the resolving power of HPLC it is unusual for co-administered drugs to interfere in assays. Watson [8] reported this to be the case with ticarcillin and clavulanic acid but did find a co-eluting endogenous plasma peak with the former compound, having an area equivalent to 3.5 mg ticarcillin/L. He also reported that the use of poor-grade imidazole as a derivatizing agent in the assay of clavulanic acid gave rise to a number of peaks which interfered in the assay of the compound.

In the case of the bioassay the co-administration of antibiotics is always a source of problems. The original bioassay for clavulanic acid described above [11] was prone to interference from co-administered penicillins. Only by increasing the benzyl penicillin in the medium to 60 mg/L with a corresponding increase in inoculum has it been possible to assay this compound without interference from amoxycillin [16] or ticarcillin [17, 18]. A further complication has been the administration of aminoglycosides with amoxycillin/clavulanic acid or ticarcillin/clavulanic acid. Here the aminoglycoside activity has been eliminated in the bioassay of the other components by incorporation of up to 1% sodium polyanethol sulphonate in the assay medium. This anionic detergent inactivates aminoglycosides by a charge effect [17]. When such assay modifications are not available, however, HPLC should always be the method of first choice when assaying antibiotic mixtures.

ACCURACY AND PRECISION

The accuracy of an assay depends greatly on the skill of the analyst. In the hands of experienced workers there is little to choose between the accuracy of HPLC and that of the bioassay for β-lactams in human serum or plasma. Examples from published data (Table 2) also show that precision expressed as C.V. (%) is often ~5% for both methods - probably because at the concentrations achieved in human serum the HPLC assays are approaching their detection limit but the bioassays are still in the middle of their range. The precision of the bioassay also depends on the number of replicate zones per sample. In this laboratory, using methods involving 4, 8 and 16 replicates, 12 assays of drug in buffer gave the following C.V. values:- amoxycillin (0.2 mg/L) - 7.01, 1.41, 1.11; ticarcillin (10 mg/L) - 3.60, 1.73, 1.15; clavulanic acid (1.0 mg/L) - 4.05, 1.19, 1.50.

Table 2. Precision of bioassay and HPLC for penicillins and clavulanic acid in human plasma samples.

Compound	Method	Concentration, mg/L	C.V. %	Ref.
Amoxycillin	Bioassay	0.25	6.35	18
Clavulanic acid	Bioassay	1.25	7.61	18
Ticarcillin	Bioassay	25.0	5.93	13
Clavulanic acid	Bioassay	1.25	5.79	13
Amoxycillin	HPLC	1.00	7.1	19
Clavulanic acid	HPLC⊗	0.67	4.02–6.23	10
Ticarcillin	HPLC	5.00	7.8 – 12.4	8
Clavulanic acid	HPLC	1.00	4.5 – 6.6	8

⊗ with fluorescence detection

SAMPLE PROCESSING AND PREPARATION

Some of the HPLC methods discussed here have relied on simple ultrafiltration to remove proteinaceous interfering substances before assay [8, 10]. Others used solid (bonded) phase extraction [19] or solvent extraction [8]. In the method used for temocillin [14], to avoid interference from endogenous constituents it was necessary to extract the protonated compound from plasma into an immiscible organic solvent, followed by back-extraction into neutral buffer before HPLC assays could be performed. The above bioassay for total temocillin has, however, been employed directly on such diverse samples as lymph [20], bile [21] and homogenates of lung [22] and other tissues [23] with little preparation apart from dilution to eliminate binding effects. This illustrates the robustness of the bioassay in coping with diverse sample types, and accounts for the short sample preparation times.

ASSAY TIME AND THROUGHPUT

Although an assay on a single sample takes minutes by HPLC whereas bioassays require several hours incubation before results are obtainable, the throughput of the two techniques depends very much on sample numbers. Consider a human study in which amoxycillin/clavulanic acid is dosed to 12 subjects who each provide 12 plasma samples. This gives 144 samples each requiring 2 analyses. Given the retention time of the two compounds in the standard HPLC methods [7], and allowing time for calibration and sample preparation, it would require >40 working hours to complete these analyses. In comparison, one person using bioassay could complete them in 1 day per compound, with a further day for plate reading. Assuming an 8-h day the total work is 24-h. The key to this high throughput is the automation of the plate-reading process.

RESOURCES

The basic equipment for a bioassay is very cheap. An otherwise well equipped microbiology laboratory needs a capital expenditure of only ~£20 to be able to perform simple bioassays. Expendable items such as media and disposable assay dishes need cost only £1 per assay. Yet the technique would be very labour-intensive, and most laboratories would attempt some form of automation. For plate reading this can be accomplished with about the same capital investment as is needed for a basic isocratic HPLC system. Beyond this the cost of a sophisticated image-analysis zone reader interfaced to a computer is of the same order as that of a computer integrator.

The ease of introducing a bioassay to a microbiology laboratory and an HPLC assay to an analytical laboratory must be roughly equivalent. Further, many microbiologists have adapted readily to the use of HPLC. The reverse, however, is not true. The techniques of microbiology are not easily acquired by scientists of other disciplines, and there are items of equipment, such as incubators and autoclaves, that are not generally available outside microbiology laboratories. Thus if the limitation on the use of HPLC is relatively high initial capital outlay, the limitation on the use of the bioassay is one of availability of technical skills and specialized equipment.

CONCLUSION

The strengths of HPLC and bioassay of β-lactam antibiotics in samples of biological origin complement each other. HPLC can provide the selectivity necessary to resolve isomeric mixtures, or mixtures of several antibiotics, or to quantify parent compounds and metabolites. The bioassay enables estimation of active compounds in large numbers of samples, at low cost, even when these samples present a wide variety of matrices. The most fortunate amongst us are perhaps those who have access to both techniques.

References

1. Garré, C. (1887) *Korrespondenzbl. Schweiz. Aertze 17*, 385.
2. Abraham, E.P., Chain, E., Fletcher, C.M., Florey, H.W., Gardener, A.G., Heatley, N.G. & Jennings, M.A. (1941) *Lancet 2*, 177-188.
3. Cooper, K.E. & Woodman, D. (1946) *J. Path. Bact. 58*, 75-84.
4. Swaisland, A.J. (1986) in *High Performance Liquid Chromatography in Medical Microbiology* (Reeves, D.S. & Ullmann, U., eds.), Gustav Fischer Verlag, Stuttgart, pp. 45-60.

5. Reading, C. (1986) as for 4., pp. 61-74.
6. Hunter, P.A., Coleman, K., Fisher, J. & Taylor, D. (1980) *J. Antimicrob. Chemother.* *6*, 455-470.
7. Foulstone, M. & Reading, C. (1982) *Antimicrob. Ag. Chemother.* *22*, 753-762.
8. Watson, I.D. (1985) *J. Chromatog.* *337*, 301-309.
9. Haginaka, J., Yasuda, J., Uno, T. & Nakagawa, T. (1983) *Chem. Pharm. Bull.* *31*, 4436-4447.
10. Haginaka, J., Yasuda, H. & Uno, T. (1986) *J. Chromatog.* *377*, 269-277.
11. Brown, A.G., Butterworth, D., Cole, M., Hanscombe, G., Hood, J.D., Reading, C. & Rolinson, G.N. (1976) *J. Antibiot.(Tokyo)* *29*, 668-669.
12. Burnett, J. & Sutherland, R. (1970) *Appl. Microbiol.* *19*, 264-267.
13. Staniforth, D.H., Coates, P.E., Davies, B.E. & Horton, R. (1986) *Internat. J. Clin. Pharmacol. Ther. Tox.* *24*, 123-129.
14. Guest, E.A., Horton, R., Mellows, G., Slocombe, B., Swaisland, A.J. & Tasker, T.C.G. (1985) *J. Antimicrob. Chemother.* *15*, 327-336.
15. Haginaka, J. & Wakai, J. (1987) *J. Chromatog.* *413*, 219-226.
16. Jackson, D., Cooper, D.L., Horton, R., Langley, P.F. & Sutton, A.J. (1980) in *Augmentin (Proc. 1st Internat. Symp., Scheveningen)* (Rolinson, G.N. & Watson, A., eds.), Excerpta Medica, Amsterdam, pp. 83-101.
17. Edberg, S.C., Bottenbley, C.J. & Gam, K. (1976) *Antimicrob. Ag. Chemother.* *9*, 414-417.
18. Davies, B.E., Boon, R., Horton, R., Reubi, F.C. & Descoedres, C.E. (1988) *Br. J. Clin. Pharmacol.* *26*, 385-390.
19. Lee, T.L.& Brooks, M.A. (1984) *J. Chromatog.* *306*, 429-435.
20. Bergan, T., Olszewski, W.L. & Engeset, A. (1985) *Drugs 29 (Suppl. 5)*, 114-117.
21. Poston, G.J., Greengrass, A. & Moryson, C.J. (1985) *Drugs 29 (Suppl. 5)*, 140-145.
22. Cowan, W., Baird, A., Sleigh, J.D., Gray, J.M.B., Leiper, J.M. & Lawson, D.H. (1985) *Drugs 29 (Suppl. 5)*, 151-153.
23. Gould, J.G., Meikle, G., Cooper, D.L. & Horton, R. (1985) *Drugs 29 (Suppl. 5)*, 167-170.

#B-2

DEVELOPMENT OF A SENSITIVE AND ROBUST ASSAY FOR THE AMINOCYCLITOL ANTIBIOTIC, TROSPECTOMYCIN

S.A. Wood and R.J. Simmonds

Pharmaceutical Research Laboratory, Upjohn Ltd., Fleming Way, Crawley, W. Sussex RH10 2NJ, U.K.

Requirement *For clinical trials and pharmacokinetic studies on the new antibiotic, microbiological assay was too insensitive and possibly unspecific; hence a new assay applicable to plasma and other biological matrices had to be developed. It had to be suitable for a high throughput and for contracting-out.*

End-step *RP-HPLC with a highly acidic mobile phase, post-column derivatization with OPA* and fluorescence detection. Pre-column derivatization was disfavoured.*

Sample handling *Selective SPE, using as i.s. a close structural analogue.*

Comments *With a 500 μl plasma sample the limit of quantification was ~5 ng/ml. In one day 35 samples could be assayed in duplicate. The method is robust, and is applicable to similar antibiotics and to matrices such as urine and solid tissues.*

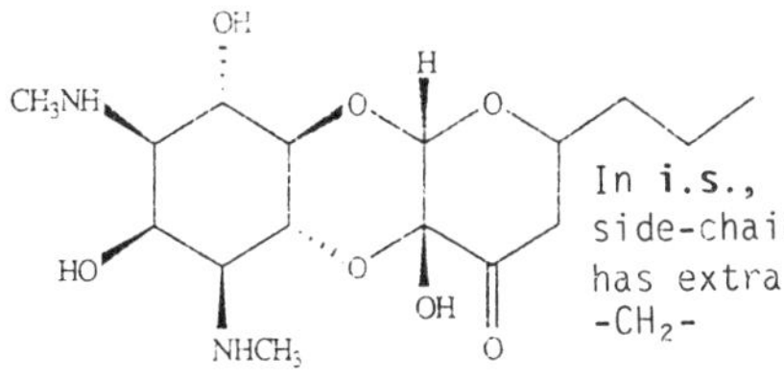

In i.s., side-chain has extra $-CH_2-$

trospectomycin

Trospectomycin is one of a new family of aminocyclitol antibiotics which is of interest to our company. They have a broad spectrum of activity against both Gram-positive and Gram-negative bacteria. Trospectomycin is being developed for use against a number of sexually transmitted diseases and is now in Phase III clinical trials.

Here we deal mainly not with analytical details but with the process by which the method was developed and the problems overcome on the way. The developed assay, in its definitive

**Abbreviations.*- OPA, *o*-phthaldialdehyde; i.s., internal standard; QC, quality-control (sample); SPE, solid-phase extraction; TFA, trifluoroacetic acid. OCL, PCRS: see text.

form, is both sensitive and robust, and suitable for the analysis of large numbers of samples at an outside contract laboratory (OCL). Microbiological assays are an obvious choice for antibiotics, and indeed were long the unique approach. However, with aminocyclitols selectivity was poor: a number of close analogues (possible metabolites) cross-reacted, and an endogenous antibiotic effect present in control plasma or serum limited sensitivity to ~500 ng/ml. A chromatographic assay was therefore required to achieve the necessary selectivity and sensitivity (~25 ng/ml).

PROBLEMS IN CHOOSING THE ASSAY APPROACH

In summary, the compound is very polar, with two basic functional groups, has no strong chromophore, and moreover we were faced with a backlog of >5000 clinical samples awaiting assay. The compound's polar nature made selective classical liquid extraction difficult; ion-pair extraction was feasible, but selectivity was poor. Its polarity and aliphatic/alicyclic nature also complicated the development of an HPLC method: poor retention, peak spreading and irreproducibility have all been associated with this class of compound.

An especial handicap was the lack of a strong UV chromophore or any particularly reactive functional groups for derivatization. Secondary amines have some electrochemical activity [1]; but we were unsuccessful using porous carbon electrodes. Our U.S. colleagues have successfully used glassy carbon electrodes for bulk drug analysis; but at the high potentials required, ~0.8-0.9 V, selectivity would be poor for bioanalysis and robustness might well be inadequate for a routine method.

Derivatization to introduce a strong chromophore or fluorophore was considered as a viable option. Classically, secondary amines can be derivatized with compounds such as dansyl chloride before HPLC [2]. Another possibility is post-column derivatization, involving oxidation of the secondary amine followed by reaction with OPA or fluorescein [3]. Post-column reaction for bulk drug assay has been tried by our U.S. colleagues but found to be very fragile. The problems, we believed, included the lack of any i.s. to compensate for system variability. However, as considered below, we did eventually adopt a post-column reaction approach.

OUR APPROACH ENTAILING A PRE-COLUMN REACTION

Either pre- or post-column reaction seemed to offer the best chance of a viable method. At the outset we did not have access to post-column reaction equipment. We therefore used the pre-column reaction approach initially.

General strategy

Firstly we had to develop a suitable HPLC system for the derivatized compound, looking for good efficiency, peak shape and adequate retention; we routinely try a number of columns in a selectivity triangle analogous to the solvent triangle, e.g. phenyl, C-8 and nitryl columns, which may differ significantly in selectivity. Then, with 2 or 3 candidate column types but deferring optimization, we settle the i.s. (speedily if there is good synthetic-chemistry back-up). Next (see below) we develop an extraction system - the choice being SPE - and then use approaches such as the solvent triangle and buffer-pH manipulation to optimize the HPLC system; the aim is to enhance selectivity whilst retaining the robustness inherent in SPE. The assay method is briefly validated in respects such as linearity, repeatability and recovery using spiked control samples and testing with actual study samples. Thereby we can formulate QC guidelines to use when the assay has been installed and validated at an OCL. If an assay is to be successfully developed rapidly, experience and judgement are cardinal. The exploration of a range of columns and SPE cartridges, using i.s. as well as the analyte, does not take long and provides a sound base for future modifications of a method to enable it to cope with new biofluids.

Selective solid-phase extraction (SPE)

In general we prefer SPE to liquid extraction, for the following reasons.

- Selectivity is easier to achieve by SPE, because of the availability of a range of over 20 different cartridge types differing in chemical nature and size. We use the 50 and 100 mg cartridges.
- Once the extraction procedure has been optimized, the cartridges give a reproducible and reliable performance. We have had no batch problems with the bonded phases.
- Safety is imperative. SPE minimizes personal contact with the sample and provides some measure of containment. The handling of tubes containing the biological material is reduced.
- SPE is quicker than solvent extraction, especially if the latter requires wash steps or back-extraction.
- SPE is easier and possibly more efficient to automate than solvent extraction.

Using dansyl chloride, pre-column derivatization on its own, without SPE, did not provide the requisite selectivity, sensitivity and reliability, although late in the investigation a pattern with fewer peaks (Fig. 1) was achieved by a column-switching stratagem: a small column furnished a 'heart-cut' fraction which was directed onto an analytical column. However, the run time was lengthy (>60 min; 15 min without heart-cutting).

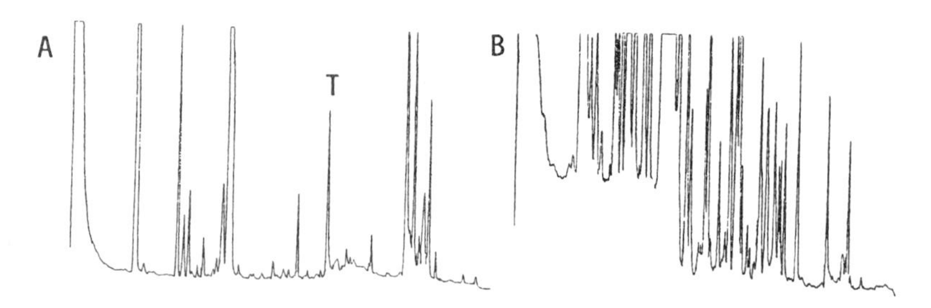

Fig. 1. HPLC patterns for a solvent extract of plasma containing trospectomycin (**T**), after pre-column derivatization with dansyl chloride: **A**, with a 'heart cut' column-switching procedure; **B**, using the main analytical column only.

An alternative strategy was to subject the sample to selective dual-cartridge SPE: reliability was good and sensitivity adequate; but this approach was discontinued because of shortness of the linear calibration range, the formation of conformational isomers in the presence of the i.s., and the lengthy derivatization procedure. The approach entailing SPE was revived with the adoption of an HPLC post-column reaction.

OUR APPROACH ENTAILING SPE AND A POST-COLUMN REACTION

The post-column reaction system (PCRS) adopted is outlined.-

→ ANALYTICAL COLUMN → MIXER → REACTION COIL → MIXER → REACTION COIL →
(↑ *Reagent A* at the first MIXER; ↑ *Reagent B* at the second MIXER)

The PRCS (Kratos 5270) mixers were specially engineered swirl-type chambers of very low dead volume, and the coils were of knitted teflon [cf. #C-5 by Nicolas & Leroy - *Ed.*]. The turbulent flow characteristics of these coils result in very little band broadening.

A post-column reaction, if a suitable scheme can be developed, offers advantages over pre-column derivatization. The separation problem was now much simpler than before, relying on the characteristics merely of the drug and not the derivative and thereby improving the selectivity of the assay. The reaction, once optimized, was reproducible, and multiple products seen with pre-column derivatization no longer presented a problem. Moreover, the system was robust and suitable for extended routine use.

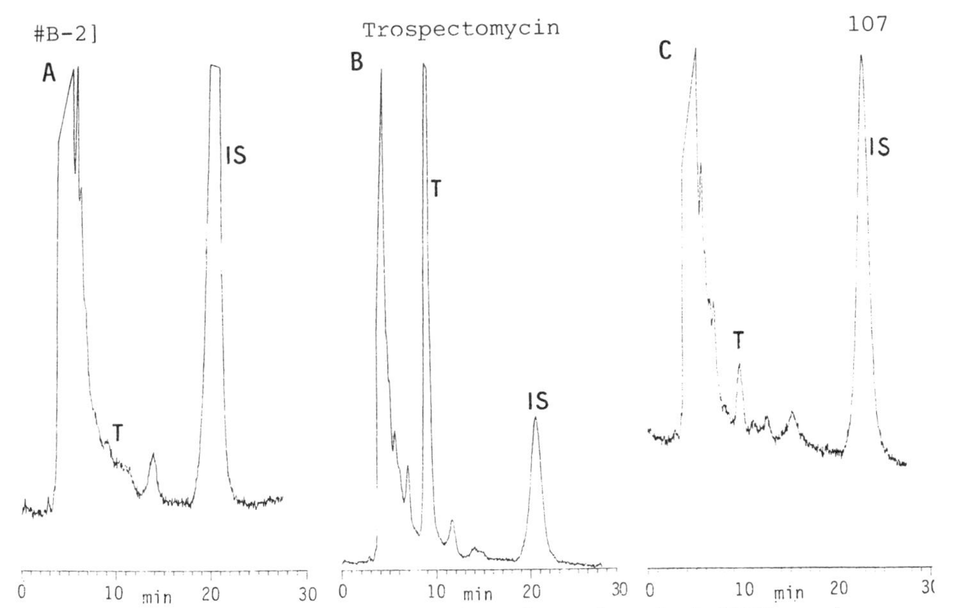

Fig. 2. HPLC patterns with the adopted method (SPE, and post-column derivatization), for plasma from a subject who received a single dose of 600 mg of trospectomycin (**T**). **A**, pre-dose; **B**, 16 h post-dose; **C**, 168 h post-dose. Internal standard peak: **IS**. The 80-100 µl injected onto the column corresponded to approx. half of the original 500 µl sample.

Steps in the method are as follows.

- Serum (0.5 ml) and i.s. are diluted to 1 ml with water and loaded onto a SPE C-18 cartridge already primed with acetonitrile and water (each 2 × 1 ml).
- Elution is effected with 0.6 ml of acetonitrile-0.5% (v/v) aqueous TFA (60:40 by vol.; hence 0.2% TFA in the mixture).
- HPLC, with a C-18 column (250 × 46 mm), is with an unorthodox eluent: acetonitrile-0.63% TFA (1.5 ml/min), which gives efficient chromatography with no need for amine modifiers and is compatible with the post-column reaction.
- This reaction entailed (1) sodium hypochlorite treatment to oxidize the secondary amine groups, via an imine intermediate; (2) reaction of the product with OPA; (3) fluorescence detection at 462 nm with excitation at 380 nm.

The method can detect <20 ng/ml using 0.5 ml of biofluid and is rapid and robust. Key features in its success are (a) the availability of a close analogue as an i.s. to compensate for any variability in extraction and post-column reaction; (b) reproducible RP-HPLC with reasonable efficiency; and (c) the use of a carefully optimized PRCR.

Results obtained.- Fig. 2 shows representative chromatograms with the PCRS method. Note the generally clean response achieved by selective extraction and selective detection from the equivalent of 250 µl plasma on-column.

The PCRS method used in-house could measure 5 ng/ml, and was successfully used at an OCL where QC samples gave Shewart plots such as that shown in a companion article from our laboratory (#ncA-2).

CONCLUDING COMMENTS

Improvements in the method, particularly the sample handling, might be achievable by introducing automated extraction with column switching (Fig. 3).

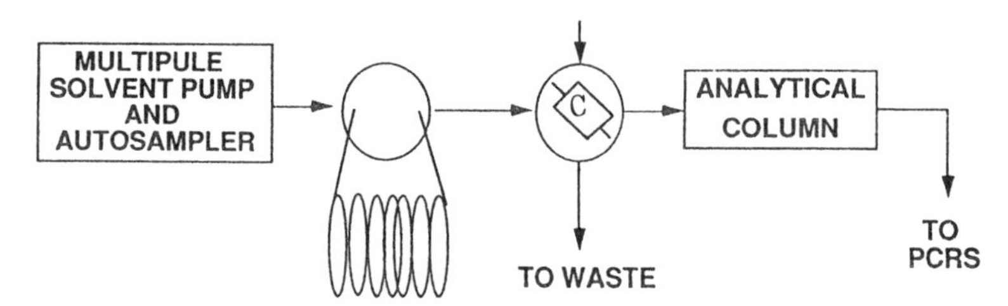

Fig. 3. Automation entailing column switching as a future improvement. C, trace-enrichment cartridge.

In summary, the following should be borne in mind when an assay like this is being developed.

- (1) Consider several options at all stages, but especially during HPLC development.
- (2) Optimize each step using the results of previous experiments.
- (3) Try all the options within reason, calling on past experience.
- (4) Most importantly, aim to develop more than one option at each stage in order to finish with two or more possible HPLC systems and two or more viable extraction schemes. Logical responses can then be made to meet changing assay requirements.

In conclusion, we believe that continuous refinement of an assay over a long period from, for example, initial animal experiments to clinical-trial analysis can follow a logical progression based upon previous experience. Given adequate sensitivity, it is rarely necessary to develop completely different assays at various phases of drug development.

References

1. Elrod, Jr., L., Bauer, J.F. & Messner, S.L. (1988) *Pharm. Res.* *5*, 664-667.
2. Tsuji, K. & Jenkins, K.M. (1985) *J. Chromatog.* *333*, 365-380.
3. Myers, H.N. & Rindler, J.V. (1979) *J. Chromatog.* *176*, 103-108.

#B-3

CEFTETRAME: HPLC ASSAY PROBLEMS AND ASSEMBLY OF RESULTS

C. Town, N. Oldfield, D. Chang and W.A. Garland

Department of Drug Metabolism,
Hoffmann-La Roche Inc., Nutley, NJ 07110, U.S.A.

Requirement	*Assay of human and animal plasma and urine, down to 0.5 µg/ml, for CFT* (a third-generation β-lactam antibiotic; its ester pro-drug, peCFT, is administered.*
End-step	*RP-HPLC (a particular phenyl type), using PenG as an i.s.; Kphos/methanol as mobile phase (with trial of a gradient); detection at 225 nm. PHR values processed by a data system, QSIMPS.*
Sample handling	*SPE (phenyl); elution by methanol.*
Comments	*See [1] for details of assay. Rapid and satisfactory, after overcoming problems:- late-eluting peaks (hence column methanol-washed between samples; HPLC column batch-variations; unsuspected CFT isomerization to an interfering isomer, iCFT; pressing deadlines in drug-development; excess manual data entry.*

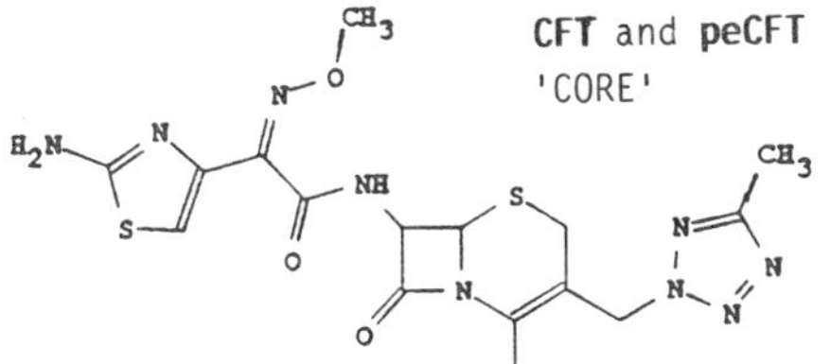

CFT and peCFT 'CORE'

R = H in **CFT**, (7*R*,8*R*)-'CORE'-3-cephem-4-carboxylic acid, where 'CORE' = 7-[(Z)-2-(2-amino-4-thiazolyl)-2-(methoxy-imino)-acetamido]-3-[(5-methyl-2H-tetrazol-2-yl)methyl]-. R = group **shown on right** → in **peCFT**, methylene-(6*R*,7*R*)-'CORE'-8-oxo-5-thia-1-azabicyclo[4.2.0]oct-2-ene-2-carboxylate pivalate.

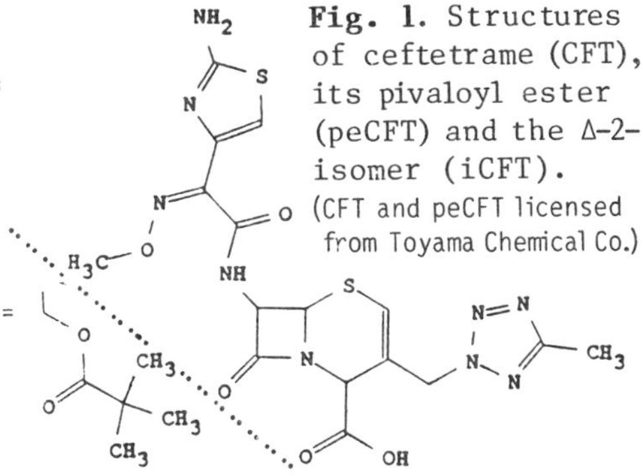

iCFT, Δ-2 isomer of CFT

Fig. 1. Structures of ceftetrame (CFT), its pivaloyl ester (peCFT) and the Δ-2-isomer (iCFT). (CFT and peCFT licensed from Toyama Chemical Co.)

**Ed.'s abbreviations (Fig. 1 explains CFT, etc.).-* PenG, penicillin G, which was used as i.s., internal (reference) standard; Kphos, potassium phosphate buffer, usually 10 mM, pH 5.2; PHR, peak height ratio for analyte *vs.* i.s. *Others:* QA, Quality Assurance; QSIMPS, an integrated set of programs (for weighted linear and non-linear curve calculation, data sorting, and calculating pharmacokinetic parameters); SPE, solid-phase extraction.

The development of a new cephalosporin, CFT [2, 3] - the major metabolite of its pivaloyl ester (peCFT) as orally dosed - entailed initial development of an HPLC assay which could detect and measure both compounds. The pro-drug (for which an i.s. was used) chromatographed at ~25 min (CFT ran at ~4 min, very close to the solvent front), but was absent in plasma from initial animal studies. Accordingly, this initial assay was discarded and efforts were directed to developing an assay optimized for measuring CFT.

Phase I: ASSAY DEVELOPMENT AND VALIDATION

A new assay was needed urgently. The decision to continue to use HPLC[⊗] seemed obvious. Another β-lactam antibiotic was sought as the reference because the β-lactam ring appeared to be the most labile part of the molecule. Penicillin G was similar to CFT in its extractability (better than Penicillins O and V), and its other physicochemical properties indicated that it would be a good match. A number of HPLC columns were tried and the best separation was found with a μBondapak Phenyl column (Waters Chromat^{y}. Divn., Millipore, Milford, MA). The extraction procedure, shown in Scheme 1 for human plasma, was worked out using similar chemistry with a Bond Elut PH (phenyl) SPE 'column' (Analytichem Internat^{l}., Harbor City, CA). The initial work with calibration standards indicated that the assay would work well with an isocratic separation. Later work with plasma extracts indicated that the baseline shifted with time due to late-eluting peaks. To keep the HPLC column unclogged and the baseline stable, it was washed with methanol every 30 samples. As samples were assayed during validation, it became clear that a methanol wash after *each* sample was needed in order to assure absolute reproducibility of the assay. At this time the scheduled time for assay development (2 months) expired, toxicology samples awaiting interpretation were accumulating, and the clinical studies had to be planned. Dog plasma was used to validate the assay.

Departmental *in-vitro* validation requires that 3 standard curves be established with QA samples on 3 different days and that the relative S.D. for the overall intra- and inter-assay precisions for any individual calibration standard not exceed 20%. The assay was validated with dog plasma from 0.47 to 9.56 μg/ml, covering the range expected in toxicological and clinical studies, and the results met the guidelines. CFT was stable in dog plasma for up to 8 h at room temperature but prolonged storage had to be at -70°. It was recommended, with no actual initial experiment, that samples be collected in a Vacutainer containing oxalate and, to prevent any pro-drug being cleaved, NaF as an esterase inhibitor.

[⊗]with Spectra-Physics integrators and, later, a Nelson Analytical 2600 data system

Scheme 1. Assay of CFT in human plasma, using SPE and HPLC with a data system which gives PHR (*vs.* PenG as i.s.)

PLASMA, 1 ml: add 2 ml pH 7.4 buffer, & i.s.
↓ *Apply to 3ml Bond Elut Phenyl cartridge pre-washed with 3 ml methanol & 3 ml buffer*
RETAINED ANALYTES
↓ *3 ml buffer (to wash), then 0.5 ml methanol*
ELUTED ANALYTES ready for HPLC
↓ *10 μm μBondapak phenyl column, 150 × 3.9 mm*
SEPARATED PEAKS, subjected to data analysis
Eluent: Kphos/methanol, 78:22, 2 ml/min

Phase II: RUNNING THE ASSAY

In outline (see [1] for more details) the assay as actually applied (with a μBondapak Phenyl column) entailed a 26-min run time with two alternate inflowing solvents:

- **A** alone from 0 to 16 min, then an inflow giving a linear change ('Curve 6'):
- **B** during 16 to 20 min (serving to wash the column); then
- **A** alone giving a sharp change over 0.1 min and continuing during 6 min, for re-equilibration prior to the next injection.

For **A** composition see Scheme 1; **B** as for **A** but 30:70 mixture.

An analytical run (a tray) consists of 5 pairs of calibration standards and (each in duplicate) a blank, a QA sample, an unextracted standard (to calculate extraction efficiency), and up to 83 experimental samples. A tray could accommodate 99 samples in all, but the number was normally fewer. Final concentration calculations were based on PHR values. For samples with concentrations higher than the upper limit of the curve, re-assay was performed using <1 ml.

Assays were performed over 1 year without significant problems on 3600 samples of plasma or urine from dogs, cynomolgus monkeys, baboons and humans. At the end of the first year the projected work-load dictated that some of the assays should be performed at an outside laboratory. Just then the characteristics of the μBondapak Phenyl columns changed: they could no longer separate CFT from an interfering endogenous component. A Nova-Pak Phenyl column was the chosen replacement; being 4 μm rather than 10 μm, the flow-rate had to be changed, as implemented by the outside laboratory under our direction; they also used a solvent gradient. The results continued to satisfy the guidelines. In continued in-house assays a gradient was likewise adopted; solvent **A** was 81:19 Kphos-methanol, and **B** was 30:70 Kphos-methanol:

- **A** alone at the start, and an inflow giving a linear change:
- **B** alone over 0-7 min and continuing with no Curve over 7-10 min; then an inflow giving a sharp linear change:
- **A** alone over 10.0-10.1 min ('Curve 11') and continuing for 10.1 to 17 min.

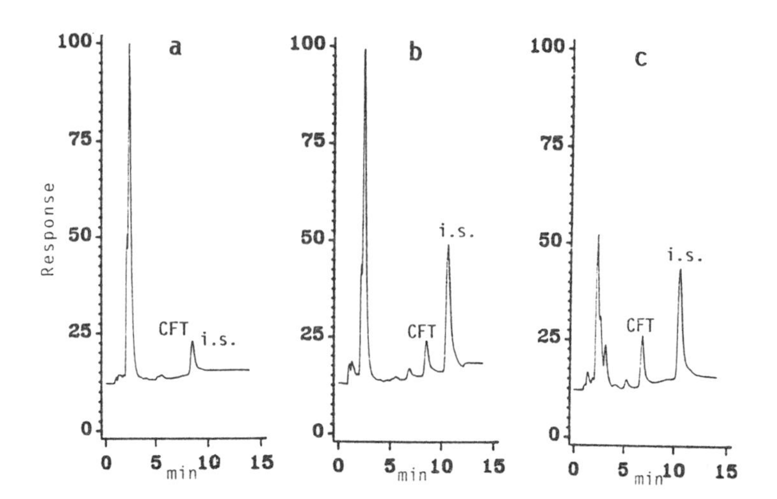

Fig. 2. HPLC patterns for control human plasma - (**a**) unspiked, (**b**) spiked with i.s. (Penicillin G) and 5 μg/ml CFT - and (**c**) for plasma from a subject 30 min after a peCFT dose, 1200 mg/kg.

The gradient system worked well, shortening the assay which continued to function properly in all other respects. Fig. 2 shows typical chromatograms for human plasma with the gradient system.

The departmental data system, QSIMPS, though having many capabilities, was the cause of some problems. It was custom-designed to perform sample injections, data collection, manipulation and sorting for MS assays [4], with report generation whose format was now to be kept. All sample definition information and relevant peak heights were manually entered. Externally produced data were also entered manually, initially by in-house analysts, and later the external analyst entered the data and review was done by an in-house analyst. It took some time to teach the external analysts to use QSIMPS.

The QSIMPS progams were designed so that the data could be packaged and electronically transferred to the corporate mainframe where they could be read and used by the clinical pharmacokineticists so that they need not manually re-enter the data. The system malfunctioned, largely due to differences in sample definition and to some hardware and software difficulties; hence the data generated from all clinical studies involving CFT had to be entered manually into the pharmacokinetic software. The pre-clinical pharmacokineticists who interpreted the animal data did not use the corporate mainframe programs and entered all the data manually. Their interpretation of these data, and release of the final reports, were thus slowed.

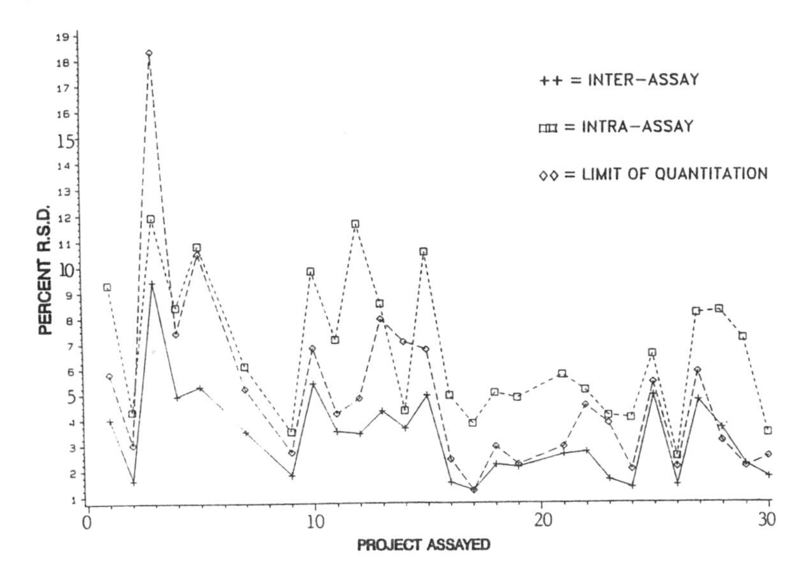

Fig. 3. Relative S.D.'s for precisions (inter- and intra-assay, and at quantitation limit), for all projects: plasma and urine samples from all 4 species (see foregoing text).

Precision.- Fig. 3 shows precisions for all 30 projects which generated samples, from humans and 4 animal species, that were assayed for CFT. Evidently intra-assay precision was always <10%, while inter-assay precision was >10% in 5 of the assay sets. The precision at the quantitation limit was always <20%, such that the guidelines were generally met.

Analyte-related observations

CFT stability in human plasma at -70° was satisfactory as shown by reproducibility checks, during 3 months and again after nearly a year, on a QA sample which was assayed in each human plasma project.

Formation of a CFT isomer [5] caused concern prior to the assays in the very last project. The inactive Δ-2 isomer (iCFT; Fig. 1) had been isolated from the faeces of dogs treated with radiolabelled CFT and conceivably might be present in plasma; it could bias CFT values if not chromatographically separate from CFT. However, using iCFT synthesized in our Medicinal Chemistry Department, no interference was found, and review of earlier chromatograms for plasma samples showed that no iCFT was detectable.

Cessation of CFT studies.- CFT was dropped from development because effective plasma concentrations could not be obtained.

Bioavailability data were crucial in making the decision, and some of the concluding studies called for – and indeed received – rapid assay and report-generation to confirm the bioavailability problem. Top-priority assays were done in-house, in time for decision-making; even for the most demanding project (a 3-way cross-over study which occupied 2 weeks and generated 648 plasma samples) an analytical report was ready within 35 days of the start of the study.

Phase III: LESSONS LEARNED, AND NEW POLICIES

In retrospect, some decisions should have been made differently for the assay of CFT. It seems clear that a gradient separation should have been used initially. More time should have been invested in the assay development phase of the project. The time lost in developing the assay for the pro-drug could have been used to develop a better assay for CFT.

The change in column chemistry.- This problem was unforeseen. The only apparent remedy would be to have purchased all of the columns from the same batch, once the assay had been established using a particular column - an impractical remedy since initially it is never clear how many samples will arise in the development of a new drug. If the compound shows unacceptable toxicity at an early stage, bulk-purchase of columns would turn out to be unwarranted. Yet if assays were still incomplete and columns ran out, the column-chemistry problem could arise. The µBondapak Phenyl column still met Waters standards, and by most other criteria the batch matched that first used; yet it would not perform this particular separation. Now more columns are tested and if a problem develops the resulting data are retrieved to furnish alternatives. Changing columns in mid-assay is never easy but at times may be unavoidable.

Data handling.- The problem of data entry and transfer has been addressed in a number of ways. A program is now in use which allows the peak-height information to be extracted from the Nelson Analytical data system and transferred to QSIMPS. Sample information is still entered manually. The mainframe computer which the clinical pharmacokineticists use has been changed, and while the programs were being moved they were rewritten to allow data to be more readily taken from QSIMPS. The pre-clinical pharmacokineticists have started using the clinical pharmacokinetics package, so that manual re-entry of data is being minimized. The ability to perform pharmacokinetic evaluation of the data has now been incorporated into QSIMPS. The system uses NONLIN to fit a curve to the data and calculate various pharmacokinetic parameters,

QUANTITATIVE SELECTIVE ION MONITORING PROCESSING SYSTEM
PHARMACOKINETIC FITTING RESULTS (12M)

PROJECT: N3156B FILENAME: N3156B MODEL: 1FOSD
SUBJECTS: ALL
TREATMENTS: C-FASTED

VD	60.2755	A	59.9994	+/-	103.608
T1/2KA	44.5330	KA	.15564E-0.1	+/-	.584089E-02
T1/2K	66.6476	K	.10400E-01	+/-	.346014E-02
AUCTH	1914.26	TLAG	24.8703	+/-	1.49844
FU		KU		+/-	

XO 1200.00

	CALCULATED	FOUND
C-MAX	8.83955	9.02832
T-MAX	102.931	120.000

CORR. COEFF. .98

AREA UNDER MOMENT CURVE 328873
MEAN RESIDENCE TIME 178.087
PLOT LU ________

TIME POINTS 13
NM CHOICE ZERO

Fig. 4. Example of calculated parameters (see text).

HOFFMANN-LA ROCHE
Q. S. I. M. P. S.
PHARMACOKINETICS PLOT
SUBJECTS: ALL
TREATMENTS: C-FASTED
COMPOUND A:
Concentration
0
4
8
144
288
min

Fig. 5. Curve for mean plasma level *vs.* time, fitted by the QSIMPS pharmacokinetic program, for 6 subjects who took orally 1200 mg of peCFT after fasting (cf.Fig. 4).

as in Fig. 4; Fig. 5 shows the corresponding graphical output, a fitted curve being drawn by the program over the mean measured concentrations of peCFT in 6 peCFT-dosed volunteers.

The rapid disappearance of the pro-drug and the formation of an inactive isomer were observed by Toyama before CFT and peCFT were licensed by Roche [6, 7]. This information was available to the clinical pharmacokineticists, but not passed on to the analysts prior to the development of the initial assay, and their written report was released much later. With improved lines of communication, such problems are unlikely to recur.

Data-handling policies.- In view of the problems with data transfer and manual entry, the programs for both analytical and pharmacokinetic data are now placed on one computer in one software package. The ultimate solution to the sample-definition problem is bar-coding or some type of tagging system that allows the computer to draw from memory the sample-identity information available in the data base. Thereby the computer can correlate the analytical results which it calculates with the sample information held in the corporate data base, both that entered when the study was initiated and that added as data become available. These data are then available to the programs which will perform the pharmacokinetic analysis. Overall, less sample information will have to be entered by the analyst. Involvement of an outside analytical laboratory should entail its using a data system of in-house type or an interface program enabling electronic rather than manual data transfer.

Metabolites, and overall retrospect.- The search for metabolites is now started earlier, so as to indicate any assay interference or the presence of active metabolites. Despite the various problems encountered, all samples (plasma, >6000; urine, >900) were assayed on time during the 18 months and results were reported to colleagues and the FDA when required. The results were crucial in the eventual decision to stop development of the drug.

In summary, assay-development complications were aggravated by the urgency, by column-batch problems, by obligatory manual entry of data, and by belated awareness of an isomer problem. Where development of a drug has to be fast, then (1) the assay for plasma and urine must be developed expeditiously; (2) actual assays must be swift, with minimal re-assay; (3) results must be available quickly, for planning subsequent studies producing reports for internal and external use.

These requirements can be met with a significant amount of effort. However, time constraints which shorten the assay development time may lead to problems later on. The difference between success and failure in a project often depends on the ability of the people in a group to overcome problems as they become apparent.

References

1. Oldfield, N., Chang, D., Garland, W. & Town, C. (1987) *J. Chromatog. 422*, 135-143.
2. Chau, P.Y., Leung, Y.K., Ng, W.W. & Arnold, K. (1987) *Antimicrob. Agents Chemother. 31*, 473-476.
3. Neu, H.C., Chin, N.X. & Babthavikul, P. (1986) *Antimicrob. Agents Chemother. 30*, 423-428.
4. Garland, W.A., Hess, J. & Barbalas, M.P. (1986) *Trends Anal. Chem. 5*, 132-138.
5. Saab, A.N., Dittert, L.W. & Hussain, A.A. (1988) *J. Pharm. Sci. 77*, 906-907.
6. Saikawa, I., Maeda, T., Yoshifumi, N., Sakai, H., Hatakawa, H., Onoda, M. & Matsutani, H. (1986) *Japanese J. Antibiotics 39*, 979-990.
7. *As for* 6., 991-995.

#B-4

DEVELOPMENT OF AN EFFICIENT HPLC ANALYSIS FOR LINCOMYCIN FOR USE IN CONTRACT LABORATORIES

C.A. James, R.J. Simmonds and S.A. Wood

Clinical Research Division, Upjohn Ltd.,
Fleming Way, Crawley, W. Sussex RH10 2NJ, U.K.

Requirement *A sensitive, inexpensive assay for lincomycin in plasma (human or porcine) and urine, sufficiently robust to be contracted out.*

End-step *Ion-pair RP-HPLC (CN or TMS for plasma, C-8 for urine) with an acetonitrile-TFA* eluent; detection at 214 nm.*

Sample handling *SPE with C-2 cartridges.*

Comments *The desired characteristics were achieved, including simplicity and selectivity for each sample type, such that >1800 samples were successfully assayed under contract. The RP-HPLC and SPE bonded phases were chosen from several tested.*

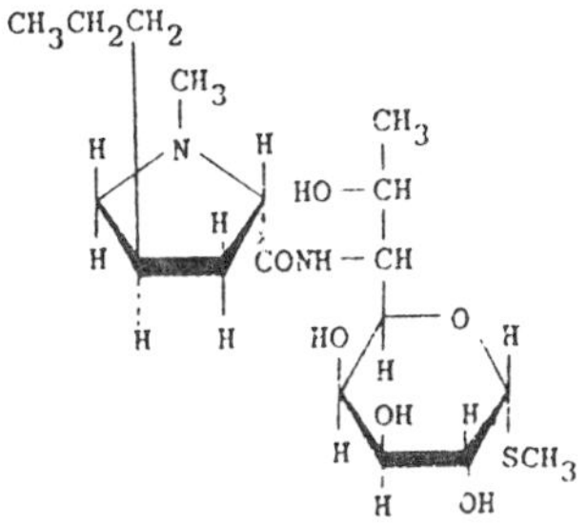

Lincomycin is an antibiotic originally discovered in the 1950's from cultures of the soil organism *Streptomyces lincolnensis*. It is used to treat serious infections in man, and in veterinary medicine for indications such as porcine dysentery. Assays were required to measure it in human plasma, human urine and porcine plasma samples, to support studies of new formulations.

Bioassay (microbiological) has traditionally been used to analyze lincomycin in biological fluids [1]. However this method lacks specificity and sensitivity, and a chromatographic technique is preferable [2]. A method developed for human plasma (J.B. Fourtillan & co-workers, pers. comm.) entails ion-pair HPLC with detection at a low UV wavelength, but it needs a relatively large sample size, and the long chromatography run-time would make it impractical for processing large numbers of samples at a contract facility. GC analysis of derivatized lincomycin is also possible, and

---

**Abbreviations*.- SPE, solid-phase extraction; TFA, trifluoroacetic acid; RP, reversed-phase.

has been used for residue analysis of lincomycin in animal tissues [3], but was too complex for our needs. As lincomycin is polar and high sensitivity was not needed, we felt that SPE followed by HPLC with low-wavelength UV would be most likely to provide a simple, robust assay.

METHOD DEVELOPMENT

Chromatography

We began by investigating the chromatography of the compound on a number of different bonded-phase silica HPLC columns. This quickly identifies usable chromatographic conditions and HPLC columns that give satisfactory retention and chromatographic efficiency for the analyte. The selectivity of different bonded phases can then be further explored and optimized with extracts of the biological matrices: we examined columns taken mostly from the Zorbax range (C-18, C-8, Ph, TMS, CN; 250 × 4.6 mm, 5 μm), these having proved robust in the past.

The potential choice of mobile phase modifiers is vast. Initially we focussed on a mobile phase containing acetonitrile-aqueous TFA (0.1-0.2% v/v in the mixture). This works well for various compound types including acids, bases and neutral drugs. Other modifiers were also considered, taking account of the chemical nature of the compound (lincomycin is basic), of published methods, and of the results of pilot experiments. Thus, we tried sulphonic acid ion-pairs, and mobile phases containing acetonitrile and 0.1 M ammonium phosphate buffer pH 5.6, already used by Upjohn for analysis of dosage formulations.

Without an ion-pair, little retention was seen with most types of RP column, although satisfactory retention was seen on a Phenyl column with ammonium phosphate or TFA mobile phases. Reasonable retention could, however, be obtained on a number of different columns in the presence of a sulphonic acid ion-pair. As expected, the retention (K') increased with increasing chain length.

We regard it as important to examine biofluids early in method development. During initial experiments with extracts of human plasma it was clear that ion-pair systems provided a selectivity not seen with mobile phases lacking such modifiers. The efficacy of an ion-pair also suggested the possibility of a cation-exchange system (Spherisorb SCX). This worked fairly well, but was significantly less efficient than the ion-pair systems and offered no other advantages such as selectivity or robustness.

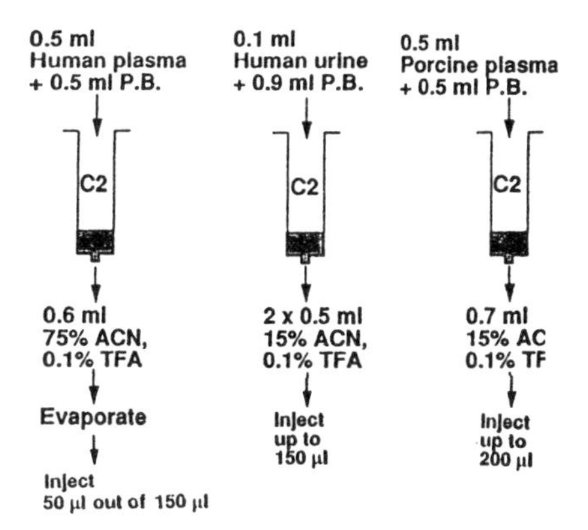

Fig. 1. SPE extraction procedure using BondElut C-2 cartridges (100 mg, 1 ml size), all solutions being drawn through using vacuum. ACN, acetonitrile; P.B., 0.1 M phosphate buffer pH 7.4. The cartridges were primed with 2 ml ACN, 2 ml elution solvent, 1 ml water and 1 ml PB, then washed with 2 ml water and 2 ml 10% (v/v) methanol.

Extraction procedure

Once suitable chromatographic conditions were obtained for lincomycin, methods for extracting from biofluids were studied. SPE, performed manually using VacElut vacuum manifolds, is the preferred sample preparation method for all of our HPLC assays. This procedure is relatively quick and easy, robust, and usable for large numbers of samples. This choice was reinforced by the polar nature of lincomycin, precluding simple liquid-liquid extraction. Moreover, SPE methods have been successfully transferred by us to contract facilities, and the equipment required is inexpensive even if the laboratory is not already equipped for SPE.

As with HPLC stationary phases, we screen the full range of Bond Elut cartridges in the RP mode (selectivity effects in particular being non-predictable) and thereby arrive at a short-list. Good retention of lincomycin was obtained on most of the non-polar types of cartridge (C-18, C8, Ph, C-2, CH), and C-2 was chosen as it provided the cleanest extracts of control human plasma. Then conditions such as wash and elution solvents and loading-solution pH were further improved.

ANALYTICAL METHOD

Extraction procedure.- The methods (Fig. 1) were similar for all three biofluids, with nearly identical priming and wash steps but slightly different elution procedures. That shown for lincomycin in **human plasma** included an evaporation step to concentrate the sample and ensure that the final extract was in a constant volume for HPLC loading. The organic content of the elution solvent was higher than required for complete elution, to allow speedy evaporation. To analyze lincomycin in **human urine** less sensitivity was needed, as reflected in the eluate composition (Fig. 1): the inconvenience of an evaporation step was avoided. For **porcine plasma** the elution procedure was slightly varied (Fig. 1), and up to 200 µl of extract injected, achieving simplification with no evident detriment to column efficiency or reliability.

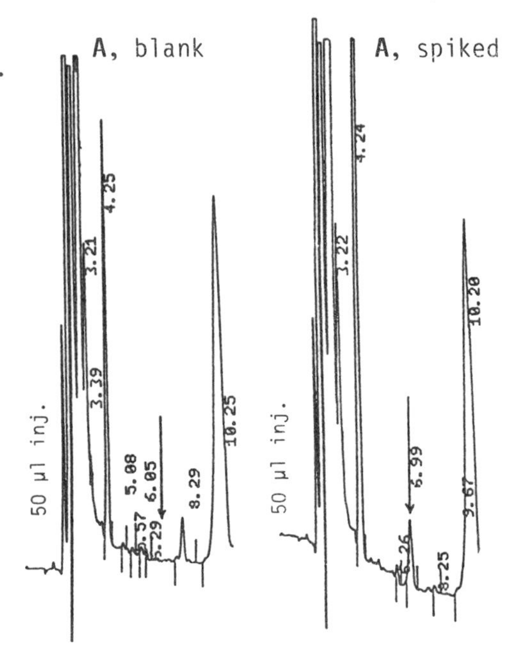

Fig. 2 *(continued opposite).* Patterns for blanks and lincomycin-spiked controls with 5 µm columns (250 × 4.6 mm) run at 35° with 1.5 ml/min flow-rate; detection at 214 nm.

A: human plasma (EDTA); 0.5 µg/ml spike.
Column: Microsorb CN.
Eluent: 8% (v/v) ACN/0.02% (v/v) TFA/0.1% (w/v) 'C-6' i.e. hexanesulphonic acid.
Chromatographic efficiency:
k' = 2.5
N (plates/column) = 4800
Skew (A_S) = 1.2
Arrows on graphs = drug position. SPE procedures as in Fig. 1.

Chromatography.- Different types of column were needed to obtain selectivity for different biofluids (Fig. 2 legends): that chosen for **human urine** as distinct from **plasma** (with similar solvent systems, both containing a 'C-6' ion-pair) gave better resolution from endogenous components of urine extracts. **Porcine plasma** was 'dirtier' than human plasma using either HPLC system: improved selectivity was unobtainable with any other type of bonded-phase column if the mobile phase contained the C-6 ion-pair, but was achieved with a 'C-7' ion-pair combined with a TMS column (Fig. 2 legend, C).

Fig. 2 shows representative chromatograms for extracts of the three biofluids assayed. It will be noted that no internal standard was included. Lincomycin was introduced in the 1950s, and unfortunately no close analogues are now available. Synthesis of a suitable compound was requested, but proved rather difficult, and none was forthcoming till the study was over.

Validation and contracting.- The following results (with µg/ml values as stated) were obtained in-house before contracting out, ***rec.*** signifying % recovery and **C.V.** (%) inter-assay variation.

- **Human plasma** (range 0.1-30): ***rec.*** *93.7* (0.25) & *92.7* (11.2);
  C.V. 11.4 (0.101) & 3.5 (11.2).
- **Human urine** (range 1-200): ***rec.*** *116* (9.9) & *104* (49.8);
  C.V. 6.2 (9.9) & 5.1 (194).
- **Porcine plasma** (range 0.2-20): ***rec.*** *99.1* (0.51) & *107* (10);
  C.V. 16.2 (0.204) & 4.0 (10).

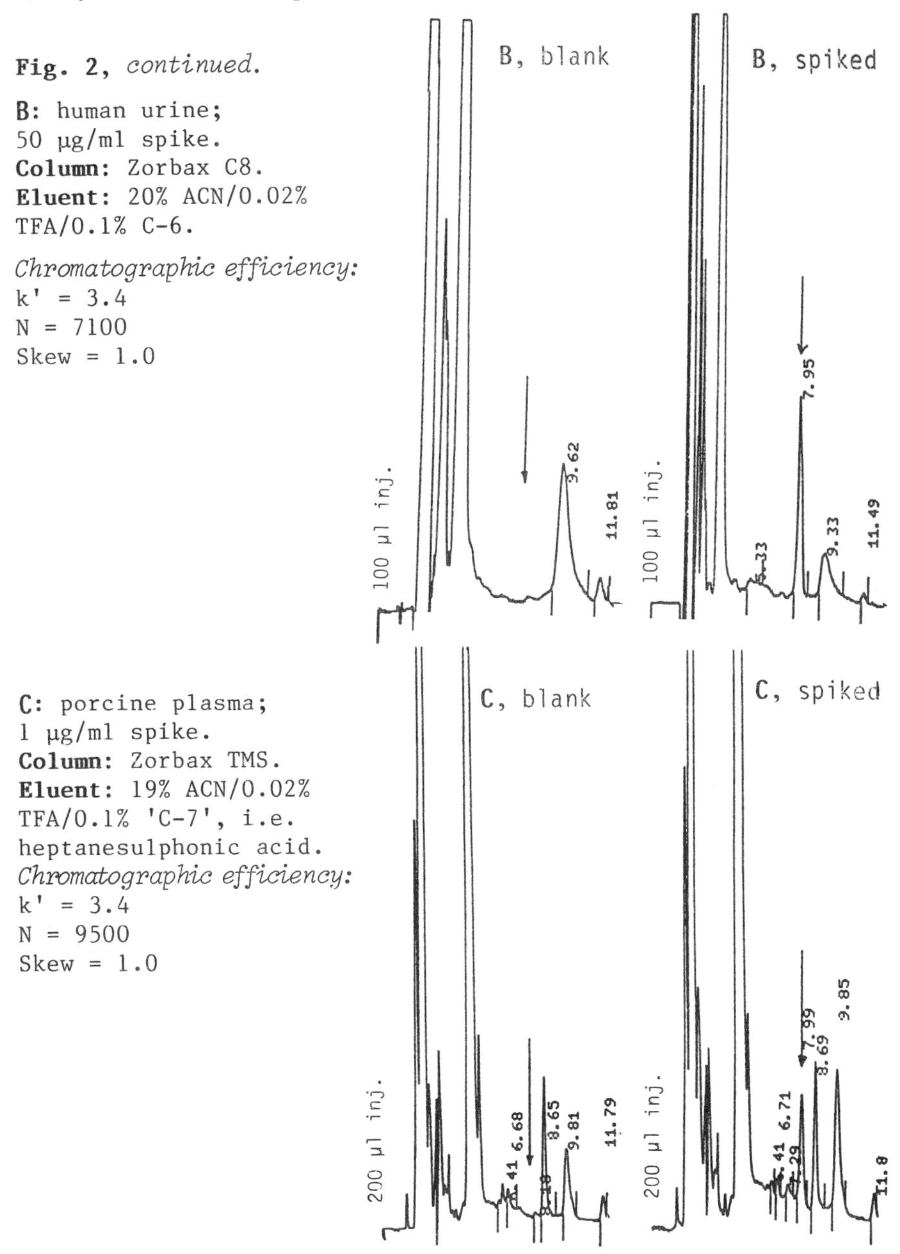

Fig. 2, *continued.*

B: human urine;
50 µg/ml spike.
Column: Zorbax C8.
Eluent: 20% ACN/0.02% TFA/0.1% C-6.
Chromatographic efficiency:
k' = 3.4
N = 7100
Skew = 1.0

C: porcine plasma;
1 µg/ml spike.
Column: Zorbax TMS.
Eluent: 19% ACN/0.02% TFA/0.1% 'C-7', i.e. heptanesulphonic acid.
Chromatographic efficiency:
k' = 3.4
N = 9500
Skew = 1.0

Linearity was good (r >0.99), and no stability problems were encountered in either biofluid samples or extracts. The assays were then transferred to two separate contract facilities and briefly validated; ~1500 human plasma and urine samples were assayed, and 300 porcine plasma samples. No major problems were encountered in either the installation or the validation of the assays at these laboratories, or in the running of the actual study samples [cf. a companion article,

#ncA-2]. Inter-assay coefficients based on quality-control (QC) results from all the contracted assays were as follows:
- **Human plasma: C.V.** 9.92 (2.2) & 8.61 (14 [μg/ml]).
- **Human urine: C.V.** 9.12 (28) & 5.25 (150).
- **Porcine plasma: C.V.** 8.8 (2) & 4.6 (12).

DISCUSSION

To achieve our aim of quickly producing robust methods for contracting-out, we screened a range of HPLC columns and extraction cartridges. We do not attempt to test every combination of columns, cartridges and conditions, but do aim to appraise the analyte's behaviour under a variety of conditions. When problems are encountered, such initial experiments may suggest alternative columns or cartridges, as proved useful here with pig plasma. Examination of a range of assay conditions also helps ensure the robustness of the method, especially non-dependence of the assay on some unique column or cartridge batch. Extraction conditions, e.g. wash solvents, can be chosen so that there is some built-in safety margin whereby small changes in reagent concentration will not affect the analysis.

The following points, in the context of contracting-out, complement pertinent points in the companion article*. #Early in method development, assay steps should be tried with the drug in its biofluid rather than water or buffer, in order to check extraction efficiency (which may differ) and chromatographic selectivity. #During development the procedures are refined as necessary, e.g. if different biofluids have to be assayed. #To avoid constraints of cost and of ineligible contract laboratories (lacking appropriate equipment or expertise), one should avoid complex extraction procedures, long run-times and use of non-standard equipment such as the Varian AASP or column switching units. #Obtaining the desired time-slot at a contract house can be a problem, as can transport of samples especially if between countries. #Familiarization gained by setting up the method in-house is vital for keeping the contracted-out work on the rails.

References

1. Barbiers, A.R. & Neff, A.W. (1976) *J. Assoc. Off. Anal. Chem. 59*, 849-854.
2. Thomas, A.H. (1987) *J. Pharm. Biomed. Anal. 5*, 319-324.
3. Farrington, W.H.H., Cass, S.D., Patey, A.L. & Shearer, G. (1987) *Food Addit. Contam. 5*, 67-75.

**Present text abridged by Editor in view of this article*

#B-5
SIMULTANEOUS HPLC DETERMINATION OF TRIMETHOPRIM, SULPHAMETHOXAZOLE AND ITS N⁴-ACETYL METABOLITE IN BIOLOGICAL FLUIDS

O. Varoquaux, P. Cordonnier, C. Advenier and M. Pays

Centre Hospitalier de Versailles,
Département de Pharmacologie et de Biochimie,
177, rue de Versailles, F-78157 Le Chesnay, France

Requirement *for 'Method* **II'**	*An assay method, simpler than the gradient method ('I') previously developed, for simultaneous assay of TMP, SMZ and its* N-*acetyl derivative (N^4SMZ; see Panel) in fluid samples, especially from patients receiving the combination TMP-SMZ (co-trimoxazole).*
End-step	*Isocratic NP-HPLC with elution by methanol-water-DEA-DCM* (6: 0.2: 0.05: 93.75 by vol.) and detection at 280 nm. 'Method* I'*, described too, uses stepwise elution.*
Sample handling	*Solvent extraction at pH 6.2: with (for* I*) chloroform-ethyl acetate, 3:1 by vol., or chloroform alone - which gives cleaner extracts and a tolerable although much lowered recovery of N^4SMZ (Method* II*).*
Comments	*The minimum quantifiable levels are 10 ng/ml for TMP and SMZ, and 50 ng/ml for N^4SMZ. The method is rapid, sensitive, precise and well suited for therapeutic monitoring; the method obviates previously encountered interferences arising from ingestion of tea or coffee. (This detailed description is novel†.)*

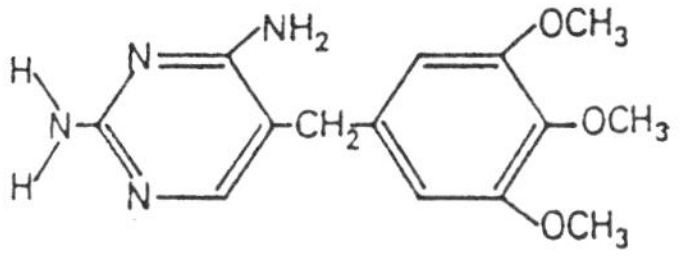

NH_2 OCH_3 OCH_3 OCH_3 CH_2

1) Trimethoprim (**TMP**)

Co-trimoxazole is a mixture of 1) and 2), 1:5 by wt.

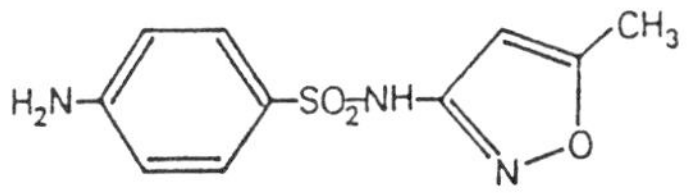

2) Sulphamethoxazole (**SMZ**)

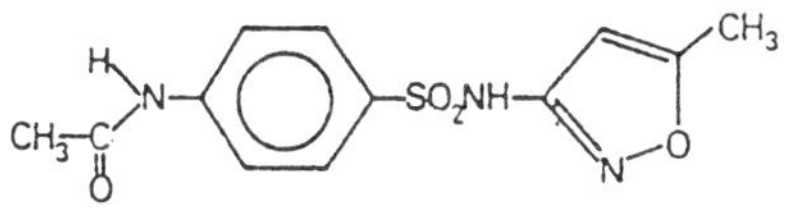

3) N-4 acetylsulphamethoxazole (**N^4SMZ**)

Abbreviations (& see above Panel).*- NP-, RP-: normal phase-, reverse phase-(HPLC); DEA, diethylamine; DCM, dichloromethane; **A & **B** signify solvent mixtures (for mobile phase); i.s., internal standard. | †Method outlined (1988) in *Ann. Clin. Biol.* *46*, 550.

Co-trimoxazole is a powerful broad-spectrum antibiotic used for treating various infections in man [1, 2]. We have re-examined methods for assaying its constituents and the metabolite N^4SMZ in biological fluids, in connection with therapeutic monitoring and clinical pharmacokinetic investigations. For SMZ there are reported methods entailing GC-FID [3-5] or HPLC-UV [6-11], and for TMP use has been made of GC with thermionic detection [12, 13] and, with greater ease and sensitivity, HPLC [14]. Determinations by GC-MS [15], polarography [16, 17] and microbiological [18, 19] or isotopic [20, 21] assay have also been described.

With co-trimoxazole, the TMP and SMZ can be determined by spectrofluorimetry [22], but the method is lengthy, requires two extractions and is not free from interferences. GC determination with thermionic detection [23] is also lengthy, requiring two extractions and preparation of a methyl derivative to separate SMZ and N^4SMZ, and plasma gives interfering peaks. More recent methods mostly use RP-HPLC with C-8 [24], C-18 [25-27] or cyanopropyl [28] phases. Some authors perform protein precipitation which, if not followed by direct injection [24, 25], is preceded [27] or followed [29] by an extraction step.

As the two compounds have different analytical characteristics, two points have to be considered: which pH is optimal for extraction and whether a single solvent or a mixture provides the best compromise that achieves simultaneous extraction and a satisfactory elution pattern. Ascalone [26] performed an ethyl acetate extraction, then C-18 RP-HPLC or, as an improvement [30], NP-HPLC which surpassed previous NP-HPLC methods [14, 31] in respect of separability and sensitivity. Despite corroborative results by van Der Steuijt & Sonneveld [27], we were unable to reproduce these results, using ethyl acetate extraction. Accordingly, for simultaneously estimating TMP, SMZ and N^4SMZ using NP-HPLC we developed two types of extraction procedure:- with **(I)** a mixed solvent (chloroform-ethyl acetate) and **(II)** chloroform alone, which optimized respectively the extractions of N^4SMZ and of TMP. These aspects have now been investigated, and the resulting new methods applied to therapeutic concentrations of co-trimoxazole. With **(I)** the 3 analytes can be determined in biological fluids, especially urines or exudates, using an elution gradient which enables traces of N^4SMZ to be determined, whereas **(II)**, optimal for TMP extraction and allowing isocratic elution, obviates possible interferences with more complex biological fluids such as plasma [32].

METHOD (I): Experimental

Chemicals.- TMP, SMZ, N^4SMZ and, as i.s. for TMP, 2,4-diamino-5-(3,5-dimethoxy-4-methylbenzyl)pyrimidine were each supplied by Hoffmann-La Roche, Basle. Sulphamoxol (Justamil®) used as i.s. for the sulphonamides was from Amphar-Rolland, Paris. The i.s. and calibration solutions were methanol dilutions of the free bases. Chloroform, ethyl acetate, and methanol Suprapur were from E. Merck, Darmstadt, and ammonia (28-30% w/v solution) was from Prolabo, Paris, which also supplied constituents of the phosphate buffer (0.2 M, pH 6.2) - a mixture (81.5:18.5 by vol.) of 2.724% (w/v) KH_2PO_4 and 7.16% $Na_2HPO_4 . 12\ H_2O$ in distilled water.

Chromatography was performed on a Varian 5000 instrument with a fixed-wavelength (280 nm) detector and a 50 µl fixed-volume injector. The stainless-steel column (250 × 4 mm) was filled with 5 µm LiChrosorb Si-60 (Merck 9388) using the balanced-density technique: the slurry (3.6 g dispersed in 15.6 ml CCl_4) was forced into the column with methanol. Alternatively, a Hibar column (Merck 50388) was used. The pre-column was a 40 × 4 mm s.s. tube filled with 25-40 µm LiChroprep Si-60 (Merck 9390). The components of the mobile phase, flowing at 2 ml/min, were chloroform-methanol-water-ammonia: **A**, 94.5:5.0: 0.25:0.19 (by vol.); **B**, 79:20:1:0.15; programming: **A** for 3 min; **A** + **B** (50:50) for 1.0 min; **B** for 8.0 min; **A** + **B** (50:50) for 1.0 min; **A** for 3.0 min (room temperature operation). As shown in Fig. 1, TMP and its i.s. were eluted between 0 and 6 min, and SMZ, N^4SMZ and their i.s. between 7 and 13 min. After ~500 injections the times were unchanged and the pressure (maximally ~180 bars) had not increased.

Sample preparation.- To 1 ml plasma, or urine (usually diluted 2- to 5-fold), in a 30 ml glass tube were added 20 µl i.s., 5 ml buffer and 12 ml chloroform-ethyl acetate (75:25). The stoppered tube was shaken for 15 min (Kahn vibrator) and centrifuged at 3000 **g** for 10 min. Then 10 ml of organic phase was taken to dryness under N_2 at 50° and the residue dissolved in 300 µl of **A** and vortexed for 20 sec; 50 µl were injected.

Calibration curves (based on peak area ratios).- To drug-free patient plasma samples (1 ml) were added 20 µl of each i.s., and 100 µl of a mixture of standards at concentrations over the range 0.5-10 (TMP), 5-100 (SMZ) or 2.5-50 (N^4SMZ) µg/ml. Diluted urines were similarly spiked (TMP-SMZ and N^4SMZ at 2.5-50 and 5-100 µg/ml respectively). Processing was as above.

METHOD (I): Results and Discussion

Choice of extraction solvent.- This was guided by previous work. TMP was extracted by Weinfeld [14] at pH 10 in chloroform and, along with the sulphonamides, by Ascalone [26, 30] at pH 6.8 in ethyl acetate. We have verified that TMP at neutral or alkaline pH is more soluble in chloroform than in ethyl acetate, in contrast to SMZ and N^4SMZ. We made mixtures of the two solvents: 3-6 extractions with each were performed on 1 ml plasma spiked with TMP and its i.s. (each 2 μg), and SMZ (40 μg), N^4SMZ (20 μg) and their i.s. (40 μg). With the usual work-up (12 ml of solvent; pH 6.2) we found that raising the proportion of chloroform or of ethyl acetate considerably improves, respectively, the extraction of TMP and of the sulphonamides (Fig. 2). Taking into account that the concentrations of TMP in samples are low, only 70% to 80% chloroform gives acceptable simultaneous extraction of 60-80% of the three analytes; this justifies the choice of a 75:25 proportion in the proposed method.

Choice of pre-extraction buffer pH.- With plasma spiked as above, extracted in quadruplicate with 12 ml 70:30 chloroform-ethyl acetate, the results shown in Fig. 3 were obtained. Extraction of SMZ is maximal at pH <5.6, and that of TMP at >8.0. The chosen pH, 6.2, is a compromise that takes account of the plasma concentration differences between TMP and SMZ. Evidently use of pH 6.8 [26] would be prejudicial to SMZ extraction.

Optimal conditions for a particular analyte.- With TMP at 1, 2 and 4 μg/ml, using pH 11 buffer and 100% chloroform for extraction, the efficiencies in 5 extractions were 96.6, 98.0 and 95.4% with, respectively, C.V.'s of 1.4, 0.7 and 0.9%. With SMZ (20, 40 and 80 μg/ml) and N^4SMZ (10, 20 and 40 μg/ml), using pH 5.4 buffer and 100% ethyl acetate, the efficiencies in 4-6 extractions were 97.0, 98.5 and 99.1% (C.V.'s 0.8, 2.4 and 1.1%) for SMZ and 99.3, 98.2 and 98.9% for N^4SMZ (C.V.'s 1.4, 2.7 and 0.9%). In view of the diverse analytical characteristics it would be illusory to envisage a quantitative yield at an intermediate pH - a compromise chosen to allow a simpler and faster assay. Each i.s. as used for validation has comparable analytical characteristics, being homologous or analogous in structure. The procedure is as used by Weinfeld [14] for TMP alone, and by Gochin & co-authors [29] for co-trimoxazole constituents with, however, an i.s. only for the sulphonamide derivatives.

Chromatographic conditions obviated an elution gradient but entailed a step whereby the proportion, in the eluent, of polar components (methanol and water) and the basicity were manipulated. With solvent **A** (as published [14] for TMP) the basicity allows the separation of TMP and SMZ (i.s.). The more polar solvent **B** accelerates SMZ and N^4SMZ elution.

Linearity was good (r = 0.999) for spiked samples with TMP (over the range 0.1-20 μg/ml), SMZ (2-150 μg/ml) and

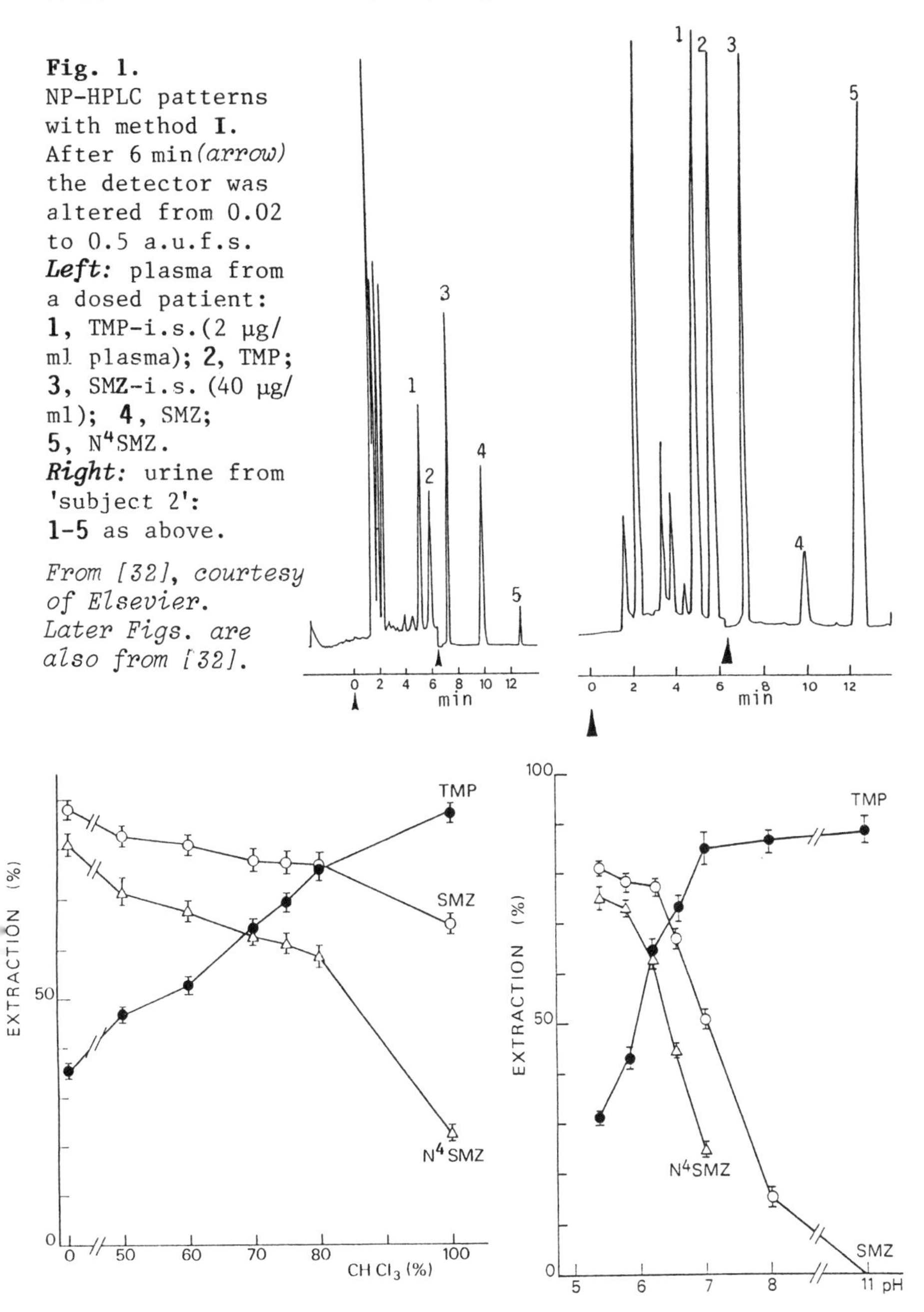

Fig. 1. NP-HPLC patterns with method **I**. After 6 min *(arrow)* the detector was altered from 0.02 to 0.5 a.u.f.s. ***Left:*** plasma from a dosed patient: **1**, TMP-i.s. (2 µg/ml plasma); **2**, TMP; **3**, SMZ-i.s. (40 µg/ml); **4**, SMZ; **5**, N^4SMZ. ***Right:*** urine from 'subject 2': **1-5** as above.

From [32], courtesy of Elsevier. Later Figs. are also from [32].

Fig. 2. Influence of the relative proportions of chloroform and ethyl acetate on % extraction mean values at pH 6.2 for analytes added to plasma (see text).

Fig. 3. Influence of pH of the phosphate buffer on % extraction (mean values) with 70:30 chloroform-ethyl acetate (see text).

N^4SMZ (range 1-100 µg/ml). The linear regression expressions (y) were respectively 0.049+0.28x, 0.089+0.025x and 0.043+0.012x. Each day's analyses included calibration curves.

Accuracy, precision, sensitivity and selectivity.- Satisfactory C.V.'s were obtained with spiked plasma samples in between-day and within-day comparisons [32]. Extraction yields, typically only ~70% for TMP, ~80% for SMZ and ~63% for N^4SMZ, are not good but are as expected in view of the pH and solvent-choice compromises. The sensitivity limit (ng/ml), taking 1 ml plasma but only 100 rather than 300 µl of the evaporation residue, is 15 for TMP and, with the detector altered from 0.01 to 0.05 a.u.f.s., 20 for SMZ and 10 for N^4SMZ. This corresponds to a signal-to-noise ratio of 3 for TMP and 2 for SMZ and N^4SMZ. The method copes with biotransformation of SMZ to N^4SMZ, usually 30% in blood and 60% in urine [33]. That of TMP is less important, the proportion of metabolites being ~10% in blood and ~20-40% in urine.

Stability is good when the evaporation residue, routinely re-dissolved just before column loading, is stored dry in the refrigerator for 2 days. **Interferences** can occur with plasma due to theophylline and an unidentified metabolite of caffeine which co-chromatograph with the TMP i.s. (caffeine itself elutes within 2 min). Hence ingestion of tea or coffee is debarred unless a different analytical method is adopted, as follows.

METHOD (II)

The following modifications were introduced: (i) a simplified isocratic elution using as mobile phase methanol-water-DEA-DCM, 6:0.2:0.05:93.75 by vol.; (ii) simplified plasma extraction by omitting ethyl acetate, so reducing extraction of possible interfering compounds; (iii) reduction of the injection volume, from 50 to 20 µl; (iv) adoption of a single i.s., sulphathiazole, for measuring all 3 analytes. The LiChrosorb column and its operation were as before. After adding the i.s. (62.5 µg) to 1 ml plasma, then 1 ml 0.5 M pH 6.2 phosphate buffer, extraction is performed with 8 ml chloroform. The organic phase is dried down under N_2, and the residue re-dissolved in 500 µl of the mobile phase; 20 µl are chromatographed. Tubes must be protected from light to avoid possible UV degradation of the compounds [34].

Retention times (min; those with Method **I** shown thus, [ ]) are 4.0 [5.8] for TMP, 6.0 [9.6] for SMZ, 7.6 for i.s. and 12.7 [12.6] for N^4SMZ (Fig. 4). The minimum quantifiable levels (µg/ml) are: TMP and SMZ, 0.01; N^4SMZ, 0.05. Accuracy and precision were determined with spiked plasma: TMP added

Fig. 4. NP-HPLC patterns with Method **II**, for 1 ml plasma samples. **C**, control; **S**, spiked with TMP, SMZ, i.s. and N^4SMZ (2.5, 25, 62.5 & 25 µg); **P**, from patient given cotrimoxazole, 9 × 960 mg tablets over 3 days (the TMP, SMZ and N^4SMZ peaks correspond to 2.2, 24 & 68 µg/ml plasma).

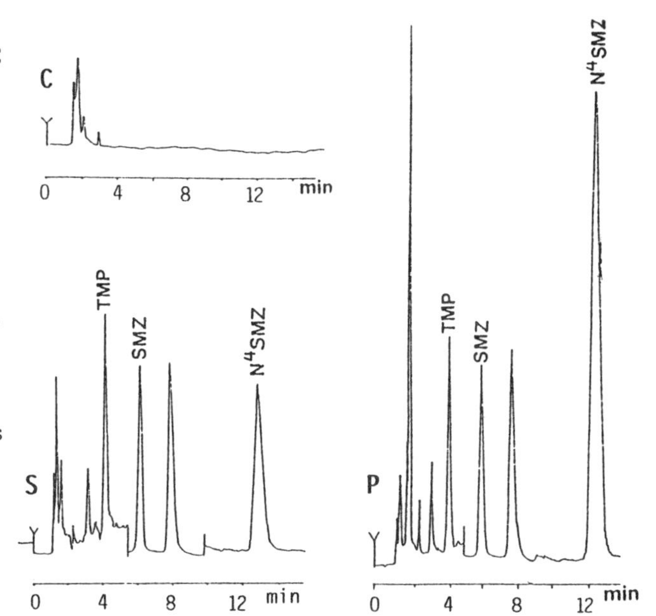

to give 1 and 5 µg/ml, SMZ and N^4SMZ 10 and 50 µg/ml (n = 10). The respective C.V.'s were 3.3 and 4.6%, 3.9 and 5.5%, and 3.3 and 1.1%. Mean recoveries were respectively 87%, 67%, and only 20% although this is acceptable, the N^4SMZ calibration curves remaining linear (r = 0.994).

The isocratic method, allowing rapid simultaneous determination of the three analytes in plasma, has already been advantageous for deciding haemodialysis frequency in a patient with a bilateral total nephrectomy, receiving co-trimoxazole because of an atypical mycobacteria osteitis.

CONCLUDING COMMENTS

The proposed NP-HPLC methods for simultaneously determining TMP, SMZ and N^4SMZ in biological fluids entail the same compromise for pH (6.2) but not for extracting solvent. They have detection limits apt for clinical pharmacokinetic studies, besides therapeutic monitoring, as published from our group [e.g. 35]. The simpler, speedier method (**II**; isocratic) minimizes endogenous interferences.

References

1. Bushby, S.R.M. & Hitchings, G.H. (1968) *Br. J. Pharmacol.* *33*, 72-92.
2. Kucers, A. & Bennett, N. McK. (1979) *The Use of Antibiotics*, 3rd edn., Heinemann Medical Books, London, pp. 687-729.
3. Daun, R.J. (1971) *J. Ass. Offic. Anal. Chem.* *54*, 1277-1282.
4. Roder, E. & Stuhe, W. (1974) *Z. Anal. Chem.* *271*, 281-283.

5. Nose, N., Kobayashi, S., Hirose, A. & Watanabe, A. (1976) *J. Chromatog.* *123*, 167-173.
6. Su, S.C., Hartkopf, A.V. & Karger, B.L. (1976) *J. Chromatog.* *119*, 523-538.
7. Peng, G.W., Gadulla, M.A.F. & Chiou, W.L. (1977) *Res. Comm. Chem. Path. Pharmacol.* *18*, 233-246.
8. Johnson, K.L., Jetter, D.T. & Clairbone, R.C. (1975) *J. Pharm. Sci.* *64*, 1657-1660.
9. Sharma, J.P., Perkins, E.G. & Bevill, R.F. (1976) *J. Pharm. Sci.* *65*, 1606-1608.
10. Penner, M.H. (1975) *J. Pharm. Sci.* *64*, 1017-1019.
11. Kram, T.C. (1972) *J. Pharm. Sci.* *61*, 254-256.
12. Lotter, A.P. & Goossens, A.P. (1977) *S. Afr. Pharm. J.* *44*, 282-283.
13. Land, G., Dean, K. & Bye, A. (1978) *J. Chromatog.* *146*, 143-147.
14. Weinfeld, R.E. & Macasieb, T.C. (1979) *J. Chromatog.* *164*, 73-84.
15. Irwin, W.J. & Slack, J.A. (1977) *J. Chromatog.* *139*, 364-369.
16. Brooks, M.A., de Silva, J.A.F. & D'Arconte, L. (1973) *Anal. Chem.* *45*, 263-266.
17. as for 16., *J. Pharm. Sci.* *62*, 1395-1397.
18. Bushby, S.R.M. (1969) *Postgrad. Med. J.* *45*, *Suppl. 10*, 18.
19. Pechere, J.C. (1970) *Path. Biol.* *18*, 343-345.
20. Schwartz, D.E., Wetter, W. & Englert, G. (1970) *Arzneim.-Forsch./Drug Res.* *20*, 1867-1871.
21. Nielsen, P. & Folke, R. (1975) *Acta Pharmacol. Toxicol.* *37*, 309-316.
22. Lichtenwalner, D.M., Suh, B., Lorberg, B. & Sugar, A.M. (1979) *Antimicrob. Agents Chemother.* *16*, 579-583.
23. Ascalone, V. (1978) *Boll. Chim. Farm.* *117*, 176-186.
24. Vree, T.B., Hekster, Y.A., Baars, A.M., Damsma, J.E. & van der Kleijn, E. (1978) *J. Chromatog.* *146*, 103-112.
25. Bury, R.W. & Mashford, M.L. (1979) *J. Chromatog.* *163*, 114-117.
26. Ascalone, V. (1980) *J. High Res. Chromatog. Chromatog. Comm.* *3*, 261-264.
27. van Der Steuijt, K. & Sonneveld, P. (1987) *J. Chromatog.* *422*, 328-333.
28. Erdmann, G.R., Canafax, D.M. & Giebink, G.S. (1988) *J. Chromatog.* *433*, 187-195.
29. Gochin, R., Kanfer, I. & Haigh, J.M. (1981) *J. Chromatog.* *223*, 139-145.
30. Ascalone, V. (1981) *J. Chromatog.* *224*, 59-66.
31. Bye, A. & Brown, M.E. (1977) *J. Chromatog. Sci.* *15*, 365-371.
32. Spreux-Varoquaux, O., Chapalain, J.P., Cordonnier, P., Advenier, C., Pays, M. & Lamime, L. (1983) *J. Chromatog.* *274*, *187-199.*
33. Patel, R.B. & Welling, P.G. (1980) *Clin. Pharmacokin.* *5*, 405-423.
34. Bergh, J.J., Breytenbach, J.C. & Wessels, P.L. (1989) *J. Pharm. Sci.* *78*, 348-350.
35. Varoquaux, O., Kasparian, P., Pieplu, C., Brion, N., Guibout, P., Pays, M. & Advenier, C. (1986) *Rev. Fr. Mal. Resp.* *4*, 213-217.

#B-6

HPLC OF RECENT QUINOLONE ANTIMICROBIALS

Klaus Borner, Ellen Borner, Hildegard Hartwig and [†]Hartmut Lode

Institut für Klinische Chemie und Klinische Biochemie, and [†]Medizinische Klinik und Polyklinik, Klinikum Steglitz der Freien Universität Berlin, Hindenburgdamm 30, 1000 Berlin 45, F.R.G.

Requirement	*An HPLC method for quinolones in biological fluids with adequate sensitivity (<10 μg/L) and specificity (metabolites to be separated, and quantified also).*
End-step	*RP-HPLC* with fluorescence detection. Alternatives: IEC; UV absorption detector; diode array detector.*
Sample handling	*Storage at -80°. Protection against light during handling. Dilution of low-protein fluids with mobile phase. Precipitation of serum proteins with acetonitrile. Special procedures for tissues.*
Comments	*Piperazine quinolone carboxylic acids are highly potent antimicrobials. For 4 examples - ciprofloxacin, enoxacin, fleroxacin and ofloxacin - and their metabolites, assay literature is surveyed. Assays serve mainly for pharmacokinetic studies and therapeutic drug monitoring. Particular problems arise (a) where metabolite analysis has to be comprehensive, (b) in patients with renal insufficiency, and (c) in the extraction of tissues.*

Quinolone antimicrobials are a new class with high potency towards Gram-negative and some towards Gram-positive pathogens. They act by inhibiting bacterial topoisomerase. This mechanism differs from that of most other antimicrobials used in human pharmacotherapy. More than 20 quinolones, sometimes also called gyrase inhibitors, are presently in various stages of development. This article describes HPLC methods for the above four representative examples and some of their metabolites. The methods have been used for pharmacokinetic studies in healthy volunteers and partly for therapeutic drug monitoring in various diseases.

**Abbreviations.*- CSF, cerebrospinal fluid; IEC, ion-exchange chromatography; i.s., internal standard; RP, reversed-phase; TBAP, tetrabutylammonium phosphate.

CHEMICAL STRUCTURES, AND PERTINENT PHYSICAL PROPERTIES AND *IN VIVO* BEHAVIOUR

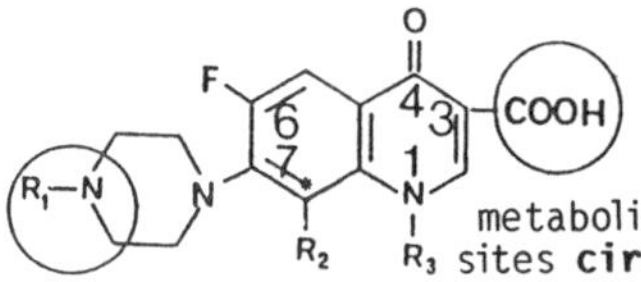

metabolic sites **circled**

The common structure comprises a dihydroquinolone carboxylic acid with, as substituents, fluorine, an aryl or alkyl group, and a piperazine or 4-methyl-piperazine. Possible variations are replacement of C by N (enoxacin) or attachment of a fourth ring (ofloxacin). The UV absorbance maxima are characteristically between 265 and 295 nm; a salient property is fluorescence with an unusually large Stoke's shift giving emission maxima between 400 and 480 nm (excitation at the absorption maximum). The nm values for absorption maximum (and second maximum) and for *fluorescence emission* are:

- ciprofloxacin: 275 (316), *450*;
- enoxacin 265 (339), *400*;
- fleroxacin: 283 (320), *469*;
- ofloxacin: 295 (327), *480*.

Solubility in water is good at extreme pH values and minimal at the isoelectric point, pH ~7-8. Solubility is very low in apolar solvents and fair in polar solvents, e.g. methanol and acetonitrile. Complex formation with Mg^{2+}, Al^{3+}, Fe^{2+} and Cu^{2+} is strongly suggested by indirect methods such as antimicrobiological activity [1], chemical [2] or pharmacokinetic data [3]. Aqueous solutions at room temperature in daylight photodecompose to a moderate extent, slightly more with fleroxacin than with the other three compounds.

Some pharmacokinetic parameters.- All four quinolones can be administered orally. For ciprofloxacin and ofloxacin i.v. preparations are also available. Serum and CSF concentrations in the range 0.01-10 mg/L are encountered in various analytical situations. Plasma protein binding is reported to be ~25% (10-55%) [4-8]. Plasma half-life varies between 3 h (ciprofloxacin) and 9 h (fleroxacin) in healthy volunteers. Volumes of distribution exceed body wt., e.g. 3.1 L/kg for ciprofloxacin and 1.34 L/kg for ofloxacin [5, 9]. The four quinolones are eliminated mainly via the kidney either as parent compound or as metabolites. Biotransformation in humans ranges from a few percent (ofloxacin) to 59% (pefloxacin) of the dose [10].

Metabolism.- Three patterns are observed (Fig. 1):
- metabolism of the piperazine side-chain, e.g. in ciprofloxacin [11, 12] and enoxacin [13];
- metabolism of the methylpiperazine side-chain, e.g. in fleroxacin and ofloxacin [5, 14]: as in the above pattern, the fluorescence quantum yield changes markedly;
- conjugation with glucuronic acid [14, 15].

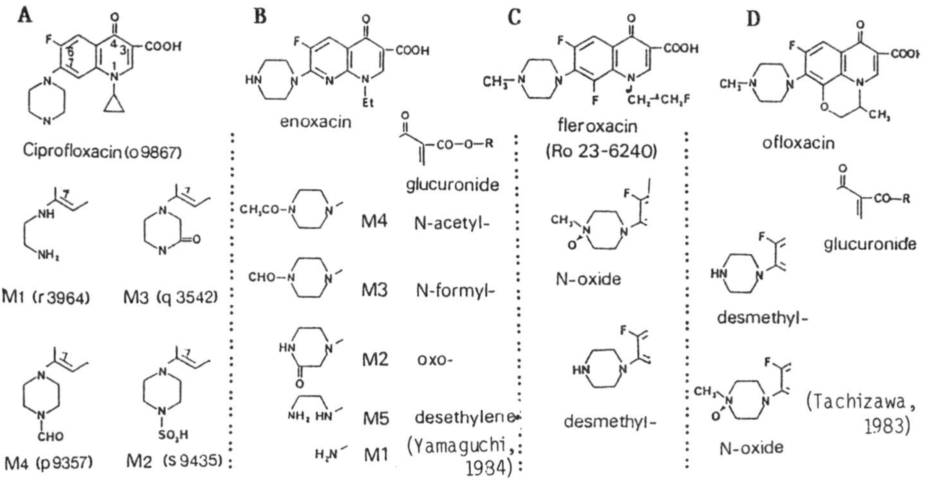

Fig. 1. Metabolites of (**A**) ciprofloxacin, (**B**) enoxacin, (**C**) fleroxacin and (**D**) ofloxacin. **M** signifies a metabolite.

SAMPLING OF BIOMATERIALS, STABILIZATION AND STORAGE

With blood, serum is preferable but heparinized plasma is also usable for analysis. Out of 15 published methods for ciprofloxacin, 11 used serum and 4 heparinized plasma (there are no published comparisons), and both specimen types have been used for the other quinolones. Serum has the advantage of stability. No stabilizer need be added to serum, CSF, urine or bile. Exposure to intense light should be obviated (see earlier); it is advisable that the containers be of brown glass or be wrapped in aluminium foil. All samples are best stored at -80°, conferring stability for at least 3-6 months if the sample is used only once after thawing. For shorter periods storage at -20° appears adequate. Tissue samples should not be allowed to take up blood before storage. Care should also be taken not to lose water by sublimation during storage. Transport on dry ice is feasible for periods no longer than 5 days; for longer periods the material might be freeze-dried.

ASSAY MATERIALS AND METHODS

Reagents, solvents, calibrators and control materials

Unless otherwise stated, our reagents of analytical purity were from E. Merck (Darmstadt). TBAP (PIC A, low-UV grade) was from Waters (Eschborn). Doubly distilled water was used for all measurements. Pure reference quinolones, with defined potencies and their metabolites were kindly supplied by the manufacturers: ciprofloxacin, Bayer (Wuppertal); enoxacin, Goedecke (Freiburg); fleroxacin, Hoffmann-La Roche (Basle); ofloxacin, Hoechst (Frankfurt).

From aqueous stock solutions of standards, working standards were prepared daily by dilution with water or buffer [16, 17] or, as is recommended for serum standards, with drug-free serum [18, 19]. For control of precision, pools of serum and urine from volunteers were made. For recovery studies blank serum and urine were spiked with reference substance.

Machery & Nagel (Düren) supplied Nucleosil 5C18 for RP-HPLC and Nucleosil 5 SA for IEC. The guard column was invariably of Perisorb RP18, 30-40 μm (Merck). Guard and main columns were respectively 30 × 4 mm and 125 × 4 mm. The range of HPLC equipment was:- pump: Perkin Elmer Series 2, LC-10 or LC-410; injector/autosampler: Rheodyne 7125, and Perkin Elmer LC-420, ISS-100 and ISS-101; fluorescence detector: Schoeffel FS 970 and Spectroflow 980, Shimadzu RF-536; absorbance/diode array detectors, Perkin Elmer LC-85 and -95/LC-480 Autoscan; integrators: Hewlett Packard 3390A and 3396A, Shimadzu C-R3A. Virtually no use was made of complex techniques such as gradient elution or column switching.

Sample preparation

Urine, and fluids with low protein content, were merely diluted with the mobile phase to a concentration suitable for chromatography and centrifuged or filtered. For serum and plasma there is a surprisingly large variety of published procedures. The original method for **ciprofloxacin** in serum recommended a simple dilution with HCl [20], phosphoric acid [16], or water and direct injection of the filtered sample. Also suggested were precipitation of serum proteins with trichloroacetic acid [8], methanol [21], acetonitrile alone [17, 22] plus chloroform [23], or perchloric acid [24], then centrifugation and injection of the supernatant.

For **enoxacin** in serum, trichloroacetic acid [25], dichloromethane [26] or acetonitrile has been used; an interesting variant is chloroform extraction entailing derivatization of the piperazine ring with ethyl chloroformate [13]. **Fleroxacin** was extracted from serum by dichloromethane-methanol-SDS; the residue from drying down the organic phase was redissolved in the mobile phase [19]. In a variation, extraction was with ethyl acetate-isopropanol [27]. Serum protein precipitation by acetonitrile can also be used, likewise for **ofloxacin** [28] which otherwise may be extracted with chloroform [29]. As **i.s.** in extraction methods with phase separation, candidates are pimemidic acid [19, 27], nalidixic acid [30], norfloxacin [21] and analogues of ofloxacin [29] or ciprofloxacin [21, 24].

Method adopted.- Serum samples (and standards, blanks and control samples) with 2 vol. of acetonitrile added are centrifuged, and the supernatant diluted 2- or 3-fold with

Table 1. Types of published HPLC methods for quinolones: **C**, ciprofloxacin; **E**, enoxacin; **F**, fleroxacin; **O**, ofloxacin. Detection: fluor(escence), UV (absorbance); sens'y, sensitivity, spec'y, specificity; accur'y, accuracy; +, high, -, low; sep., separation.

Type	Sep. mode	Deriva-tization	Detec-tion	Analyte	Merits and demerits	Meta-bolites
a	RP	no	fluor	C E F O	sens'y +, spec'y +, accur'y ?	some
b	RP	no	UV[⊗]	C E F O	sens'y -, spec'y -, accur'y +	yes
c	IE + RP	no	fluor	F O	sens'y +, improved sep.	yes
d	RP	post-col.[†]	fluor	C	sens'y +	yes
e	RP	pre-col.	UV	E	improved sep.; recovery -	some
f	RP	pre-col.	fluor	O	sep. of O enantiomers	

[⊗]Diode array has been used also. [†]Photochemical reactor.

the mobile phase buffer. Treatment of CSF or bile is similar. The extraction of soft tissues required treatment with a Potter-Elvehjem homogenizer or a 'stomacher'[⊗]. Processing was then as for serum. Bone usually required several successive extractions. In general, extractions entailing phase separation, e.g. with chloroform, normally gave low recoveries and thus needed a numerical adjustment to correct, or an i.s. [30].

HPLC separation methods

Table 1 surveys published approaches. Isocratic RP-HPLC seemed the preferable approach, as exemplified in Fig. 2 for ciprofloxacin. Addition of TBAP or another amine appeared essential in order to avoid tailing. Stationary phases with a high degree of end-capping, such as Nucleosil 5 C18, were preferred. Methylpiperazine quinolones were difficult to separate from their desmethyl metabolites by RP-HPLC. But IEC combined with a RP pre-column produced a different elution pattern with complete separation of fleroxacin and ofloxacin from their *N*-oxide and desmethyl metabolites (Fig. 3). With fluorescence detectors (see below), separation from endogenous compounds was generally good. Several other drugs, e.g. salicylic acid, produce fluorescent peaks, and need checking for possible interference under the individual chromatographic conditions used.

Most methods allow the simultaneous detection of some metabolites similar in polarity to the parent drug. For other metabolites, e.g. oxo-ciprofloxacin, *N*-sulpho-ciprofloxacin and oxo-enoxacin, separate chromatographic runs with modified mobile phases were warranted (cf. Table 2). The separation of ofloxacin enantiomers (cf. Table 1, *f*) is achievable by derivatization or by use of a column containing bovine serum albumin bound to silica (K-H. Lehr & P. Damm; #ncB-4, this vol.).

[⊗]tissue squeezer (A.J. Seward, London)

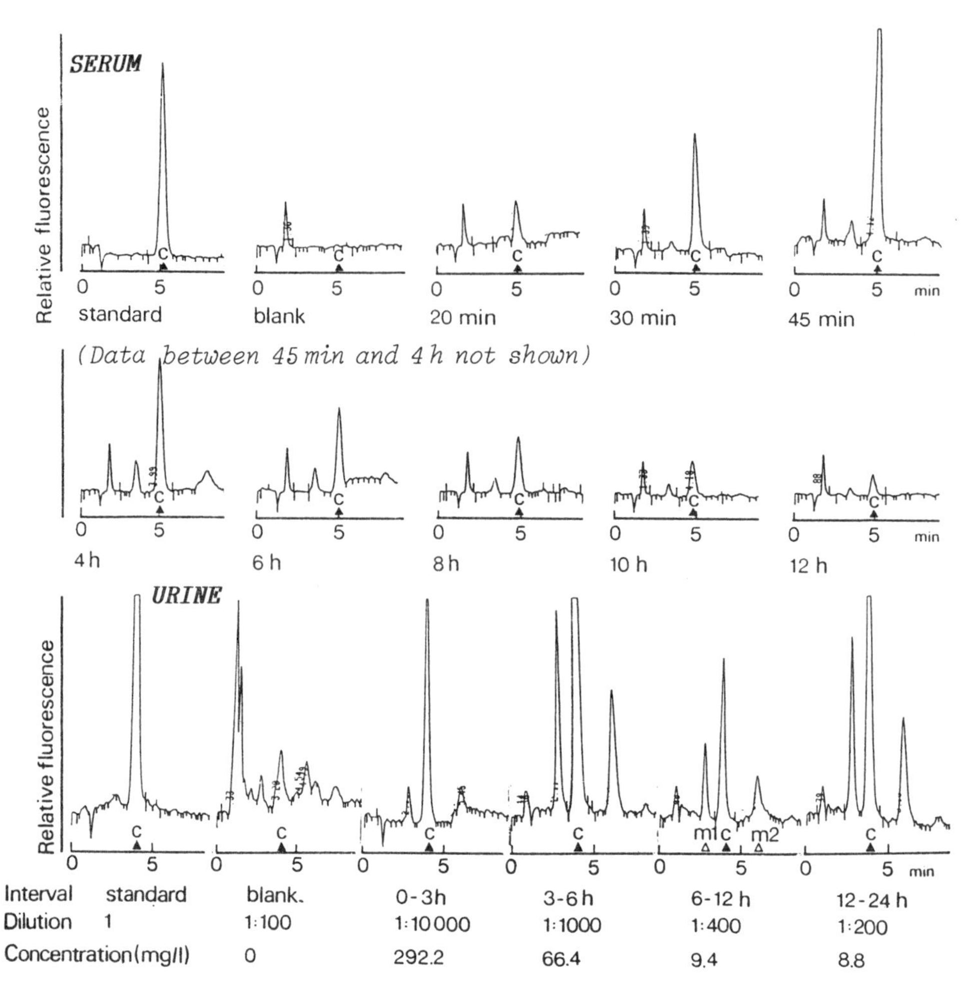

Fig. 2. Patterns for ciprofloxacin: in serum after a single oral dose of 750 mg, and in urine after a single i.v. dose of 100 mg. For serum the patterns at intervals of 60, 75, 90 & 105 min were as for 45 min except for doubling of the pre-**C** peak, slightly diminished at 3 h when **C** was back on-scale.

Methods of detection

The intrinsic fluorescence of the quinolones makes them good candidates for fluorescence detection - as used in 18 out of 23 published methods; 5 used UV absorbance. Fluoresence detection was highly sensitive and specific, and also gave good precision and a wide range of linear detector response; it is obligatory if only small sample volumes are available and for clinical specimens where interferences are likely to be present. Although in most studies apparent recoveries from 90 to 110% were reported, the accuracy of

Fig. 3. Patterns for fleroxacin in serum and urine after an oral dose of 400 mg. Analysis was by Method 8 in Table 2 (below). For serum the patterns at 30 & 45 min and 1,2 & 3 h were similar to those at 20 min and 4 h.

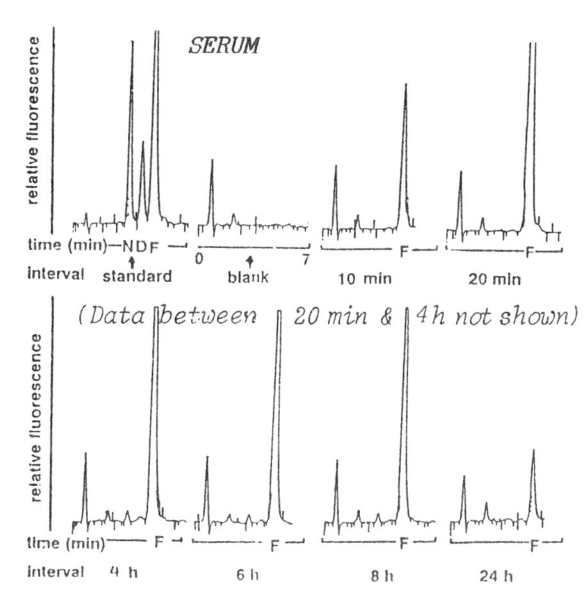

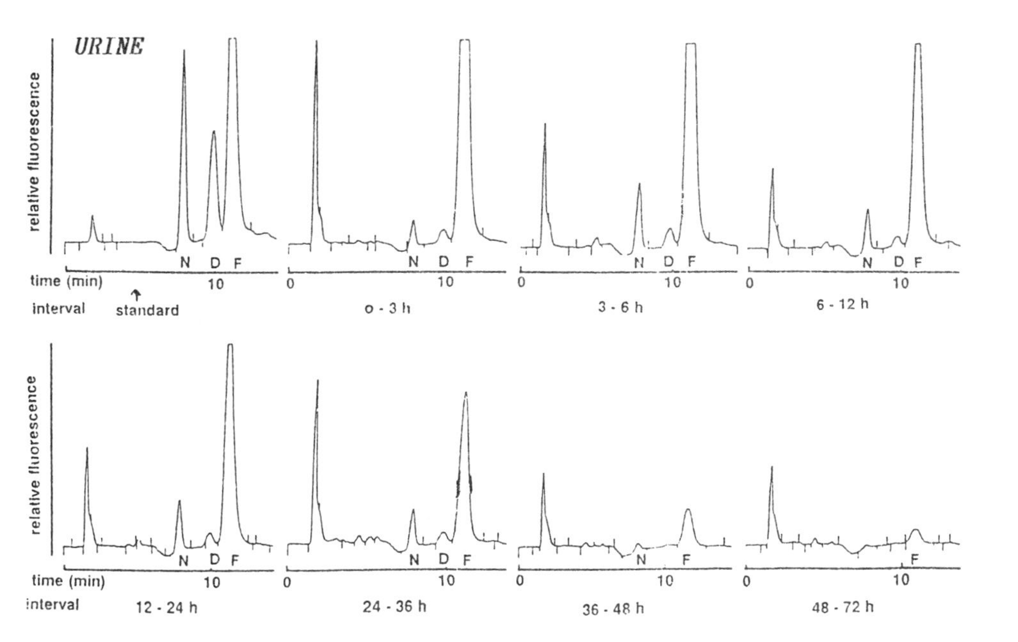

fluorimetric determination of quinolones in serum remains a critical issue. In a comparative study the reported recoveries from serum were 101-110% for two HPLC methods and 125-109% for two microbiological methods [31]. Fluorescence yield is strongly influenced by the matrix, which if slightly different between standards and samples may cause systematic errors from -20% to +20%.

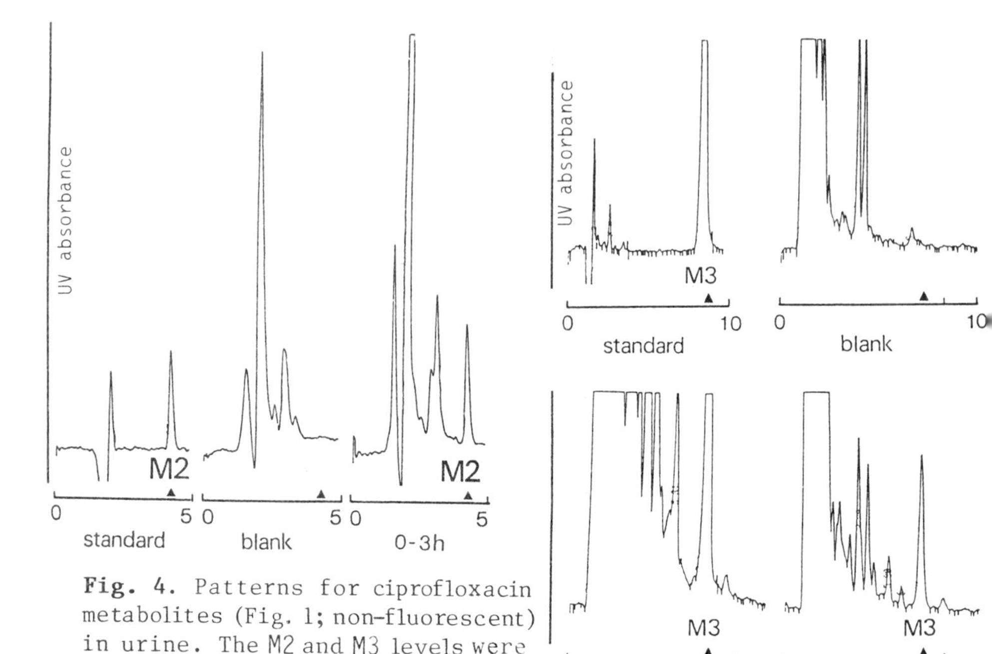

Fig. 4. Patterns for ciprofloxacin metabolites (Fig. 1; non-fluorescent) in urine. The M2 and M3 levels were falling after 3 h; low in 12-24 h samples (not shown).

The fluorescence yield of metabolites with a modified piperazine ring is altered, usually downwards. Thus fluorescence detection does not allow the determination of all metabolites, although it is achievable for all known ciprofloxacin metabolites by photothermal derivatization (IV, Table 1) [32]. The UV absorbance detector serves mainly for non-fluorescent metabolites (Fig. 4), although its sensitivity is much lower (40-fold for ciprofloxacin itself) [22]; it is also less specific since many endogenous compounds absorb in the 260-320 nm range. It has been advocated [30] for the parent compound also, but the sample size was rather large and the apparent recovery from serum was only 55%. Whilst absorbance detection appears poorly suitable for therapeutic monitoring, the diode-array detector seems apt for identifying metabolites, as shown for oxo-enoxacin in urine (Fig. 5).

ANALYTICAL RELIABILITY (Summary: Table 3)

Sensitivity.- With fluorescence detectors ~0.01 mg/L serum is detectable, but the limits for urine were generally higher due to higher background noise. With absorbance detection the limits were in the range 0.5-2.0 mg/L urine for the parent compounds and the oxo and sulpho metabolites.

Precision.- C.V.'s depended markedly on concentration: 1.0-7.2% within-batch and 2.2-9.2% between-batch for urine. For serum the respective ranges were 0.8-5.5% and 2.2-9.3%.

Table 2. HPLC (usually RP) conditions. Abbreviations besides those at start of article and (including drugs: **C**, **E**, **F**, **O**) in Table 1 (*Note:* TBAP is aqueous).- MeCN, acetonitrile; phos. = phosphate, **or** phosphoric (in #**10**, where **B** is phos. acid taken to pH 2.92 with 4 mM NaOH); to = 'taken to', in context of taking to a stated vol. or pH; r.t., room temperature (20-26°); **se** = serum, **ur** = urine; λ values (nm) following 'fluor' are excitation/emission or, if 'cut' added, excitation/cut-off for emission measurement [relative values are in comparison with the parent drug as unity, on a weight basis].

Method and analytes	**1**: **C**, des-ethylene-**C** (**I**), unknown m2 (**II**)	**2**: oxo-**C**	**3**: sulpho-**C**	**4**: **O**, des-methyl-**O** (**I**), **O** *N*-oxide (**II**)	**5**: **E**	**6**: **E**	**7**: **E** oxo-**E** (**I**)	**8**: **F**, des-methyl-**F** (**I**), **F** *N*-oxide (**II**)	**9**: **F** and as for **8**	**10**: **O**, des-methyl-**O** (**I**), **O** *N*-oxide (**II**)
µl injec'd /sep. mode	**se** 20-50, **ur** 20/RP	**ur** 20 /RP	**ur** 20 /RP	**se** 5-50, **ur** 5-50/RP	**se** 100 /RP	**ur** 5 /RP	**ur** 20 /RP	**se** 10, **ur** 5 /RP	**se** 20, **ur** 10 /RP + IE	**se** 25, **ur** 25 /RP + IE
Mobile phase (& *ml/min*; MPa pressure; temp.)	125 ml MeCN, + 5 mM pH 2 TBAP to 1 L (*1.0*; 15; r.t.)	220 ml MeCN, + 5 mM pH 2 TBAP to 1 L (*1.5*; 12; r.t.)	400 ml MeCN, + 5 mM pH 2 TBAP to 1 L (*1.0*; 14; r.t.)	140 ml MeCN, + 5 mM pH 2 TBAP to 1 L (*1.0*; 10; r.t.)	140 ml MeCN, + 10 mM pH 2.2 TBAP to 1 L (*1.0*; 15; r.t.)	120 ml MeCN, + TBAP as in **5** (as in **5**)	200 ml MeCN, + 10 mM pH 2.5 TBAP to 1 L (*1.5*; 15; 30°)	140 ml MeCN, + TBAP as in **5** (*0.8*; 10; r.t.)	**se** (**ur**) 500 (750) ml MeCN + 500 (250) ml 0.1 mM pH 2.66 Na phos. (*1.5*; 12; 30°)	To 750 ml MeCN add **B** (see heading) to 1 L; pH to 3.82 (*2.5*; 20; r.t.)
Detection & peak quantit^n. basis	fluor: 275/ cut 418 [**I**, 7.94]; areas	UV: 280; areas	UV: 330; areas	fluor: 295/ cut 418 [**I**, 0.95; **II**, 0.77]; areas	fluor: 265/400; heights	as for **5**; heights	UV: 340 (diode array: 240-430); areas	fluor: 285/ 460 [**I**, 0.38; **II**, 0.65]; areas	fluor: 283/ 460 [**I**, 0.38; **II**, 0.65]; areas	fluor: 295/ 480 [**I**, 0.86; **II**, 1.17]; areas
Retention time, min⊗	**C** 3.5 **I** 2.5 **II** 5.9	**C** 2.4 oxo-**C** 8.6	sulpho-**C** 4.1	**O** 3.7 **I** 1.9 **II** 2.4	**E** 3.4	**E** 4.0	**E** 1.9 **I** 5.5	**F** 10.5 **I** 11.9 **II** 15.6	**F** 1.67, 11.5 **I** 13.2, 9.9 **II** 11.8, 7.9 (footnote)	**O** 5.2 **I** 4.8 **II** 3.9

⊗ In **9**, values are for **se** and **ur** respectively, with different mobile phases

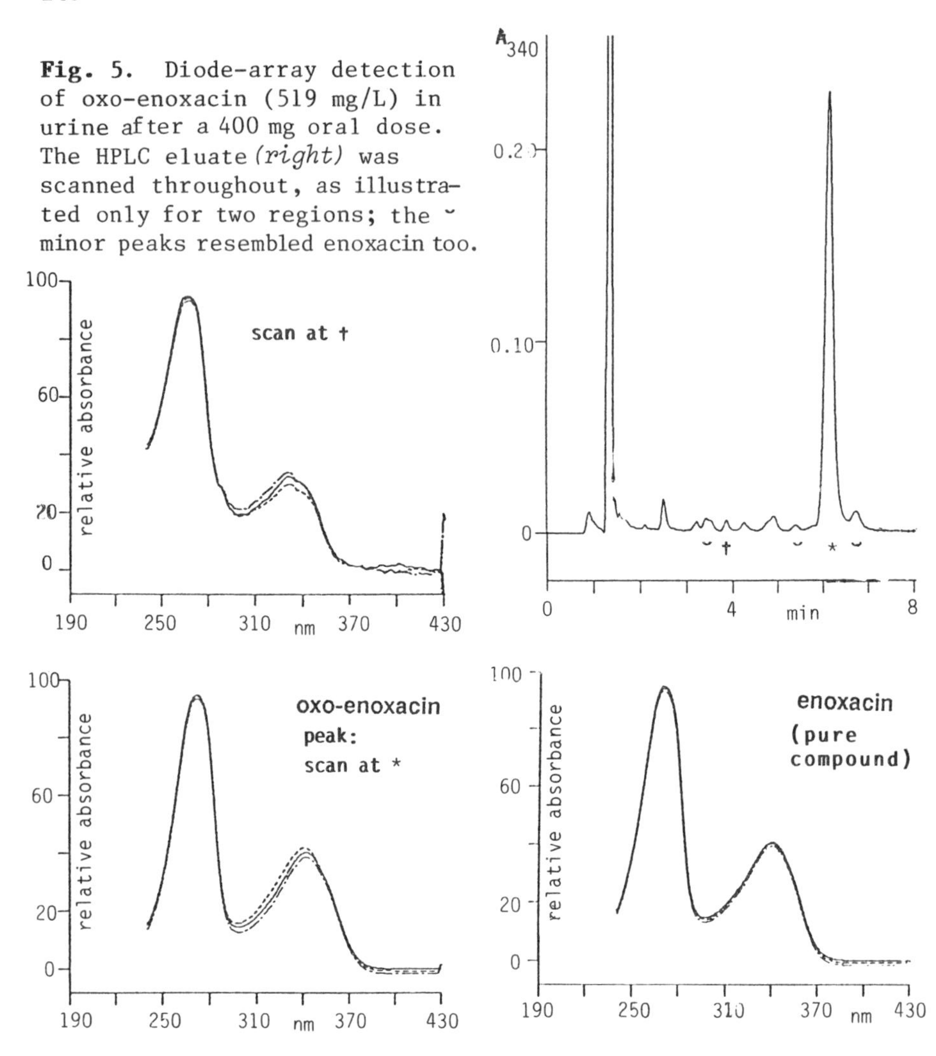

Fig. 5. Diode-array detection of oxo-enoxacin (519 mg/L) in urine after a 400 mg oral dose. The HPLC eluate *(right)* was scanned throughout, as illustrated only for two regions; the ˘ minor peaks resembled enoxacin too.

Recoveries of reference compounds from spiked urine were 98-109%. With fresh serum from volunteers, the mean recovery of ciprofloxacin and ofloxacin was ~95%, and values from patients were corrected accordingly. In volunteer studies the basis for correcting results was the recovery of spiked drug from the same individual's blank serum. For ciprofloxacin spiked into a commercial human control serum (Boehringer Mannheim) the recovery was 113%; such serum, maybe differing in matrix from fresh material, should be used with discrimination. Animal serum apppears less suitable for preparing standards [24].

Specificity.- With Method **1** (Table 2; fluorimetry), a number of endogenous and exogenous substances were tested and shown not to interfere in the assay of ciprofloxacin,

Table 3. Validation of analytical methods. For **C**, **E**, **F** and **O** (drug analytes) see heading to Table 1; **se**, serum; **ur**, urine. Sens(itivity) is expressed as mg/L, and prec(ision) as C.V. %. Method nos. are as in Table 2; where two were used, values with the second are given in *italics*. Values after ± are S.D.

Drug; method	Material	Sens.	Prec. within-series	Prec. between-series	Recovery, %
C;*1	**se**	0.01	0.8-2.4	4.8-9.3	96 ±2
	ur	0.2	1.7-2.1	2.4-7.2	99.8
E; 5	**se**	0.025	3.9-5.5	7.3-9.2	86 ±7
	ur	0.1	2.8-5.2	3.7-14.6	100-101
F; 8,†*9*	**se**	0.03 *0.05*	1.5-4.3 *3.2-6.7*	3.3-6.2 *2.5-7.4*	94-99 *98-106*
	ur	0.5 *1.4*	*4.6-10.4*	2.3-3.6 *2.6-6.7*	*100-109*
O; 4,⊗*10*	**se**	0.02 *0.05*	1.2-2.6 *1.1-5.1*	2.2-5.2 *2.3-4.3*	97 ±1 *96-99*
	ur	0.2 *0.05*	1.0-1.1 *1.9-7.2*	2.9-3.3 *1.9-2.2*	100-101

REGRESSION ANALYSIS (**b**, slope; **a**, intercept, mg/L):
*se: **b** 1.107-1.309, **a** -0.036 to +0.023; **ur**: **b** 1.042-1.556, **a** -3.3 to +9.2
†**se**: **b** 1.010, **a** -0.01; **ur**: **b** 1.039, **a** -1.6
⊗**se**: **b** 1.011, **a** -0.034; **ur**: **b** 1.572, **a** 31.3
ALTERNATE METHODS FOR COMPARISON: * & ⊗, bioassay; †, RP-HPLC.

relative to which (as unity) the retention times were: tryptophan, 0.40; **M1**, 0.71; tyrosine, 1.08; **m2**, 1.59; paracetamol, 1.76; doxycycline, 2.25; metamizole, 7.39; salicylic acid, 9.67. With methods 1 and 10 of Table 2, several hundred clinical specimens have been analyzed for ciprofloxacin and ofloxacin without any observations of serious interferences. In fleroxacin assay 21 drugs have been stated not to interfere [27].

Method comparisons

Comparison was mainly with microbiological assay. Accord between HPLC and bioassay depends on the absence of active metabolites; thus desmethylofloxacin is still microbiologically active. However, in healthy volunteers the two methods gave identical results for ofloxacin in serum and urine, which contain very low concentrations of metabolites. For ciprofloxacin (Fig. 6), which might give active metabolites in urine, bioassay results were higher - by ~20%. In the case of serum, wherein metabolite concentrations are very low, other factors evidently operate; the discrepancy has been observed also by other authors [8, 31]. In serum from patients with renal disease, bioassay may yield higher results than HPLC for ofloxacin and ciprofloxacin: we have found that these give rise to high metabolite concentrations in serum (Fig. 7).

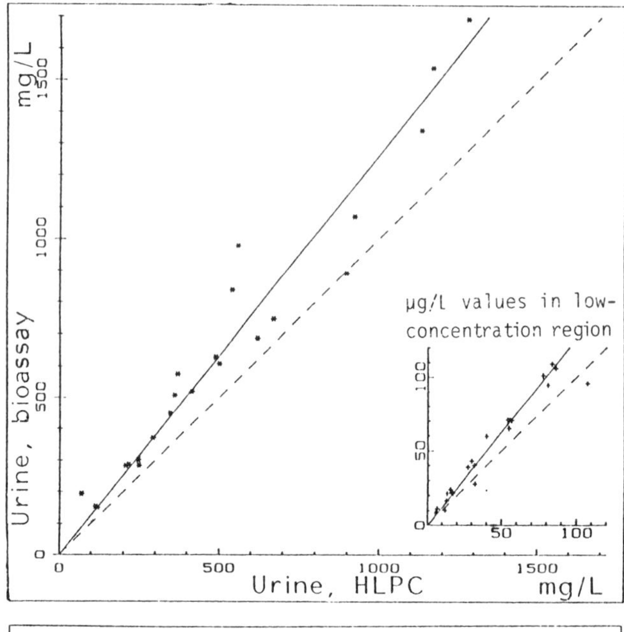

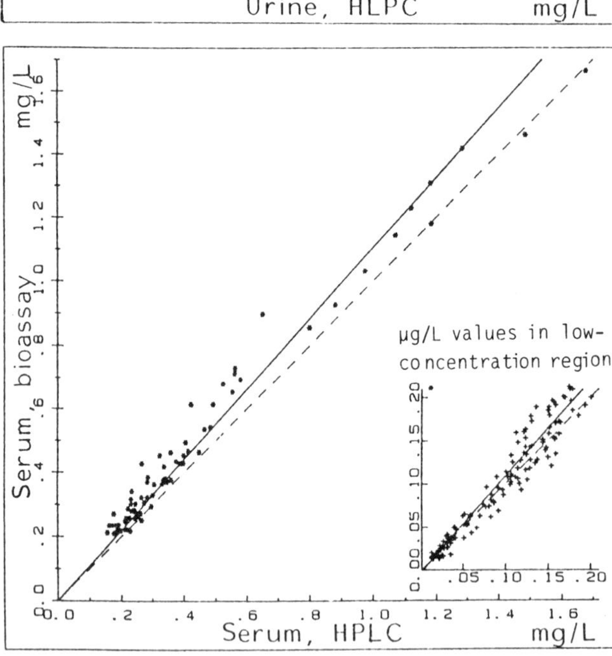

Fig. 6. Ciprofloxacin method comparison, for urine and serum from volunteers.

With enoxacin and fleroxacin a similar discrepancy may occur. Now that such important quinolone antimicrobials can be assayed by good HPLC methods with fluorescence detection, it appears highly desirable that an external quality assurance scheme be set up on a regular basis.

Fig. 7. Patterns for serum (method 10, Table 2), after oral oxfloxacin (**O**) administration to a healthy volunteer and *(right)* a patient with terminal renal insufficiency. **N** = **O** *N*-oxide; **D** = desmethyl-**O**.

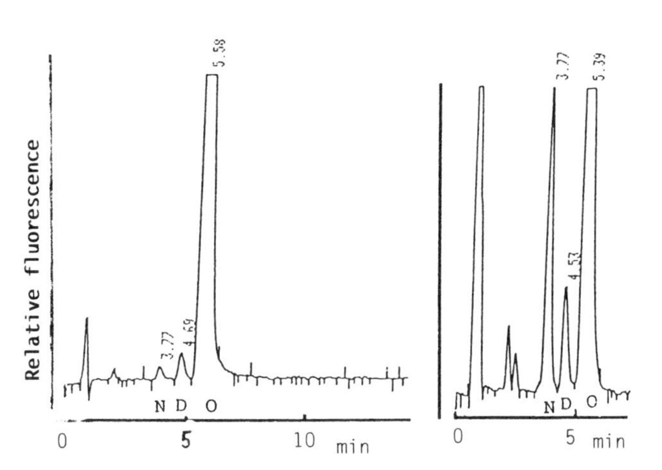

Acknowledgements

The authors are indebted for financial support to Bayer AG (Leverkusen), Goedecke GmbH (Freiburg), Hoechst AG (Frankfurt) and Hoffmann-La Roche (Basle). The provision of pure reference materials and of metabolites is also thankfully acknowledged.

References

1. Ratcliffe, N.T. & Smith, J.T. (1984) *F.A.A. Chemo.** *3-5*, 563-569.
2. Marolt-Gomiščeк, M., Veber, M., Veber, A., Smerkolj, A. & Gomiščeк, S. (1987) in *Progress in Antimicrobial and Anticancer Chemotherapy* (Berkada, B. & Kuemmerle, H-P., eds.), Ecomed Verlagsgesellschaft, Landsberg/Lech., pp.1792-1794.
3. Höffken, G., Borner, K., Glatzel, P., Koeppe, P. & Lode, H. (1985) *Eur. J. Clin. Microbiol. 4*, 345.
4. Höffken, G., Lode, H., Prinzing, C., Borner, K., & Koeppe, P. (1985) *Antimicrob. Agents Chemother. 27*, 375-379.
5. Lode, H., Höffken, G., Olschewski, P., Sievers, B., Kirch, A., Borner, K. & Koeppe, P. (1987) *Antimicrob. Agents Chemother. 31*, 1338-1342.
6. Yamaguchi, T., Suzuki, R. & Sekina, Y. (1984) *Chemotherapy 32 S-3*, 109-116.
7. Weidekamm, E. & Dell, D. (1987) *F.A.A. Chemo.** *6-10*, 1615-1622.
8. Joos, B., Ledergerber, B., Flepp, M., Bettex, J-D., Lüthy, R. & Siegenthaler, W. (1985) *Antimicrob. Agents Chemother. 27*, 353-356.
9. Borner, K., Höffken, G., Lode, H., Koeppe, P., Prinzing, C., Glatzel, P., Wiley, R., Olschewski, P., Sievers, D. & Reinitz, D. (1986) *Eur. J. Clin. Microbiol. 5*, 179-186.
10. Borner, K. & Lode, H. (1986) *Infection 14, Suppl. 1*, 54-59.

**Fortschr.Antimikrob. Antineoplast. Chemotherapie*

11. Gau, W., Kurz, J., Petersen,U., Ploschke, H.J. & Wuensche, L. (1986) *Drug Res. 36 (II)*, 1545-1549.
12. Borner, K., Lode, H. & Höffken, G. (1986) *Eur. J. Clin. Microbiol. 5*, 476.
13. Nakamura, R., Yamaguchi, T., Sekina, Y. & Hashimoto, M. (1983) *J. Chromatog. 278*, 321-328.
14. Sudo, K., Okazaki, O., Tsumura, M. & Tachisawa, H. (1986) *Xenobiotica 16*, 725-732.
15. Tanimura, H., Tominaga, S., Rai, F. & Matsumoto, H.(1986) *Drug Res. 36 (II)*, 1417-1420.
16. Gau, W., Ploschke, H.J., Schmidt, K. & Weber, B. (1985) *J. Liq. Chromatog. 8*, 485-497.
17. Borner, K., Lode, H., Höffken, G., Prinzing, C., Glatzel, P. & Wiley, R. (1986) *J. Clin. Chem. Clin. Biochem. 24*, 325-331.
18. Nilsson-Ehle, I. (1987) *J. Chromatog. 416*, 207-211.
19. Dell, D., Partos, Ch. & Portmann, R. (1988) *J. Liq. Chromatog. 11*, 1299-1312.
20. Wingender, B., Graefe, K.H., Gau, W., Förster, D., Beermann, D. & Schacht, P. (1984) *Eur. J. Clin. Microbiol. 3*, 355-359.
21. Kees, F., Naber, K.G., Meyer, G.P. & Grobecker, H. (1989) *Drug Res. 39,(I)*, 523-527.
22. Weber, A., Chaffin, D., Smith, A. & Opheim, K. (1985) *Antimicrob. Agents Chemother. 27*, 531-534.
23. Myers, C.M. & Blumer, J.L. (1987) *J. Chromatog. 422*, 153-164.
24. Nix, D.E., De Vito, J.M. & Schentag, J. (1985) *Clin. Chem. 31*, 684-686.
25. Vree, T.B., Baars, A.M. & Wijnands, W.J.A. (1985) *J. Chromatog. 343*, 449-454.
26. White, L.D., Bowyer, H.M., McMullin, C.H. & Desai, K. (1988) *J. Antimicrob. Chemother. 21*, 512-513.
27. Awni, W.M., Maloney, J.A. & Heim-Duthoy, K.L. (1988) *Clin. Chem. 34*, 2330-2332.
28. Borner, K., Hartwig, H., Lode, H. & Höffken, G. (1986) *Fresenius' Z. Anal. Chem. 324*, 355.
29. Ichihara, N., Tachisawa, H., Tsumura, M., Uno, T. & Sato, K. (1984) *Chemotherapy 32 S-1*, 118-149.
30. Vallée, F., LeBel, M. & Bergeron, M. (1986) *Therapeutic Drug Monitoring 8*, 340-345.
31. Weber, A., Smith, A.L., Wong, K., Painter, B. & Krol, G. (1988) *Meths. Findings Exptl. Clin. Pharmakol. 10*, 123-127.
32. Scholl, H., Schmidt, K. & Weber, B. (1987) *J. Chromatog. 416*, 321-330.

#B-7

QUANTITATIVE ANALYSIS OF THE ANTIFUNGAL AGENT, FLUCONAZOLE BY CAPILLARY GAS CHROMATOGRAPHY WITH ION-TRAP DETECTION

P.V. Macrae

Pfizer Central Research, Sandwich, Kent CT13 9NJ, U.K.

Requirement — *An automated method, sufficiently specific and sensitive to allow the determination of fluconazole in small volumes of human plasma (<50 µl) and vaginal secretions (<10 mg).*

End-step — *Capillary GC with ITD*.*

Sample handling — *Fluconazole and a structurally similar i.s. extracted from basified plasma into ethyl acetate; back-extraction into acid, then re-basification and re-extraction into ethyl acetate. Final extract evaporated to dryness, re-constituted in a small volume of ethyl acetate and injected onto the GC-ITD using an auto-injector.*

Comments — *The GC-ITD proved to be reliable and sensitive (<2 ng of analyte per sample detectable). Modifications to the assay procedure were needed to take into account the uniqueness of the ion trap's design. High-purity solvents and injector deactivation proved critical to the reliability of the method.*

R = -H, Fluconazole (I)
R = -F, Internal standard (II)

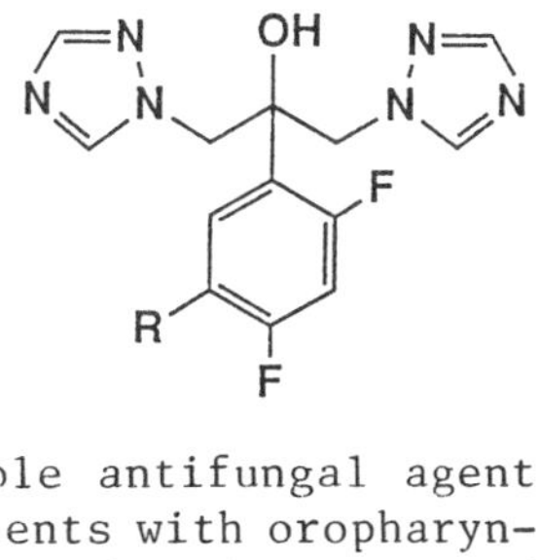

Fluconazole (**I**) is a new bis-triazole antifungal agent which has shown excellent activity in patients with oropharyngeal or vaginal candidiasis [1-3]. Besides having good absorption after oral administration (>90% bioavailability), a long plasma half-life (30 h) and low plasma-protein binding (11%), fluconazole is rapidly distributed in fluid tissue

**Abbreviations.*- MS, mass spectro-meter/metry/metric; ITD, ion-trap detector; i.s., internal standard; SPE, solid-phase extraction.

compartments [4]. Typically, maximum plasma concentrations at normal therapeutic doses are <3 μg/ml. Sample size may be limited to <50 μl of plasma from neonates or <10 mg of vaginal secretions. Bioassay, HPLC or GC-electron capture [5] methods do not have the required sensitivity when sample size is limited, whereas an MS end-point such as the ITD has far greater inherent sensitivity and specificity.

The aim of this work was to develop an automated assay method for small samples of plasma and vaginal secretions with a sensitivity greater than the typical minimum inhibitory concentrations (~0.6 μg/ml) against most strains of *Candida albicans*. The assay was therefore required to be able to quantify fluconazole in the low-ng range. A complementary goal was to determine the suitability of the Finnigan ITD for use in the field of biomedical analysis where different constraints are put on the instrument compared to analysis in the petrochemical and flavour industries, where this instrument is used extensively.

SAMPLE PREPARATION

Plasma (50 μl), after the addition of i.s. (**II** in panel; 100 ng) and 1 ml 5 M NaOH, was extracted with ethyl acetate (4 ml) on a rotary mixer (30 rpm, 10 min). With vaginal secretions (pre-weighed; ~10 mg) these steps were preceded by digestion in 1 ml 5 M NaOH for 1 h at room temperature. After centrifugation and transfer of the ethyl acetate to a clean tube the analytes were extracted into 2 ml 1 M HCl on the rotary mixer, centrifuged, and the organic layer aspirated and discarded. The acid layer was basified with 1 ml 5 M NaOH and the analytes extracted with 4 ml ethyl acetate on the mixer. After centrifugation the ethyl acetate layer was transferred to a clean tube, evaporated to dryness under N_2, reconstituted in 35 μl ethyl acetate and transferred to an autosampler vial. A 2 μl sample was injected onto the GC using a Hewlett Packard 7673A autosampler.

GC-ITD ANALYSIS

Capillary GC analysis was carried out on a Hewlett Packard 5890 in the splitless injection mode. The column, HP Ultra 1 (12 m × 0.2 mm i.d., 0.33 μm film thickness) was temperature-programmed from 200°, holding for 0.7 min, to 255° at 25°/min. The injector temperature was 300° and the column head pressure was 70 kPa (helium). The GC was directly coupled to a Finnigan ITD 810 (running V.4.00 software) via a 1 m heated transfer line (250°). Acquisition was performed over the mass range 210-249 at 16 microscans/sec. The ion traces at m/z 224 (fluconazole) and 242 (i.s.) were created using selected ion chromatograms and the peak areas measured automatically using the ITD calibration software.

Calibration curves were constructed using selected ion peak area ratios against concentration using non-weighted linear regression. Sample concentrations were obtained by interpolation of peak area ratios from the calibration curve using the ITD quantitative software.

SENSITIVITY, PRECISION AND ACCURACY

The detection limit for fluconazole on the ITD at m/z 224 was 50 pg for a 10:1 signal-to-noise. This amount was equivalent to 0.02 µg/ml from plasma based on analysis of a 50 µl aliquot. Precision of the assay was ascertained at 0.04, 0.2 and 0.4 µg/ml: the C.V.'s based on selected ion peak area ratios were 8%, 10% and 7% (n = 6) respectively. Accuracy was checked by analyzing independently prepared samples of plasma containing fluconazole at concentrations unknown to the analyst. Mean accuracy, expressed as the ratio of amount found to amount added, was 1.02 ±0.115 S.D. (n = 12), giving a mean C.V. of 11% over the whole calibration range.

These values are as might be expected for a sensitive GC-MS assay using a conventional magnetic-sector or quadrupole instrument [6], suggesting that ITD in the electron-impact mode offers similar sensitivity, precision and accuracy. The mass spectrum of fluconazole on the ITD was essentially identical to that obtained on a conventional quadrupole MS with extensive fragmentation. The molecular ion (m/z 306) was absent and the base peak (m/z 224) used for quantitation was due to loss of one of the methylene triazole side chains. The sensitivity of the assay was limited by extensive fragmentation in the electron-impact mode and the relatively low mass of the ion used for quantitation, and not by any instrumental shortcomings. This was confirmed by comparison with the sensitivity obtained on a conventional quadrupole MS (VG 12-250): it gave a signal-to-noise ratio only two-fold greater than with the ITD for a 50 ng injection of fluconazole.

ION TRAP OPERATION

The detection limit on the ITD could be improved to approach that of the quadrupole with careful instrument tuning using pure fluconazole. However, sensitivity was not improved, and indeed was dramatically reduced, when plasma extracts were analyzed. This was rationalized by consideration of how the ITD ion-generation and storage function operates. The small size of the ion trap limits the amount of ions that can be stored in stable orbits without space-charging effects. To achieve the optimum amount of ions a software algorithm reduces the ionization time during the scan. When pure drug is injected, the ionization time is determined solely by the amount of fluconazole present. However, with a plasma extract the ionization time is governed by the amount of drug and the amount of any other component eluting at the same GC retention time. An increasing amount of co-eluting components would decrease the ionization time and hence decrease sensitivity. It is possible to bypass this automatic ionization time, but the drawback to this approach is that the dynamic range is severely limited due to the limited ion-storage capacity of the trap.

SOLVENT PURITY

Initially, the sensitivity of the assay was significantly reduced when plasma extracts were analyzed. This was shown to be due to a decrease in the ionization time in the ion trap at the retention time of fluconazole, and not due to poor extraction efficiency. The inherent high sensitivity of the ITD in full-scan mode was utilized to identify the co-eluting components. Library searching using the NBS library on the datasystem revealed the identity of the interfering components as phthalate plasticizers. Their source was traced to the ethyl acetate used at the final extraction stage. Several grades of ethyl acetate were analyzed on the ITD for the presence of plasticizers by evaporating 6 ml aliquots to dryness under N_2 and reconstituting in 50 µl methyl t-butyl ether and injecting 5 µl. General-purpose and Analar-grade ethyl acetate from several suppliers all contained significant amounts of plasticizer. Redistillation of Analar ethyl acetate also proved unsatisfactory. As the phthalate esters were still present, it was thought azeotroping had occurred. The use of Distol-grade ethyl acetate (FSA Lab. Supplies, Loughborough, U.K.), which has been specially checked for purity, solved this problem. This grade of solvent, significantly more expensive than other grades, was the only one to show no detectable levels of plasticizers under the analytical conditions used. Hence, when plasma extracts were prepared using Distol-grade ethyl acetate, sensitivity was maintained as no reduction in ionization time occurred.

This problem with low-grade solvents was not found on the quadrupole MS as the sample capacity of the ion source was not exceeded. This effect, specific to the ITD, would manifest itself with any analyte and co-eluting component, either endogenous to the sample or introduced in the extraction procedure. This suggests that more careful optimization in sample work-up is required than with a conventional GC-MS analysis. In particular, SPE methods should generally be avoided. The amount of matrix and background contamination produced from an SPE method is usually substantially higher than from a multi-stage solvent extraction procedure and will significantly degrade the ITD performance.

DROPPING NEEDLE INJECTOR

An all-glass solid injector (dropping needle) was fitted to the GC's injection port. It is ideal for biomedical analysis of high-boiling compounds as there are no solvent effects, no column contamination by non-volatiles and no septum bleed. It is also easy to clean and, being all-glass, has low catalytic activity. A further advantage is the ability to concentrate dilute samples so that a much larger proportion of the final extract can be analyzed compared with any other injector. Using this injector peak shape was good and the required sensitivity was achieved. Unfortunately, this injector has two major disadvantages. Being all-glass, it is fragile. Secondly, and of importance to this analysis, it is not amenable to automation. To enable this assay to have a high throughput, automatic sample injection was considered essential; hence conventional injectors capable of automation were investigated.

SPLITLESS INJECTOR

Initially the splitless injector was set up in accordance with the manufacturer's instructions. However, the peak shapes were very broad and tailing on the capillary column; peak shape and sensitivity were in fact worse than with the published packed-column method [5]. This result suggested that fluconazole was being retained in the splitless injector. The injector was reassembled using a variety of quartz inserts which had been deactivated with different silanizing reagents. In particular, the Grob method [7] using hexamethyldisilazane was unsuccessful. The optimum conditioning was achieved using an insert which had been immersed in HCl (s.g. 1.18) for 16 h, then washed with distilled water, methanol and ethyl acetate. The clean insert was then immersed in diphenyltetramethyldisilazane for a further 16 h, allowed to dry in a stream of dry helium and then fitted into the GC injection port. The injector, with helium flowing through at 35 ml/min, was heated to 300° for 16 h before connecting the capillary column. Under these conditions the peak shape approached that obtained on the dropping needle injector, although the sensitivity was at least an order of magnitude worse.

Fluconazole is a relatively polar drug (log D = 0.6) and is more soluble in solvents with a high polarity index such as ethyl acetate (P' = 4.3) than in aprotic solvents such as benzene (P' = 3.0). It was proposed that fluconazole, when injected in ethyl acetate, would have a greater affinity for the non-evaporated solvent droplets than the deactivated walls of the injector and would flood through the injector and column with the solvent front, leaving only a small proportion retained at the head of the column. Part of the classical concept of splitless injection was the use of a non-retaining empty insert in the injector to minimize retention power. However, non-retention of fluconazole was occurring to such an extent that the sample was lost for the analysis. A compromise was achieved by loosely packing the injector port with glass wool silanized with dimethyl dichlorosilane. The glass wool had potentially two effects on the injection. Firstly, it split the large solvent droplets into smaller ones which easily released the analyte. Secondly, the methyl silane groups on the glass wool had greater retention power than the phenyl silane groups of the insert and stopped the sample flooding. In consequence, a compromise was reached where most of the solute was retained with only a slight broadening of the peaks. The glass wool also acted as a trap for non-volatile material which would otherwise have caused column degradation.

ON-COLUMN INJECTION

The autosampler used for the analysis had the facility of on-column injection. This, via a coupled length of megabore tubing, should have eliminated the problem of active sites for adsorption, but the non-retention problems occurred again. Several different types of megabore tubing, including a deactivated retention gap, were tried; but solute retention was poor. Chromatographic performance also rapidly deteriorated when plasma extracts were analyzed, suggesting non-volatile components were being coated onto the walls of the analytical column. These effects appear to limit the usefulness of the on-column injector not only for analyzing fluconazole but also with any biomatrix extraction procedure where extracts are often contaminated with non-volatile components.

TRANSFER LINE

The ITD is coupled to the GC *via* a 1 m heated transfer line. To ensure chromatographic parity with conventional detectors the ITD was decoupled from the end of the analytical column by use of an open-split interface. The analytical column was held in a glass liner in close proximity to the deactivated inlet line of the ITD. Peak broadening was observed, presumably because fluconazole was being adsorbed onto this coupling. Deactivation of this open-split interface was attempted by procedures similar to those adopted for the splitless injector, but peak shape was not improved. The remedy was to remove the open-split interface and pass the column directly through the transfer line into the IDT housing. Unfortunately this approach had a major disadvantage. To replace a column required venting the ITD to atmosphere: the ITD needs 12-24 h pumping to stabilize after a shutdown, compared to a 15 min turnround time on a fast-pumping quadrupole instrument.

ANALYSIS OF SAMPLES

The goal of a high-throughput, high-sensitivity assay for fluconazole on an ITD was achieved by adopting an extraction procedure that uses high-purity solvent combined with automatic splitless injection onto a capillary GC column. Fig. 1 shows typical selected ion current profiles for fluconazole and the i.s. extracted from human plasma. Fig. 2 shows

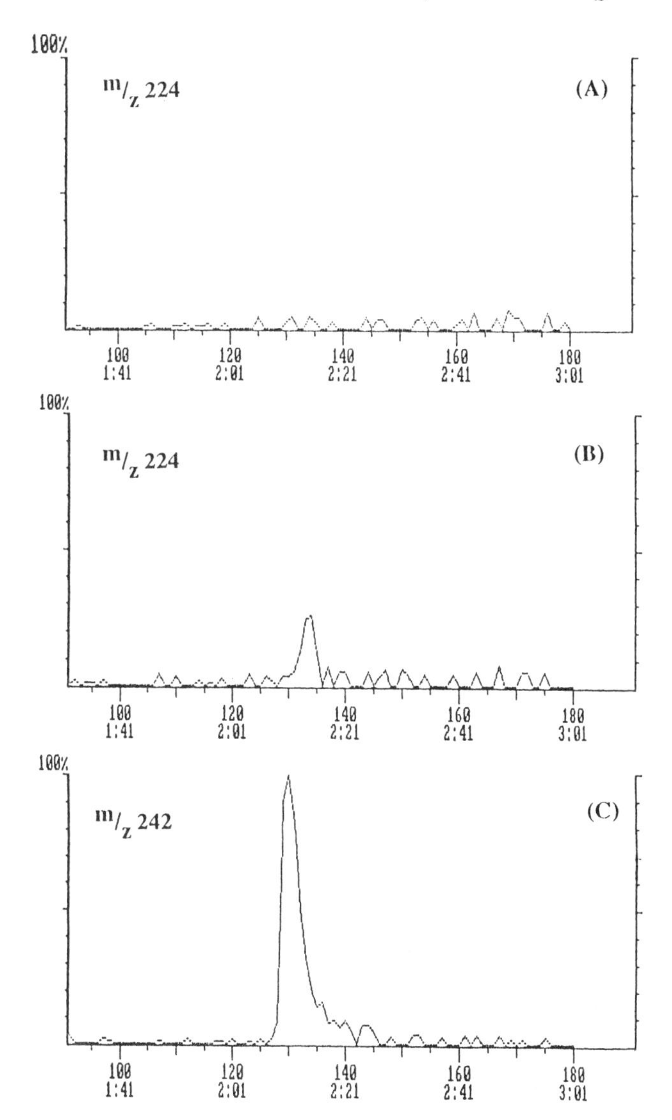

Fig. 1. Selected ion profiles (fragment ions: m/z 224, fluconazole; 242, i.s.) for extracts from 50 µl plasma samples: **A**, blank; **B**, fluconazole, 0.07 µg/ml plasma: ~200 pg put on the column; (**C**), i.s.: ~1 ng put on the column.

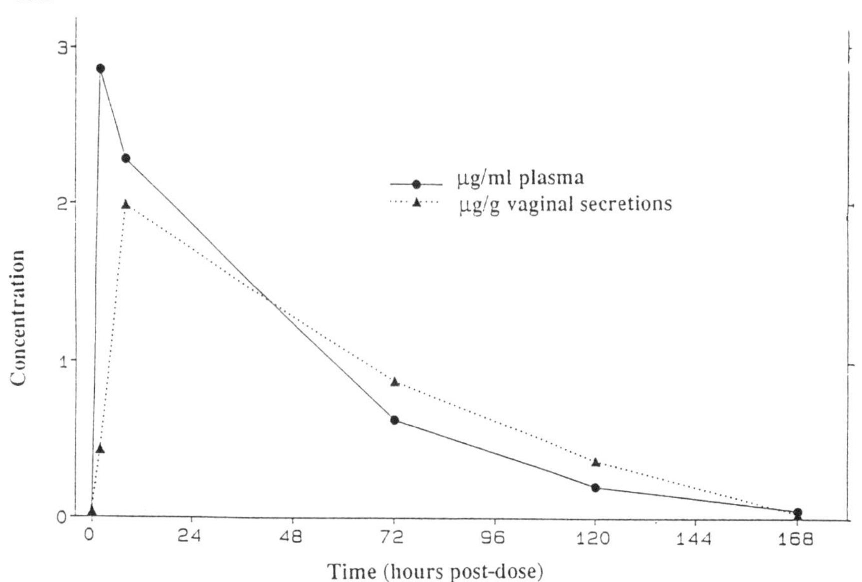

Fig. 2. Typical pharmacokinetic profile of fluconazole in plasma and vaginal secretions after a single oral dose (150 mg) in a female subject, showing therapeutic levels in the vagina for at least 72 h.

a typical profile of fluconazole in plasma and vaginal secretions in a female subject, obtained using this method, after a single oral dose of 150 mg.

References

1. Dupont, B. & Drouhet, E. (1988) *J. Vet. Med. Mycol. 26*, 67-71.
2. Kutzer, E., Oittner, R., Leodolter, S. & Brammer, K.W. (1988) *Eur. J. Obstet. & Gyn. & Rep. Biol. 29*, 305-313.
3. Brammer, K.W. & Feczko, J.M. (1988) *Ann. N.Y. Acad. Sci. 544*, 561-563.
4. Brammer, K.W. & Tarbit, M.H. (1987) in *Recent Trends in the Discovery, Development and Evaluation of Antifungal Agents* (Fromtling, R.A., ed.), J.R. Prous Science Publ., Barcelona, pp. 141-149.
5. Wood, P.R. & Tarbit, M.H. (1986) *J. Chromatog. 383*, 179-186.
6. Garland, W.A. & Powell, M.L. (1981) *J. Chromatog. Sci. 19*, 392-434.
7. Grob, K. & Neukon, H.P. (1985) *J. Chromatog. 323*, 237-246.

#B-8

ASSAY METHODS FOR IMIDAZOLE- AND TRIAZOLE-LIKE ANTIFUNGALS AND SOME ANTIHELMINTHICS

R. Woestenborghs and J. Heykants

Department of Drug Metabolism and Pharmacokinetics, Janssen Research Foundation, B-2340 Beerse, Belgium

Bioanalytical experience with a number of Janssen antifungals and antihelminthics has been gained over 15 years. Whilst initially much useful information was gained with bioassays, chromatographic techniques and RIA's have become indispensable for studying the pharmacokinetics and toxicokinetics and for clinical monitoring of new anti-infective compounds.

The antifungals of lower mol. wt. (miconazole, econazole, isoconazole, imazalil/enilconazole, azaconazole, penconazole, parconazole and propiconazole) can easily be assayed using GC-ECD. Detection limits are in the lower ng/ml range, making the methods very useful for determining residues of the compounds used for veterinary, agricultural or industrial applications (e.g. imazalil and azaconazole). On the other hand, the antifungals of higher mol. wt. (ketoconazole, itraconazole and saperconazole), having better UV-absorption characteristics, are typically analyzed by HPLC with UV detection.*

Standard assay procedures for the antihelminthics are GC with NPD (levamisole) and RIA or HPLC-UV (mebendazole and flubendazole).

The anti-infective compounds originally synthesized and developed by the Janssen Research Laboratories include agents considered in this article, and also the following types (not considered here): a salicylanilide antihelminthic, closantel (synthesized in 1974); a nitroimidazole antiprotozoal, carnidazole (1972); aryltriazine anticoccidials, clazuril and diclazuril (1984); a pyridazine antiviral (ref. no. 79975; 1988).

For antihelminthic and antifungal agents various analytical methods have been described, mainly entailing bioassay, chromatography or immunoassay. Here we summarize the analytical

---

* *Abbreviations include* ECD, electron-capture detector; NPD, nitrogen-phosphorus detector; IAA, isoamyl alcohol; MS, mass spectrometry; i.s., internal standard; RP, reversed phase.

strategies used in the Janssen Bioanalytical Laboratory for the imidazole antifungals (Fig. 1a), the triazole antifungals (Fig. 1b) and some antihelminthics (Fig. 1c).

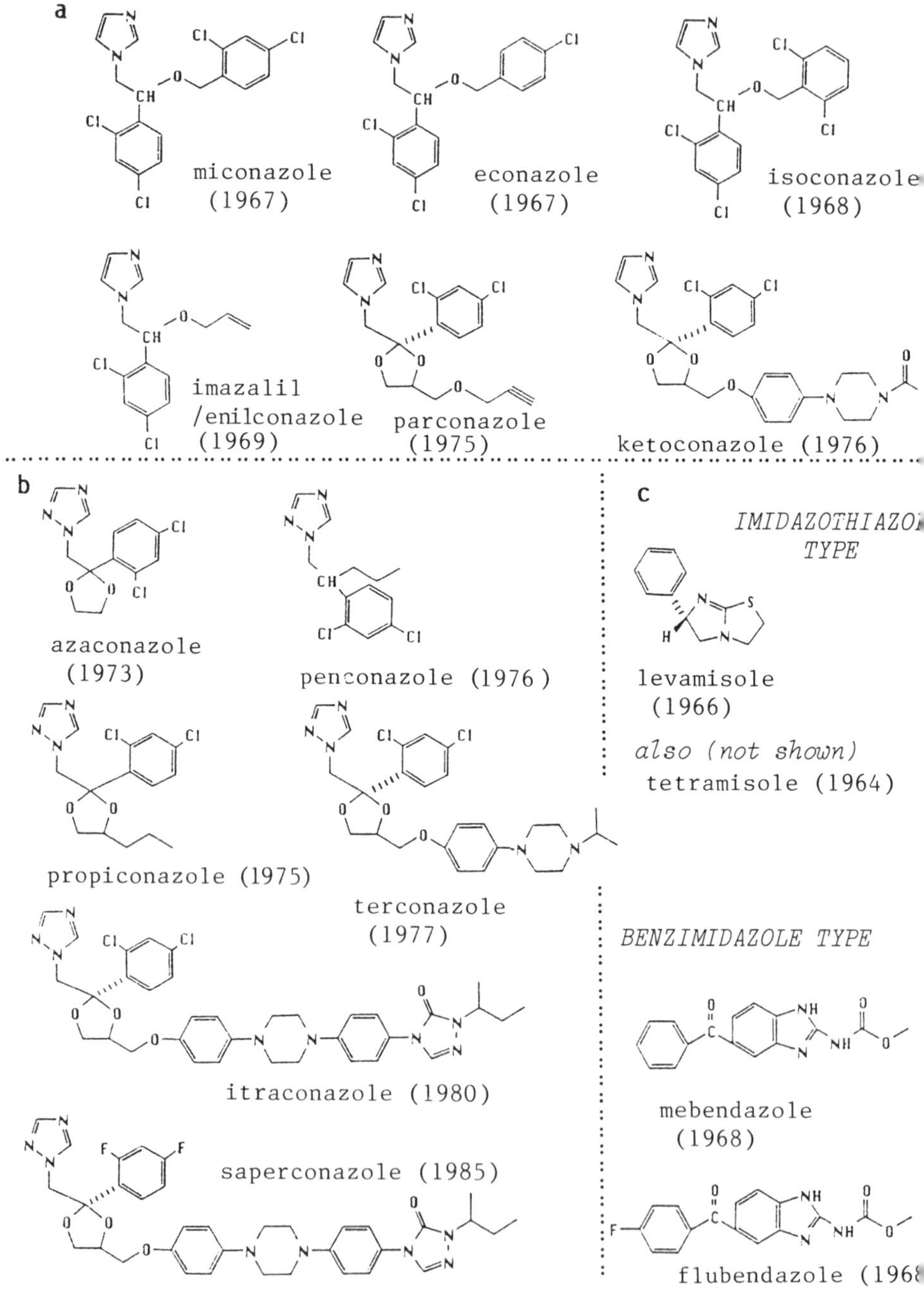

a

miconazole (1967)

econazole (1967)

isoconazole (1968)

imazalil /enilconazole (1969)

parconazole (1975)

ketoconazole (1976)

b

azaconazole (1973)

penconazole (1976)

propiconazole (1975)

terconazole (1977)

itraconazole (1980)

saperconazole (1985)

c

IMIDAZOTHIAZO TYPE

levamisole (1966)

also (not shown) tetramisole (1964)

BENZIMIDAZOLE TYPE

mebendazole (1968)

flubendazole (196

Fig. 1. Structures (with year of synthesis) of (**a**) imidazole and (**b**) triazole antifungals, and (**c**) some antihelminthics.

ANTIFUNGALS

General strategy.- Although bioassays are very useful tools for investigating and developing antifungal agents, they are rather time-consuming, often lack specificity and are of limited use in pharmacokinetic studies where plasma drug concentrations are needed rather than antifungal activity levels in plasma. Chromatographic methods have therefore become increasingly important for pharmacokinetic studies with antifungals. The choice between GC and HPLC is governed more by the mol. wt. of the investigated compound than by its chemical structure (imidazole or triazole). Hence for the antifungals of lower mol. wt. (≤500) GC is used; for those of high mol. wt., HPLC is the method of choice.

GC method.- All lower mol. wt. imidazoles and triazoles (miconazole, econazole, isoconazole, imazalil/enilconazole, azaconazole, penconazole, parconazole and propiconazole) can be successfully extracted at alkaline pH using 5% (v/v) IAA in n-heptane. A preceding acid extraction step was found to be necessary for imazalil in order to 'liberate' it from vegetable or animal tissues [1]. Sample clean-up is usually accomplished by back-extraction and re-extraction and is necessary only when low detection limits are needed. Overall extraction recoveries are typically >90%.

GC analysis is performed on packed columns (OV-17, SP-2250-DB) or capillary columns (OV-101) using ECD or NPD. With the ECD and adequate sample clean-up, detection limits are <1 ng/ml. Problems encountered with the general assay method outlined above were mostly due to adsorption phenomena. For method improvement, therefore, all anti-adsorption stratagems are warranted, e.g. silanization of the glass columns, use of deactivated GC-packing (SP-2250-DB), and even the injection of polyethylene glycol at regular intervals.

HPLC method.- The higher mol. wt. antifungals (ketoconazole, itraconazole and saperconazole) have superior spectroscopic characteristics and are typically analyzed by HPLC with UV detection [2]. The compounds are extracted at alkaline pH with 1.5% IAA in heptane and, if lower detection limits are needed, are back- and re-extracted. Keratinous material (hair, nails) should be chemically hydrolyzed prior to extraction. Extraction recoveries for the metabolically unaltered compounds are >90%.

HPLC is performed on RP (C-18) columns, using acetonitrile-water-diethylamine as eluent. UV detection is generally used, although comparable sensitivities can be obtained by using fluorescence or electrochemical detection. Detection limits are ~1 ng/ml or ~1 ng/g for all three compounds.

The use of more polar extraction solvents (5% or 10% IAA in heptane) is indicated where hydroxylated metabolites have to be assayed as well as the parent compound. Attention should be paid to the low solubility of the compounds: polar organic solvents should be used to prepare standard solutions, and evaporated extracts should immediately be reconstituted in the HPLC elution mixture and be quickly processed to avoid adsorption onto the glassware.

ANTIHELMINTHICS

For levamisole, GC-NPD is the generally used approach [3] although TLC, polarography and HPLC have also been used. The compound and its i.s. are extracted from the alkalinized samples using 5% IAA in heptane. Clean-up of the crude extracts is accomplished by back extraction and re-extraction. Overall extraction recoveries are >80% for plasma and the main animal tissues. GC is performed on packed or capillary columns as for antifungals (above) using NPD in the routine method or MS detection in the confirmatory work; detection limits were <10 ng/ml or <10 ng/g. The method was applied to measure plasma concentrations in bioequivalence studies or animal-tissue concentrations in residue-decline studies. The use of Vacutainer® tubes and stoppers should be avoided for the preparation of plasma samples as it was shown that rubber-stopper constituents displaced the drugs from the plasma proteins, resulting in an underestimation of the levamisole plasma levels by ~20%.

For the benzimidazoles, HPLC-UV [4] is typically employed, or else RIA, performed directly on plasma [5]. For HPLC the compounds (e.g. mebendazole, flubendazole) are extracted at alkaline pH with 5% IAA in heptane, giving ~75% recovery from animal tissues after back- and re-extraction. HPLC is performed on RP columns (C-18), using ammonium formate-methanol-acetonitrile (68:16:16 by vol.) as eluent. Detection limits of ~10 ng/ml or ~10 ng/g are attainable using 254 nm detection. Problems often encountered with the bioanalysis of the benzimidazoles, reflecting their low solubility or high polarity, are peak broadening or decreased extraction recoveries.

References

1. Woestenborghs, R., Michielsen, L., Pauwels, C., Van Leemput, L. & Heykants, J. (1988) *Med. Fac. Landbouww. Rijksuniv. Gent 53/3b*, 1425-1432.
2. Woestenborghs, R., Lorreyne, W. & Heykants, J. (1987) *J. Chromatog. 413*, 332-337.
3. Woestenborghs, R., Michielsen, L. & Heykants, J. (1981) *J. Chromatog. 224*, 25-32.
4. Karlaganis, G., Muenst, G.J. & Bircher, J. (1979) *High Resol. Chromatog. Chromatog. Comm. 2*, 141-144.
5. Michiels, M., Hendriks, R., Heykants, J. & Van den Bossche, H. (1982) *Arch. Int. Pharmacodyn. Thér. 256*, 180-191.

#B-9

EXPERIENCES WITH ASSAYING HYDROXYNAPHTHOQUINONES - A PROBLEM OF SOLUBILITY?

M.V. Doig and A.E. Jones

Department of Bioanalytical Sciences,
Wellcome Research Laboratories,
Beckenham, Kent BR3 3BS, U.K.

Here the analytical methods developed to analyze two related antimalarials 58C80 and 566C80 in plasma are described. Once the drug development program for 566C80 looked as if it might progress to a commercial product, it was sought to improve the analytical method and to develop or modify it for assaying the drug in other biological matrices. The problems encountered and experiences gained are highlighted. These are likely to be of interest to any analyst trying to assay lipophilic agents in biological fluids.

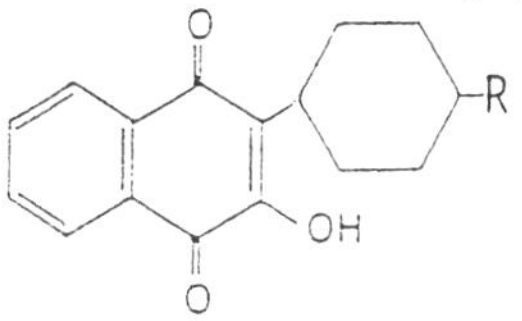

R in the cyclohexyl-substituted hydroxynaphthoquinone

in 58C:	t-butyl
in 568C:	1,1-dimethylpropyl
in 566C:	4-chlorophenyl
in 59C:	phenyl

The history of the development of hydroxynaphthoquinones as antimalarial agents has been reviewed [1]. Compounds 58C80 and 566C80 are two from a series of 2-substituted 3-hydroxy-1,4-naphthoquinones synthesized here in the hope of finding a new chemical entity to replace existing antimalarial therapies. Their action involves inhibition of the mitochondrial electron-transport cytochrome complex bc_1 which is linked to pyrimidine biosynthesis via ubiquinone [2]. Unlike mammalian systems, plasmodia are unable to salvage pre-formed pyrimidines, so inhibition of this pathway causes death. The potencies of compounds in the series were evaluated using *in vitro* and *in vivo* bioassays [3]. The results obtained for 58C80 played a primary role in its selection as a compound for further development, support of which called for an analytical method for quantifying 58C80 in plasma; 568C80 was used as i.s.*

METHOD FOR 58C80

Plasma (1.0 ml) acidified with 0.5 ml of 0.5 M acetic acid was extracted using 2 × 5 ml of IAA/HEX. The solvent was removed under N_2 and the 'acidic' 3-hydroxy group derivatized

**Abbreviations.*- i.s., internal standard; IAA/HEX, 2% (v/v) isoamyl alcohol in hexane; SPE, solid-phase extraction.

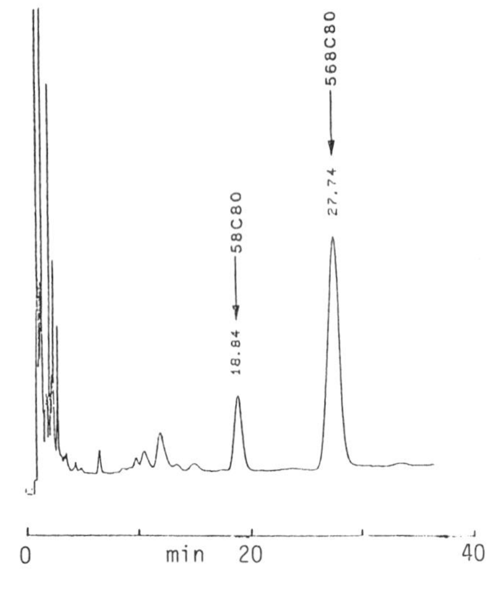

Fig. 1. Packed-column GC pattern for 58C80 in human plasma (~150 ng/ml).
GC: HP 5880 with ECD.
Column: 2 m × 2 mm, 10% OV-17 on Chromosorb W (HP) 100-120 mesh, at 245° (injector 275°, ECD 300°).
N_2 carrier gas at 35 ml/min.
Injection vol. 1 µl.

with ethereal diazomethane. An example of a GC trace is shown in Fig. 1 (conditions in legend).

Unfortunately when 58C80 was administered to humans it was rapidly metabolized. One metabolite was dark-red and had a long half-life; when excreted, it changed the colour of the urine to a dark claret-red. This and other problems halted the development of 58C80 as an antimalarial.

While performing the assays for 58C80, two essential points were noted. The use of plastics during sample preparation had to be avoided because plasticizers caused chromatographic interferences. The use of a close analogue as an i.s. was essential because absolute peak areas varied from day to day and sample to sample but peak-area ratios remained constant.

The next in the series to be developed as an antimalarial was 566C80. It was selected because it displayed good antimalarial potency when examined in a variety of *in vitro* and *in vivo* bioassays and was not metabolized when incubated with human liver microsomes.

INITIAL METHOD FOR 566C80

Extraction and derivatization were performed as for 58C80, and the i.s. was 59C80 (shown on title p.). Fig. 2 shows an example of a GC trace, the use of a bought fused-silica capillary column reflecting the time-gap since the 58C80 assay for which hand-packed glass columns were used. The method was robust, with linearity from 10 ng/ml to 10 µg/ml plasma (r = 0.997977; peak-height ratio: slope 0.559, intercept 0.101) and with good within-day and (Fig. 3) between-day

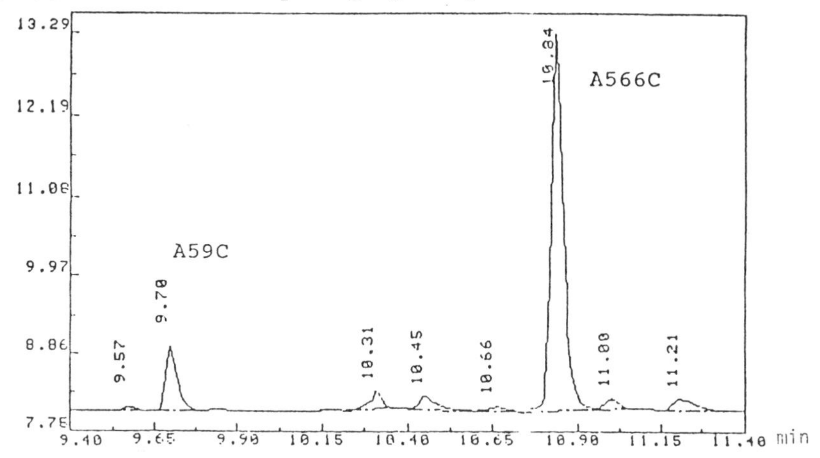

Fig. 2. Capillary GC pattern for 566C80 in dog plasma (9.6 μg/ml). GC: HP 5880 with ECD. Column: 25 m × 0.25 mm i.d. 100% methyl silicone fused silica; oven at 85° for 1 min then 25°/min to 275° and 5 min at 300°. Injector: 300°; ECD: 300°. Splitless injection, 0.5-4 μl; valve time 0.5-0.7 min.

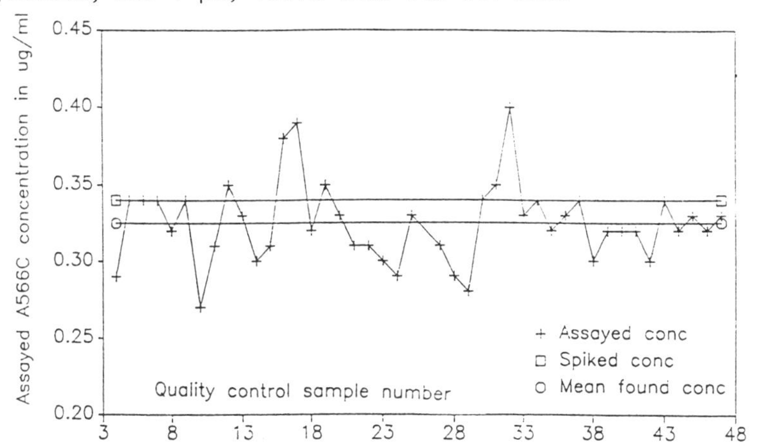

Fig. 3. Graphical representation of the QC results from a single study (BPAD 87/17).

reproducibility. The main problems with the method for large-scale routine analysis were the removal of the extraction solvent and the presence of a small peak that would interfere with the 566C80 peak if chromatographic resolution were not maintained at the highest level.

EVALUATION OF SPE

Using ^{14}C-labelled 566C80, various Bond-Elut cartridges (C-18, C-2, phenyl, cyano, amino) were first examined in the reversed-phase (RP) mode; but regardless of pH or the dilution of the plasma, none of these phases retained >50% of the applied radiolabel. Then C-2, cyano, diol, amino and straight-silica cartridges were evaluated in the normal-phase (NP) mode. The best cartridge for retaining 566C80 from non-aqueous solvents was the amino cartridge, but solvent extraction followed by SPE was even less practical for large-scale routine assays than the original solvent extraction method. Accordingly, the following 'double-decker' method was developed. Plasma (0.5 ml), i.s. and 0.5 ml of 0.1 M acetic acid were mixed and applied to a Chem-Elut liquid-liquid extraction cartridge, The 566C80 and i.s. were then eluted using 15 ml of IAA/HEX, applied directly to an amino Bond-Elut cartridge pre-washed with 2 ml methanol and then 2 ml IAA/HEX. The loaded cartridge was then washed with 2 ml ethyl acetate followed by 2 ml methanol. The 566C80 was then eluted using 2 ml 0.1 M acetic acid in methanol. The eluate was concentrated under N_2 and derivatized using ethereal diazomethane.

This SPE method produced notably clean chromatograms and avoided the need to remove 10 ml of extraction solvent under N_2. Unfortunately within-day reproducibility and recoveries for this method deteriorated in analytical runs when >10 samples were extracted in a day and when the batch of Chem-Elut cartridges was changed. When these problems occurred we had no time to try to solve them, so the SPE method was abandoned and we returned to the original solvent extraction method.

EXTRACTION OF 566C80 FROM URINE

SPE was again tried and abandoned due to variable results. Solvent extraction with ethyl acetate produced the best recoveries but dirty chromatograms. An adequate method was developed by washing the ethyl acetate with pH 8 buffer. Then we started validation of the method, including measuring the stability of added 566C80 in urine stored at -20°. The results appeared to indicate that 566C80 was unstable in urine. This was strange because 566C80 was stable in plasma for months or even years. Repeating the experiments with the spiked urine samples in glass rather than plastic containers indicated that the 566C80 was not degrading but was gradually adsorbing onto the plastic containers. As the latter were the same as those used for evaluating the stability of 566C80 in plasma, this phenomenon is dependent on the matrix. This phenomenon could be partially suppressed by acidifying the urine to pH 3, and appeared to be dependent on the urine's salt concentration.

EXTRACTION OF 566C80 FROM FAECES

The extraction method was essentially the same as that used for urine. Prior to assaying real samples, the method was developed and validated by using spiked aqueous control homogenates (1% and 10%) of human faeces. The method appeared to be robust, and extraction recoveries were >95% regardless of the faecal content of the homogenate. In addition the homogenates, unlike urine, could be stored in plastic containers. Then we assayed real samples and found that the 566C80 concentration increased when the faecal content of the homogenate increased: thus a 1% homogenate contained a total of 103.5 mg and a 10% homogenate of the same sample contained 52.4 mg. This trend was consistent and not dependent on the consistency of the faecal sample. This means that spiked samples do not accurately reflect the extraction of 566C80 from real samples. Optimizing the assay and obtaining some indication of the true extraction recoveries must await our analyzing samples from a radiolabel study in man and comparing the values we obtain with those obtained from combustion analysis.

CONCLUSIONS

The antimalarial agents 58C80 and 566C80 are highly lipophilic and their extraction from biological samples containing ng/ml concentrations can produce surprising results.

Solvent extraction was far more reproducible than SPE. This is only because a close structural analogue, used as an i.s., could correct for losses at any stage of the solvent extraction, provided that clean glassware was used. The ability of the i.s. to correct for losses during SPE is more difficult as the analyte has a tendency to adsorb to plastic especially when it is in an aqueous matrix. This phenomenon became apparent only when conducting stability studies on urine samples stored in plastic containers, since plastic had been avoided during sample preparation because plasticizers caused chromatographic interferences.

Another problem was seen when analyzing faecal samples: when analyzing different dilutions of real samples it was discovered that extraction recoveries from spiked samples did not accurately reflect the recovery from real samples. This is the ultimate problem for an analyst because an analyte can hardly be quantified without utilizing spiked samples and assuming that the recoveries from spiked and real samples are the same.

Acknowledgements

The authors thank R. Casey, I. Fraser, S. Hannan, D. Jardine and G. Ridout for conducting some of the practical work included in this article.

References

1. Hudson, A.T. (1984) in *Handbook of Experimental Pharmacology*, Vol. 68/II (Peters, W. & Richards, W.H.G., eds.), Springer-Verlag, Berlin, pp. 343-361.
2. Hammond, D.J., Burchell, J.R. & Pudney, M. (1985) *Mol. Biochem. Parasitol. 14*, 97-109.
3. Hudson, A.T., Randall, A.W., Fry, M., Ginger, C.D., Hill, B., Latter, V.S., McHardy, N. & Williams, R.B. (1985) *Parasitology 90*, 45-55.

#B-10

ASSAY OF FAMCICLOVIR AND ITS METABOLITES, INCLUDING THE ANTI-HERPES AGENT PENCICLOVIR, IN PLASMA AND URINE OF RAT, DOG AND MAN

C.F. Winton, S.E. Fowles, R.A. Vere Hodge and †D.M. Pierce

Beecham Pharmaceuticals Research Division,
Coldharbour Road, The Pinnacles,
Harlow, Essex CM19 5AD, U.K.

Requirement	*A method for specifically determining famciclovir (BRL 42810), penciclovir (BRL 39123) and 4 intermediate metabolites in plasma and urine with reliability down to ~1 and ~10 μg/ml respectively.*
End-step	*RP-HPLC on an Apex 1 ODS 3 μm column, with gradient elution by pH 7.0 phosphate buffer/methanol and UV detection with programmed wavelength changes.*
Sample handling	*After TCA* addition to precipitate proteins and inactivate viruses, extraction by an SCX cartridge and elution by pH 11.0 phosphate buffer/methanol.*
Comments	*Lower limit of sensitivity 0.5 μg/ml for plasma and, depending on the compound, 2-50 μg/ml for urine. HPLC run time maximally 35 min, depending on species, for plasma and 50 min for urine in all species.*

Famciclovir is the inactive diacetyl 6-deoxy analogue of the antiviral agent penciclovir, 9-(4-hydroxy-3-hydroxymethyl-but-1-yl)guanine.⊗ This pro-drug is under evaluation as an oral therapy for the treatment of Herpes Simplex (HSV types 1 and 2) and Herpex Zoster infections [1,2]. So as to evaluate the kinetics of the *in vivo* conversion of famciclovir to penciclovir and subsequent disposition in animals and man, a sensitive and specific assay for biological fluids was required for both these compounds and intermediates in the biotransformation pathways (Fig. 1): BRL 43594 (monoacetyl, 6-deoxy-BRL 39123), BRL 42359 (6-deoxy-BRL 39123), BRL 39913 (diacetyl-BRL 39123) and BRL 42222 (monoacetyl-BRL 39123). A gradient HPLC-UV method which fulfils these requirements was developed for plasma and urine of rat, dog and man. Though specific details of the method vary from species to species and between biological fluids, the essential features are common throughout.

---

†addressee for any correspondence. ⊗BRL nos. : *see* Requirement.

**Abbreviations.*- i.s., internal standard; SPE, solid-phase extraction; TCA, trichloracetic acid; MS, mass spectrometry (FAB, fast atom bombardment).

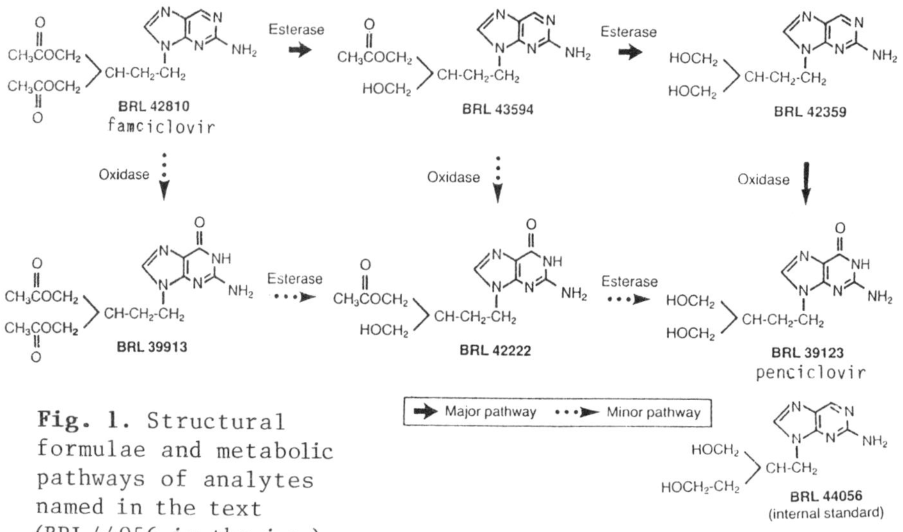

Fig. 1. Structural formulae and metabolic pathways of analytes named in the text (BRL 44056 is the i.s.).

MATERIALS

Solutions of all analytes and the i.s. (above) were made up in ultra-pure water from an Elgastat Spectrum System (Elga Products, High Wycombe). Organic solvents were of HPLC grade (Rathburn Chemicals, Walkerburn, Scotland). Other reagents were of Analar grade (Fisons, Loughborough, and BDH Chemicals, Poole). SPE cartridges (Bond-Elut) were from Jones Chromatography (Hengoed, Mid-Glamorgan).

SAMPLE TREATMENT

Blood samples were collected on ice into plastic EDTA tubes with added NaF or KF, to inhibit hydrolysis of famciclovir and its esterified metabolites by esterases in plasma and red cells. Samples were centrifuged within minutes of collection, and the plasma frozen (dry ice) for transport to the laboratory. In samples kept frozen at appropriate temperatures over many months, there was no evidence of significant analyte degradation. Just before analysis, samples were thawed and aliquots (typically 0.5 ml plasma or 100 µl urine) with 1 vol. of 16% (w/v) TCA added were kept for 10 min. This step precipitated proteins from plasma and was also expected to inactivate Herpes virus and HIV by virtue of its very low pH (<1).

EXTRACTION

SPE was adopted, with future automation in mind. A variety of cartridges were tested and compounds were found to be effectively retained on C-8, C-18 and SCX (sulphonic acid cation-exchange) cartridges; SCX gave the most complete elution. (Pre-conditioning: 1 ml each of methanol, water, pH 7 Na_2HPO_4)

Optimal conditions for sample clean-up involved initially washing the loaded SCX cartridge with 1 ml methanol/1 mM Na_2HPO_4 adjusted to pH 7.0 (1:4 by vol.), which removed the majority of endogenous components, leaving the analytes on the cartridge. Analytes were subsequently eluted under strongly alkaline conditions, using 1 ml methanol/100 mM K_2HPO_4 adjusted to pH 11.0 (1:3). The switch from low-molarity Na_2HPO_4 in the wash to higher-molarity K_2HPO_4 in the eluent was critical for the control of elution. Na^+ is a weaker counter-ion than K^+. At low ionic strength in the wash, there was minimal risk of its displacing the bound analytes, whereas K^+, a stronger counter-ion used at higher molarity, was adsorbed to the cation-exchanger in preference to the analytes, thus promoting elution.

CHROMATOGRAPHY

A gradient HPLC approach was adopted because of the wide differences in polarity between the acetyl esters and the de-esterified analogues. The HPLC system configuration consisted of a dual-pump gradient solvent delivery system with high-pressure mixing. The system (Millipore/Waters) comprised two Model 510 pumps and a 680 gradient controller, with a WISP 712 or 710B autoinjector; the sensitive UV detector (Kratos SF783) was of variable wavelength.

An Apex 1 ODS 3 µm column of i.d. 4.6 mm was used, 5 cm or, for dog plasma and for urine of all species, 15 cm; it was protected by a Brownlee RP-18 Newguard guard column. For the gradient elution a methanol/phosphate buffer mobile phase was used; the proportions of the two components in the inflowing solution were optimized, for each assay, during method development, as were the buffer pH and ionic strength.

For **rat and human plasma** the gradient programme was as shown in Fig. 2a, the solutions delivered by pumps A and B having the compositions given in the legend. Endogenous components in **dog plasma** necessitated some modification of conditions: those used were a compromise between those used for plasma and urine of other species. The longer column (15 cm) was required, together with a mobile phase containing more methanol (10%) in solution A, and a tailor-made gradient programme, which with 1 ml/min flow-rate resulted in a run-time of 35 min.

With **urine** of all species, separation of analytes from endogenous contaminants again necessitated the 15 cm column, and a different gradient programme (Fig. 2b); as is indicated

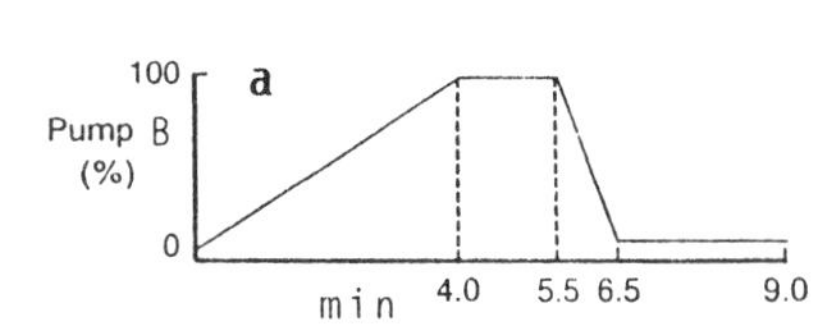

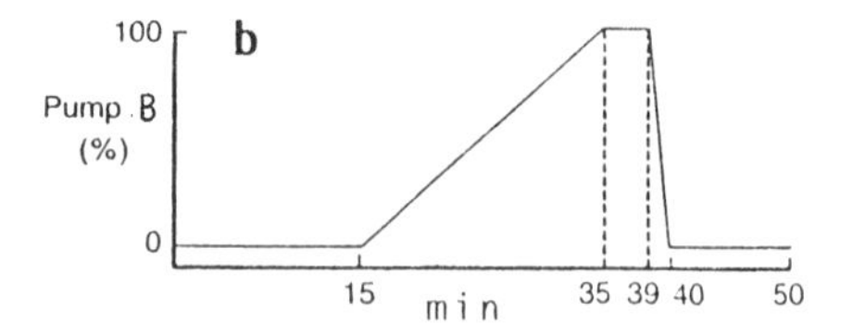

Fig. 2. Gradient conditions for separation of famciclovir (BRL 42810) and metabolites in (**a**) human plasma, and (**b**) human urine. Methanol/10 mM Na_2HPO_4 (pH 7.0) mixtures were in the following proportions: **a**, Pump A, 7:93 and Pump B, 35:65; **b**, Pump A, 5:95 and Pump B, 35:65. Flow-rates (ml/min) differed between **a** (1.0) and **b** (2.0), also column lengths as stated in the text which gives other details.

in the legend, the proportion of methanol in Solution A was lowered (to 5%). With the 1 ml/min flow-rate the run-time was 50 min.

DETECTION

Detection was by UV absorbance, with programmed wavelength changes to optimize the sensitivity and specificity for individual analytes. Penciclovir and its 6-oxy precursors (BRL 39913 and BRL 42222) demonstrated the characteristic guanine spectrum, with absorption maxima at 214 and 254 nm and a shoulder at ~270 nm. In contrast, the absence of the 6-oxy moiety from famciclovir and the other deoxy compounds (BRL 43594, BRL 42359 and the i.s. BRL 44056) resulted in significant shifts of absorbance maxima to longer wavelengths, 220 and 305 nm, the latter band being reduced in intensity compared with that of the 254 nm maximum for the guanine analogues.

Thus the simplest wavelength-change programme, as used for human and rat plasma assay, involved merely appropriately timed alternations between 254 nm for determining the guanines and 305 nm for monitoring the 6-deoxyguanines. For assays in dog plasma and in urine of all species, these programmes were complicated by the need to minimize endogenous interferences, met by monitoring both the guanine BRL 42222 and the deoxyguanine BRL 43594 at 245 nm instead of 254 and 305 nm respectively. Similarly, the guanine BRL 39913 was monitored at 270 nm. Thus, in these cases lower sensitivity was accepted as a trade-off against increased specificity. The programming details are given in the inserts to Figs. 3a and 3b, which show typical chromatograms for extracts of human (and rat) plasma and of urine from all three species. As expected, penciclovir (the most polar compound) eluted first and famciclovir (the least polar) last; for the other compounds, side-chain free OH's reduced retention, outweighing the influence of the guanine 6-oxy substituent.

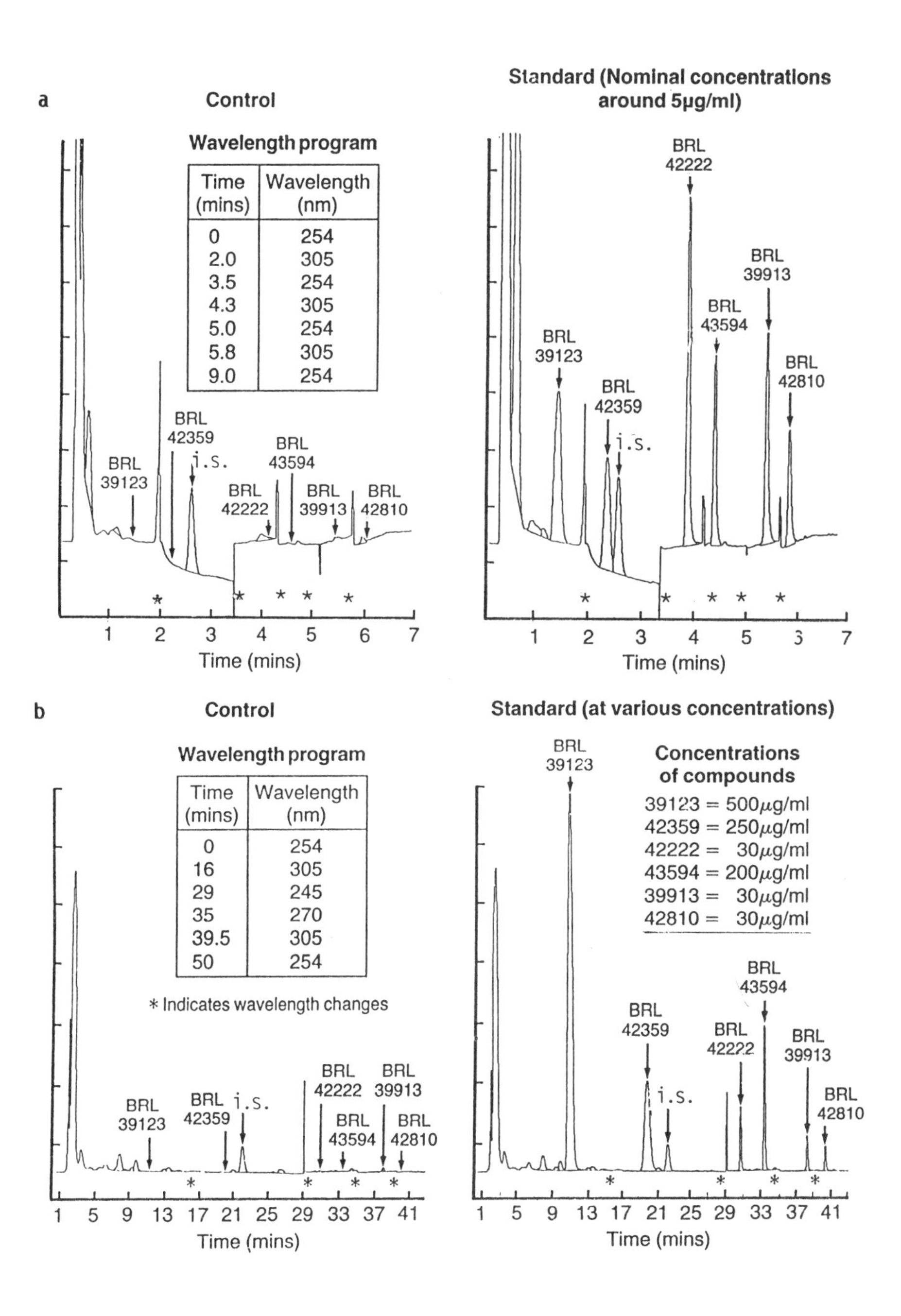

Fig. 3. Typical chromatograms of famciclovir (BRL 42810) and its 5 metabolites in (**a**) human plasma, and (**b**) human urine. Wavelength changes denoted *. 'Control' was from undosed volunteers, spiked with i.s. 'Standard' likewise, but spiked with *all* analytes.

VALIDATION

Extraction efficiencies from plasma were only ~50% with means (n = 5) ranging from 42% to 63% for individual analytes. In general they were better than 70% from urine, with mean values ranging from 71% to 91% for all compounds except penciclovir (BRL 39123), which was somewhat lower (54%). Although the recoveries of famciclovir (BRL 42810) and metabolites from plasma were low, they were not subject to extensive variation between runs. The losses appeared to occur predominantly during the precipitation step.

Linearity.- Standard curves relating peak-height ratio to analyte concentration were linear for all analytes up to 80 µg/ml in plasma of all three species, with little scatter (R.S.D. typically <5%). In urine, linear ranges were established for individual analytes, taking account of concentrations observed *in vivo* early in the study. Thus for penciclovir the calibration was linear up to 2000 µg/ml, and for its deoxy precursor (BRL 42359) up to 1600 µg/ml; for BRL 43594 it held up to 200 µg/ml and for famciclovir and the other metabolites up to 40 µg/ml.

Sensitivity.- The lower limit of reliable determination (LRD) for all analytes in plasma from each of the three species was routinely 0.5 µg/ml, although lower LRD values have been achieved using appropriately low calibration ranges. In urine, the typical LRD for penciclovir was 50 µg/ml, though values as low as 12.5 µg/ml were achieved over a limited calibration range. For BRL 42359 the corresponding values were 40 and 10 µg/ml urine. For BRL 43594 the LRD was 5 µg/ml urine, and for all other analytes 2 µg/ml.

Specificity.- Chromatographic peaks in the spiked calibration standards and authentic samples were analyzed by on-line frit-FAB-MS on a Jeol DX-303 mass spectrometer with a DA5000 data system. Each peak contained the appropriate $(M+H)^+$ ion, as follows: penciclovir, 254; BRL 42359, 238; BRL 42222, 296; BRL 43594, 280; BRL 39913, 338; famciclovir, 322.

In addition, peak homogeneity was examined by substituting a Water 990 photodiode-array detector for the standard UV detector. The penciclovir peak was free from co-eluting compounds in plasma and urine from all three species. Although interferants co-eluting with famciclovir, BRL 43594, BRL 42359 and BRL 42222 were observed, these did not give significant peaks at the wavelength chosen for their detection. Only BRL 39913 suffered interference from a small but detectable peak which limited the specificity of the method with respect to this analyte.

Table 1. Within-day variation in detected concentrations (μg/ml) in human plasma and urine spiked at 3 different concentrations. Precision is defined as the C.V., and accuracy as the % deviation of the mean value from the nominal.

	Level		BRL 39123	BRL 42359	BRL 42222	BRL 43594	BRL 39913	BRL 42810
PLASMA (n = 6)	Low	Mean concentration	1.0	1.0	1.0	1.0	0.97	1.1
		Precision (%)	3.1	1.7	3.5	0.7	7.6	4.3
		Accuracy (%)	−0.06	1.8	1.0	0.0	2.1	−3.7
	Med	Mean concentration	8.1	8.1	8.3	8.1	8.2	8.2
		Precision (%)	3.1	1.6	3.6	2.4	5.1	4.0
		Accuracy (%)	−0.1	0.3	−1.4	−0.5	−2.6	−1.9
	High	Mean concentration	84.9	80.4	85.4	80.9	84.1	83.2
		Precision (%)	3.5	2.1	3.9	2.4	7.1	7.9
		Accuracy (%)	−4.8	1.1	−4.9	−0.7	−5.6	−2.8
URINE (n = 5)	Low	Mean concentration	194.5	146.7	3.8	19.5	3.1	3.9
		Precision (%)	3.1	5.0	5.8	27.0	27.0	4.7
		Accuracy (%)	1.9	5.7	6.1	4.7	18.4	3.1
	Med	Mean concentration	637.5	506.9	12.5	63.9	11.2	12.2
		Precision (%)	3.0	3.6	2.4	3.5	2.7	4.2
		Accuracy (%)	−7.2	−8.6	−4.1	−4.3	0.6	−0.1
	High	Mean concentration	1974	1426	39.7	201.1	37.7	40.7
		Precision (%)	1.1	2.6	2.2	1.8	2.8	3.0
		Accuracy (%)	0.4	1.8	0.7	1.5	−0.1	−0.4

Accuracy and reproducibility.- The within-day variation of assay accuracy was tested by analyzing replicate samples in each fluid containing each of the analytes at 3 different concentrations (low, medium and high with respect to ranges found in plasma and urine samples). The different concentration ranges used for the various analytes in urine reflected the much lower concentrations of some (e.g. famciclovir, BRL 42222, BRL 39913) than of others (penciclovir, BRL 42359, BRL 43594) which were detected. The first group of compounds was rarely quantifiable in plasma.

The results (Table 1) demonstrated that the assay for all compounds in human plasma was reproducible within-day with C.V. values <8%, and accurate, with deviations from set values <6% in each case. Comparable results were obtained in other species. In human urine the assay was likewise good for most compounds, except that the interference with BRL 39913 resulted in lower reproducibility (C.V. 27%) and accuracy (18.4% deviation from nominal) at the lowest concentration.

Comparable results were obtained with rat urine. However, both precision and accuracy were somewhat less with dog urine, the C.V. values and deviations being 10-29% for some of the analytes.

Results of a similar experiment conducted by repeating analyses on 5 different days demonstrated almost as good between-day reproducibility, the C.V.'s for plasma being ~11% for most compounds although slightly higher (14%) for BRL 39913. In urine, C.V.'s were 12% or lower for all compounds.

APPLICATION

The method has now been applied extensively to studies in animals and man, and has contributed greatly to understanding metabolism and supporting toxicology studies, as well as to some early pharmacokinetic studies in man. One example involved the assay of famciclovir and its metabolites, including the antiviral agent penciclovir, after single and repeated administration of the pro-drug to 6 healthy human subjects. Fig. 4 shows a typical plasma concentration-time curve for the major detectable compounds in plasma (penciclovir and BRL 42359) after single and repeated 1000 mg doses of famciclovir. Assay sensitivity is sufficient to allow the penciclovir plasma concentration-time curve to be followed for up to 6 h (~3 half-lives).

Unchanged famciclovir was not detected in plasma, and BRL 43594 (its monoacetyl analogue) only at the first sampling time, 0.25 h after dosing. Neither of the 6-oxy metabolites in the minor pathway (Fig. 1), BRL 39913 and BRL 42222, were detected in plasma.

In urine, 50-60% of the dose was recovered as the active antiviral agent penciclovir and a further 4-15% as its 6-deoxy precursor, BRL 42359. The monoacetyl analogue of famciclovir, BRL 43594, contributed only trace amounts (<0.5% of the dose). Of the minor metabolites BRL 42222 (6-oxy monoacetyl) was detected in an occasional urine sample, though its diacetyl analogue, BRL 39913, was not detected at all.

COMMENTS

SPE in conjunction with a short HPLC run-time makes the human (and rat) plasma assay quite rapid, with a throughput of up to 70 samples/day. The longer run-times required for the chromatography of dog plasma extracts, and of urine extracts from all species, coupled with a lower stability of analytes in these extracts, limits the throughput in these cases to ~27 and ~24 samples/day respectively. Column life exceeds 1000 injections.

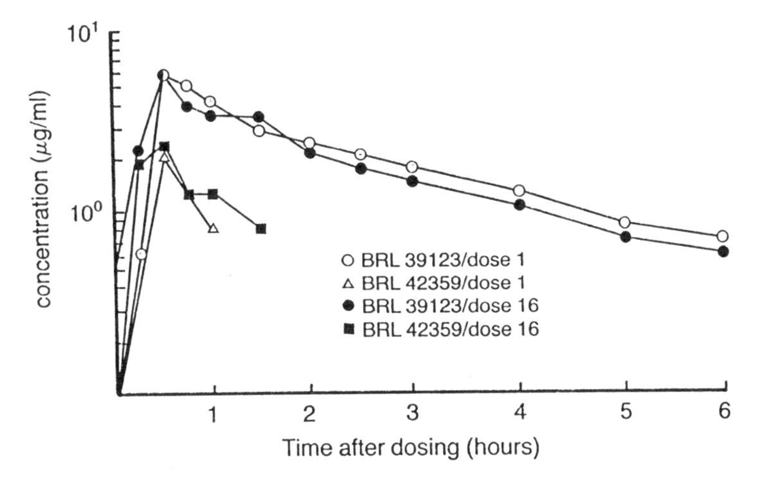

Fig. 4. Typical profile of plasma concentrations of BRL 39123 and BRL 42359 detected in a healthy subject following the first and last of 16 oral administrations of 1000 mg BRL 42810 (famciclovir) every 8 h.

The presence in urine of an endogenous interferant which cannot be chromatographically separated from BRL 39913 limits the sensitivity of the assay with respect to this analyte. However, this is not a serious limitation, since in practice this compound is hardly present among the compounds excreted in urine.

CONCLUSION

The complex gradient-HPLC method described, with UV detection encompassing programmed wavelength changes, allows accurate and precise determination of the pro-drug famciclovir, the antiviral agent penciclovir and four intermediates (Fig. 1) in plasma and urine of rat, dog and man.

Acknowledgements

The authors thank Mr G.D. Allen for MS analysis and Miss A. Fairless for excellent technical assistance.

References

1. Boyd, M.R., Bacon, T.H., Sutton, D. & Cole, M. (1987) *Antimicrob. Agents Chemother. 31*, 1238-1242.
2. Boyd, M.R., Bacon, T.H. & Sutton, D. (1988) *Antimicrob. Agents Chemother. 32*, 358-363.

#B-11

THE ANALYSIS OF ZIDOVUDINE AND ITS GLUCURONIDE METABOLITE BY HPLC

S.S. Good, D.J. Reynolds and P. de Miranda

Burroughs Wellcome Co., 3030 Cornwallis Road, Research Triangle Park, NC 27709, U.S.A.

Requirement	*Procedures for simultaneous assay of the anti-HIV agent AZT* and its glucuronide (GAZT) in human serum.*	
End-step	***Procedure (1):*** *RP-HPLC with an acetonitrile gradient.*	***Procedure (2):*** *isocratic RP-HPLC, with an i.s.*
Sample handling	*Ultrafiltration.*	*SPE, with C-18 cartridges.*
Comments	*During optimization of **(2)**, which is faster, the effects of varying mobile phase pH and buffer strength were investigated, along with different sample preparation procedures including ultrafiltration, SPE and, more recently, acid precipitation. Both **(1)** and **(2)** have been used for the quantitation of AZT and GAZT in serum and other biological matrices in support of the pre-clinical and clinical development of Retrovir®. Procedure **(2)** is being used in various laboratories for therapeutic drug monitoring and studies on pharmacokinetics and disposition.*	

The thymidine analogue AZT has been shown in clinical trials to be an effective agent in the therapy of AIDS and ARC. More recently, the drug has been shown to slow significantly the progression of the disease in patients with early ARC and in HIV-positive, asymptomatic patients. AZT displays a short half-life in humans, and is extensively metabolized to GAZT, its 5'-*O*-glucuronide conjugate. Although the latter has no antiviral activity and is chemically and biologically stable, it is often important to determine blood levels of both drug and metabolite. The development of two HPLC methods for their simultaneous determination in serum is now described.

**Abbreviations.- (See Fig. 1 in next article for formula)* AZT, zidovudine (azidothymidine, ZDV, Retrovir®; 3'-azido-3'-deoxythymidine); GAZT, 5'-glucuronide of AZT (5'-β-D-glucopyranuronosyl substituent); RP, reversed-phase; SPE, solid-phase extraction; i.s., internal standard - A22U [1-(3-azido-2,3-dideoxy-β-D-*threo*-pentofuranosyl)thymine]; AIDS, acquired immune deficiency syndrome; ARC, AIDS-related complex.

ULTRAFILTRATION FOLLOWED BY GRADIENT ANALYSIS

To support the unusually rapid pre-clinical and clinical development of AZT, an RP-HPLC method for the quantitation of AZT and GAZT in serum and urine was quickly devised [1]. For safety as well as simplicity, a sample preparation procedure was selected which used ultrafiltration to remove macromolecules and particulates, including potentially infectious HIV particles.

Samples (0.25-0.5 ml) were placed in the top chamber of Centrifree micropartition systems (Amicon Corp., Danvers, MA) and the units were centrifuged according to the manufacturer's instructions (2000 **g**, 45 min, fixed angle rotor). The ultrafiltrates were then analyzed by HPLC: 0.1 ml samples were injected onto an Adsorbosphere C-18 guard column (5 μm, 10 × 4.6 mm i.d.) linked to a C-18 analytical column (5 μm, 250 × 4.6 mm; Alltech Assoc., State College, PA) which were preceded by an in-line filter (0.5 μm; Scientific Systems Inc., State College, PA). Samples were eluted using a buffered aqueous mobile phase (25 mM ammonium phosphate, pH 7.2) with a 35-min linear acetonitrile gradient from 0% to 30% and 1.0 ml/min flow rate. Effluent monitoring was at 267 nm, the absorbance maximum for AZT. In order to avoid build-up of non-polar material with a resulting degradation in column performance, it was necessary to purge the system between individual analyses with buffer containing 60% (v/v) acetonitrile for 10 min, followed by a 15 min re-equilibration to initial conditions. Including ramping times between the different mobile phases, the total time required per analysis was 66 min.

Known-concentration AZT and GAZT standards were prepared in bulk by adding both analytes to pooled normal human serum. Careful serial dilutions in normal serum gave a 0.1-20 μM range for AZT and 0.2-40 μM for GAZT. Each bulk standard was split into 1.1 ml replicates and stored at -20°. Replicate standards were shown to be stable for at least 5 months [2]. Prior to depletion of a set of replicates, new standards were prepared as described and checked against the old set in the same assay. For each analysis, 5 to 8 serum standards were thawed, ultrafiltered and analyzed in parallel with unknowns. Standard curves were generated from unweighted least-squares linear regression analysis of peak areas corresponding to prepared concentrations.

The typical chromatograms in Fig. 1 (p. 178) show separation of AZT and GAZT in human serum and urine. A minor interference with GAZT in urine was occasionally observed (Fig. 1B), but due to the relatively high levels of urinary GAZT (usually >4 times the AZT concentration) this was of little significance.

Table 1. Determination of AZT and GAZT by ultrafiltration and HPLC: measured concentrations (µM) in serum at different spike levels. Replicates were processed individually.

Analyte	Conc^{n}. *Measured*:-	Mean	Range	C.V., %	
Within-assay (n = 4)					
AZT	9.50	9.28	9.22-9.35	0.7	
	2.38	2.35	2.27-2.41	2.5	
	0.950	0.93	0.92-0.95	1.6	
	0.238	0.24	0.23-0.25	3.5	
GAZT	44.0	44.0	43.9-44.1	0.2	
	11.0	11.2	11.1-11.2	0.6	
	4.40	4.47	4.42-4.51	1.0	
	1.10	1.11	1.11-1.12	0.4	
Between-assay (n = 5)					(% deviation)
AZT	9.50	9.17	9.02-9.28	1.2	(-3.5)
	2.38	2.35	2.31-2.37	1.0	(-1.2)
	0.950	0.948	0.925-0.962	1.7	(-0.3)
	0.238	0.244	0.166-0.244	16.4	(-5.6)
GAZT	44.0	43.7	43.1-44.4	1.3	(-0.6)
	11.0	11.1	11.0-11.2	1.1	(+0.6)
	4.40	4.41	4.30-4.49	1.7	(+0.2)
	1.10	1.06	1.11-0.98	5.1	(-3.5)

Accuracy and precision were excellent, as shown by evaluating assay parameters for AZT and GAZT added to serum: standard curves were determined on 5 different days, with sets of replicates for each of the 4 spike levels listed in Table 1. For the standard curve the mean values were as follows:

0.0286 (range 0.0282-0.0289; C.V. 0.9%) for the slope;
0.0020 (range 0.0010-0.0046; C.V. 73.9%) for the y-intercept, expressed in A_{267} units;
1.0000 (range 0.9999-1.0000; C.V. zero) for the correlation coefficient, r.

Other validation observations are shown in Table 1. The within-assay C.V.'s for AZT and GAZT were 3.5% or less, and the between-assay values generally <6% except for the lowest AZT standard (16.4%).

Drawbacks of using ultrafiltration and a gradient.- Although ultrafiltration is a simple and rapid way to prepare serum samples for HPLC analysis, and may provide a measure of safety by removing virus from samples obtained from HIV-infected patients, the method is based on the assumption that analytes in prepared serum standards bind to macromolecules to the same extent as in unknowns. This is a reasonable assumption for compounds which display little or no serum binding. However, for analytes having moderate or strong binding the % of the total serum concentration present in the unbound state may depend on the health and nutritional status of the patient as well as on the relative time and composition of the patient's last meal. In addition, since the level of analyte(s) in the ultrafiltrate reflects the original unbound concentration in the sample, the sensitivity of the technique (which attempts to quantitate total analyte levels, i.e. bound + unbound), is decreased to the extent to which the analyte is bound. Although the serum binding of AZT is relatively low (~35% in human serum), because of these uncertainties and because of the long analysis time required per sample using the HPLC gradient described (66 min) an alternative method for determining AZT and GAZT in serum was desired.

SOLID-PHASE EXTRACTION FOLLOWED BY ISOCRATIC ANALYSIS

Samples of serum standards and unknowns (0.5 ml) were individually mixed with 0.1 ml of 25 μM A22U (the 3'-β-azido isomer of AZT) as i.s. and allowed to equilibrate at room temperature while the SPE columns were prepared. These columns (high-hydrophobic C-18, 500 mg; J.T. Baker, Phillipsburg NJ; or Bond Elut C-18, 500 mg; Analytichem International, Harbor City CA) were placed in a vacuum manifold. Plastic stopcocks were mounted between the columns and manifold so that the flow of liquid through individual columns could be stopped just as the meniscus reached the top of the column. Thus air was prevented from entering the columns between additions of solvent or sample. Each column was pre-wetted with methanol (1 reservoir vol.) and rinsed with phosphate-buffered saline (PBS, pH 7.40:, 2 reservoir vols.). Sample mixtures were drawn into individual columns and allowed to equilibrate for at least 2 min. Each column was then washed with 1.0 ml PBS and pulled dry for 3-5 min. Extracts were eluted into test tubes with 3 sequential applications of 0.35 ml methanol. Columns were drained completely after each elution. The extracts were evaporated to dryness, re-dissolved in 0.2 ml 15% aqueous acetonitrile, centrifuged briefly to remove particulates and analyzed by HPLC with the same hardware as stated in the legend to Fig. 1.

Fig. 2A shows the resolution of AZT, GAZT and A22U, during 15 min, using the mobile phase stated in the legend (1.0 ml/min). Even when viewed at high sensitivity, the baseline resolution of AZT and A22U from endogenous serum components is apparent. Occasionally a minor endogenous interference with GAZT peak was seen, but it corresponded to <0.1 μM GAZT.

Validation.- Standard curves were constructed from peak areas relative to that for A22U, and unknown concentrations were calculated using the curve parameters obtained after unweighted least-squares linear regression analysis. Data describing the precision and accuracy of the assay have been published elsewhere [2]. The detection limits (concentrations with C.V. <15%) are 0.1 µM for AZT and 0.1-0.5 µM for GAZT. Between 0.03 and 0.1 µM, signal-to-noise ratios were between 3 and 8 but precise quantitation of AZT and GAZT was unobtainable. Perhaps by using a lower concentration of A22U, or 2 i.s.'s at different concentrations, the limits of accurate detection could be improved.

HIV inactivation.- With discontinuance of ultrafiltration, a method for inactivating HIV in serum was sought. Although detergents and organic solvents appeared to be effective inactivators [3], it was felt that these reagents would interfere with the binding of AZT and GAZT (or of endogenous interferants) to the SPE media. Following a report that HIV could be inactivated in serum by heat-treatment [4], the heat-stabilities of AZT (0.44-44 µM) and GAZT (1.5-150 µM) in serum were investigated. Under the incubation conditions selected for inactivation of HIV in serum samples (58° for 1 h), no degradation of either AZT or GAZT was observed. After heat treatment, the concentrations compared with controls were 101.0 ±0.6% for AZT and 101.7 ±1.1% for GAZT (mean ±S.D.; n = 4). Heat-stability (55°, 5 h) in serum and plasma has been confirmed independently (N. Klutman, N. Hinthorne & C. Riley, 1989®).

Mobile-phase optimization.- The effects of various modifications were investigated in the course of optimizing the chromatographic conditions. When methanol was used as the organic modifier in place of acetonitrile, incomplete resolution of GAZT, A22U and/or AZT was observed using a variety of buffer strengths and pH values. Resolution was found to be highly pH-dependent. A mobile phase of acidic pH was selected to suppress the ionic character of GAZT (pKa ~3.5) and permit a longer retention of the metabolite. Not surprisingly, then, the capacity factor (k') and resolution of GAZT were particularly sensitive to mobile phase pH. Even more sensitive, however, was the k' for an unidentified, highly retained compound. When the pH was raised to 3.0, its retention time shifted from ~720 sec (later than AZT, Figs. 2A & 3A) to ~420 sec, resulting in co-elution with GAZT (Fig. 2B). With a pH decrease to 2.4, acceptable resolution of all 3 analytes was obtained but the retention of the unidentified compound increased to 1140 sec, causing an unacceptable increase in the required analysis time (Fig. 2C). In addition, when the strength of the buffer was increased

® *Am. Soc. Microbiol. 29th Interscience Conf.* (Houston) *on Antimicrobial Agents and Chemotherapy.*

[*Continued on p. 180*

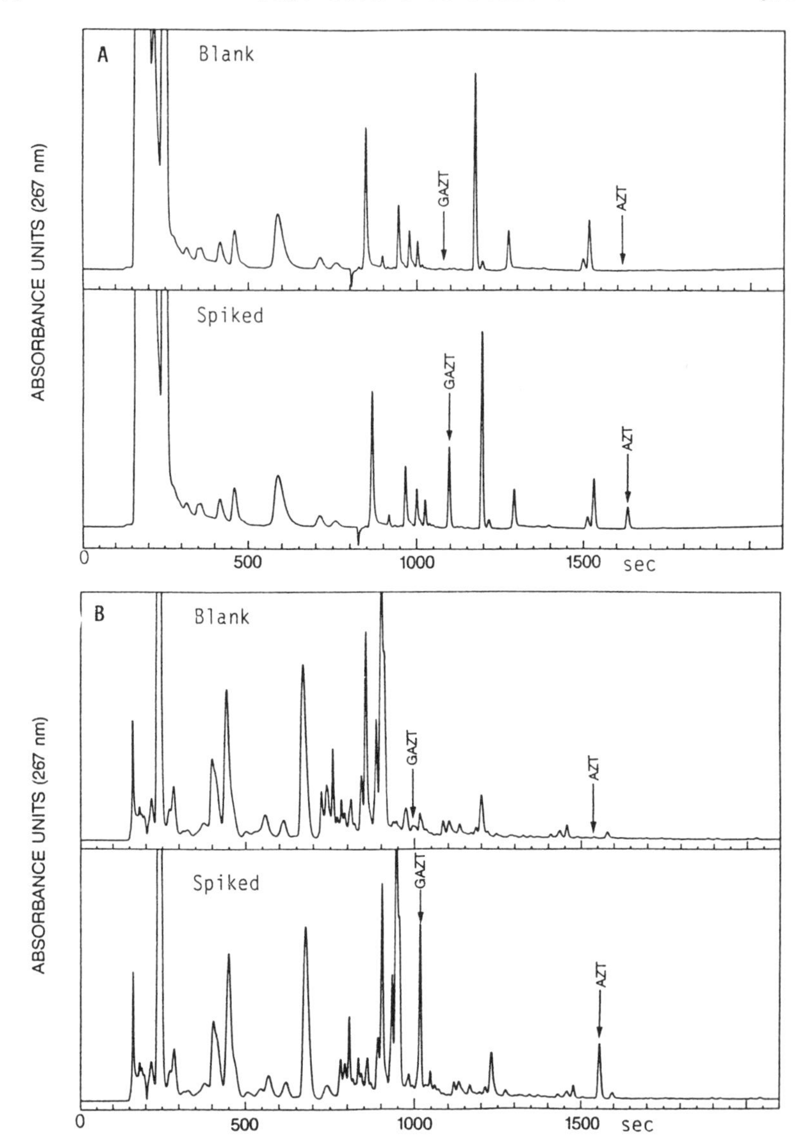

Fig. 1. HPLC analysis of AZT and GAZT by gradient elution using ultrafiltration for sample preparation. **A**, normal human serum: blank or *(below)* containing 2 μM GAZT and 1 μM AZT (positions marked; 0-0.06 aufs). **B**, normal human urine: blank or *(below)* containing 30 μM GAZT and 15 μM AZT (0-0.4 aufs). Equipment: WISP 712 Autoinjector (Waters); Constametrics pumps (2), Dynamic Mixer & SpectroMonitor 3000, variable wavelength (Milton Roy); gradient control and, for integration, DS-80Z microcomputer (Digital Specialties, Chapel Hill, NC); column, pre-column and filter as described in text.

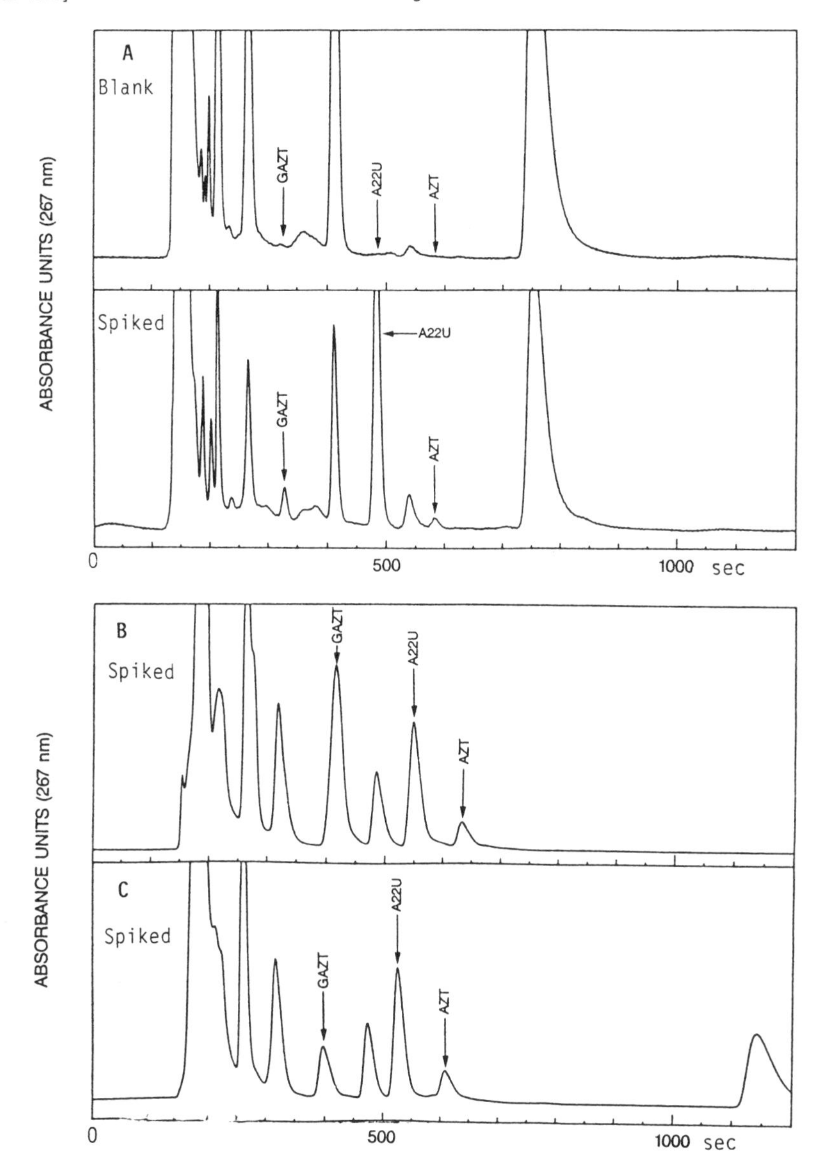

Fig. 2. HPLC analysis of AZT and GAZT by isocratic elution using SPE for sample preparation. The mobile phase was a 15:85 (by vol.) mixture of acetonitrile and 0.085% H_3PO_4, pH-adjusted with conc. ammonia: **A**, to pH 2.7; **B**, to pH 3.0; **C**, to pH 2.4. Compounds added to the serum (retention times marked): **A**, *upper:* nil (blank); *lower:* AZT to 0.2 μM, GAZT to 0.5 μM and A22U (i.s.) to 12.5 μM; **B** & **C**: AZT, 1 μM; GAZT, 2.0 μM; A22U to 12.5μM. Absorbance scale: 0-0.015 aufs in blank, 0-0.03 aufs in other chromatograms including later Figs.

from 12.4 mM (Fig. 3A; 0.085% ammonium phosphate) to 25 mM (Fig. 3B) while maintaining the pH at 2.7, the unidentified compound eluted just prior to A22U, resulting in incomplete resolution. It is unclear whether the shift reflects buffer strength, ionic strength or a combination of the two.

Column purging and survival.- When samples from AIDS patients were analyzed, highly retained compounds from previous injections occasionally interfered with subsequent analyses. The assumed cause was the numerous concurrent medications used by this patient population, although no correlations with specific therapies were discernible. The frequency of the interferences warranted performing, 10 min after each injection, a short column purge (4 min with 80% aqueous acetonitrile) and re-equilibration (12 min) with the analysis buffer, to clean the column of any highly retained materials. To prolong column lifetime, the changes in mobile phase composition were done in 4-min linear steps, resulting in a total HPLC analysis time of 34 min/sample.

Although the pH of the mobile phase was close to the manufacturer's recommended limit for silica-based columns (pH 2.0), as many as 3000 samples could be analyzed using one column, provided that the pre-column was changed at the earliest sign of deterioration (usually peak tailing or poor resolution) and the entire system was purged with 80% acetonitrile before shut-down.

Inter-laboratory comparison.- With the growing use in other laboratories of the sample-preparation and HPLC procedure, an attempt was made to determine the reproducibility of the analyses at various sites. This entailed distribution of sets of samples (4 spiked, 1 blank) as summarized in Table 2. There was surprisingly good agreement among the replicates (Table 2). For none of the 10 laboratories did the reported results include false positives; C.V.' s were <20% for all but the lowest GAZT concentration, whose C.V. was 31%.

ULTRAFILTRATION FOLLOWED BY ISOCRATIC ANALYSIS

It was of interest to determine whether good resolution of GAZT, A22U and AZT from compounds found in ultrafiltrates of normal human serum could be obtained using the isocratic analysis method. For serum containing AZT and GAZT, the profile (Fig. 3C) was as for SPE-prepared samples except that the peak corresponding to GAZT ran earlier and was no longer resolved from endogenous material. Serum ultrafiltrates are alkaline (pH 9) and probably contain enough buffering capacity to affect the retention of relatively quickly eluting acids such as GAZT. Accordingly, when the ultrafiltrates were acidified prior to injection (10 µl 85% H_3PO_4 added to 120 µl of ultrafiltrate) retention times for both GAZT and AZT were like those in the SPE method (Fig. 3D). However, due to the intrinsic limitations of using ultrafiltration for sample preparation (discussed above), this technique was not pursued further.

ACID PRECIPITATION FOLLOWED BY ISOCRATIC ANALYSIS

Sample preparation using acid precipitation prior to isocratic HPLC has also been investigated. Trichloroacetic acid (TCA) was selected as the precipitating agent since 10% TCA (w/v) has been shown to remove >99% of plasma proteins when used in acid:plasma ratios as low as 0.2 [5]. Because the supernatant of an acid-treated serum would be expected to have significant buffering capacity, the choice was a ratio of 0.25 which provided, after removal of precipitated proteins by centrifugation, a supernatant of pH 2.7. Accordingly, 200 µl aliquots of serum standards and unknowns were individually mixed with 20 µl of 50 µM aqueous A22U (i.s.) in 1.5 ml polypropylene microfuge tubes; after a 15-min equilibration at room temperature and then adding 50 µl of 10% TCA, the samples were re-mixed and similarly equilibrated. They were then centrifuged (microfuge; 12000 **g**, 15 min), and the clarified supernatants were analyzed by the isocratic HPLC method. Fig. 3E shows a representative profile.

No decreases in GAZT or AZT concentrations (0.2-5 µM) in supernatants of TCA-precipitated serum samples could be detected by repeated HPLC analysis for up to 24 h, indicating that the analytes are stable enough to permit acid precipitation with no need for neutralization - a labour-intensive procedure to be avoided. Preliminary comparisons, using isocratic HPLC, with identically prepared aqueous standards have shown that the recoveries of GAZT and AZT from TCA-precipitated serum samples are consistently >85% in the 0.2-5 µM concentration range. Within-assay precision was also shown to be good, the C.V.'s for this concentration range being 2.5-11.4%.

Although still at the pilot stage, the preparation of serum samples by TCA precipitation followed by isocratic HPLC appears promising. Analyte recoveries are slightly less than with SPE, but identical chromatographic resolution from endogenous materials has been observed. The acid precipitation method could greatly shorten sample preparation time compared to use of SPE and would eliminate the cost of the extraction columns.

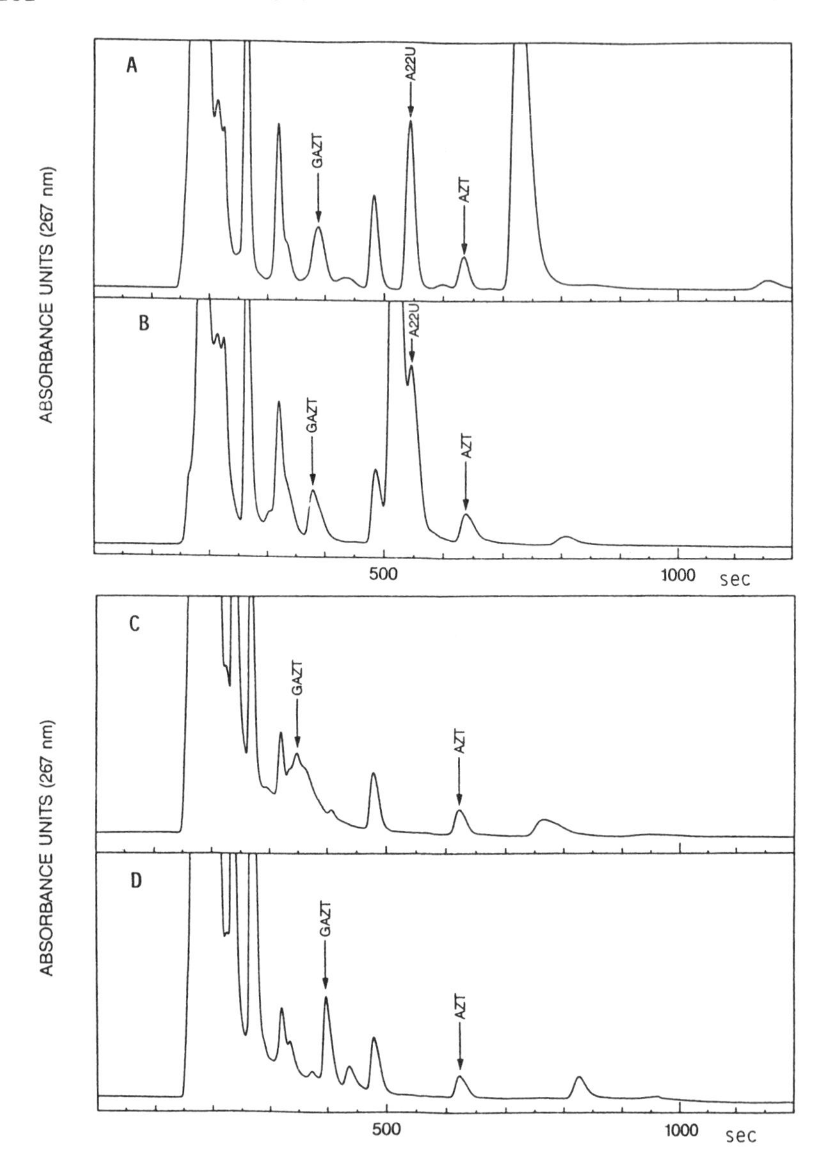

Fig. 3. Isocratic HPLC analyses of AZT and GAZT (+ A22U in some; positions marked) in human serum, with variations in sample-preparation or mobile-phase (always pH 2.7) conditions. AZT 1.0 or (**E**) 0.2 µM; GAZT 2.0 or (**E**) 0.5 µM; A22U 12.5 µM in **A** & **B**, 3.6 µM in spiked **E**. Mobile phase as in Fig. 2A except in **B** where the H_3PO_4 was raised to 25 mM. Sample preparation was by the usual SPE (**A**, **B**), by ultrafiltration with analysis directly (**C**) or after acidification (**D**), or by TCA precipitation (**E**, ***on right***; *upper chromatogram* = unspiked blank).

Table 2. Site-to-site reproducibility for the determination of AZT and GAZT by SPE and isocratic RP-HPLC. Normal human serum was prepared in bulk to contain the indicated concentrations, split into 1.1 ml replicates and stored frozen. Coded samples of each replicate were sent to participating analytical sites (altogether 10) for analysis by the described method.

Analyte	Conc^{n}., µM	*Measured:* (N.D. for blank) Mean	Range	C.V., %
AZT	13.1	12.3	11.0-13.6	6.0
	5.35	5.26	4.87-5.90	6.6
	1.36	1.33	1.06-1.58	9.8
	0.27	0.28	0.22-0.37	15.8
GAZT	18.4	19.7	17.4-21.5	7.5
	6.29	7.13	5.62-8.28	14.2
	2.10	2.30	1.75-2.91	18.3
	0.38	0.41	0.22-0.54	30.7

References

1. Blum, M.R., Liao, S.H.T., Good, S.S. & de Miranda, P. (1988) *Am. J. Med. 85*, 181-194.
2. Good, S.S., Reynolds, D.J. & de Miranda, P. (1988) *J. Chromatog. 431*, 123-133.
3. Resnick, L., Veren, K., Salahuddin, S.Z., Tondreau, S. & Markham, P.D. (1986) *J. Am. Med. Ass. 255*, 1887-1891.
4. McDougal, J.S., Martin, L.S., Cort, S.P., Mozen, M. & Heldebrant, C.M. (1985) *J. Clin. Invest. 76*, 875-877.
5. Blanchard, J. (1981) *J. Chromatog. 226*, 455-460.

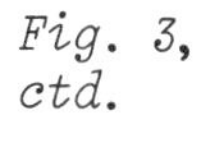

Fig. 3, ctd.

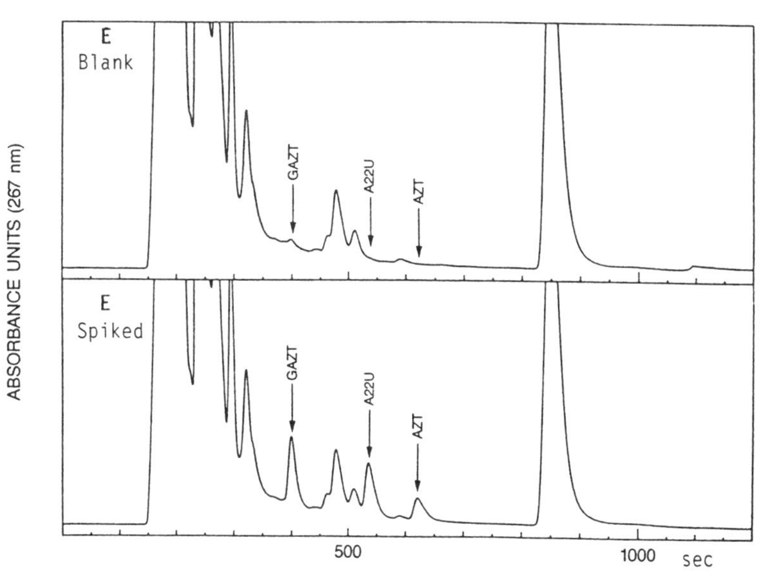

#B-12

IMMUNOASSAY PROCEDURES, INCLUDING RIA, FOR THE DETECTION OF ANTIVIRAL COMPOUNDS: ZOVIRAX® AND RETROVIR® AS EXAMPLES

Richard P. Quinn, Sarvamangala Tadepalli, Barbara S. Orban and †Leroy Gerald

Burroughs Wellcome Co., 3030 Cornwallis Road, Research Triangle Park, NC 27709, U.S.A.

A variety of IA procedures are considered for the support of antiviral compounds in the U.S.A.: RIA, ELISA, EIA and TR-FIA. Initially several different immunogens were prepared to elicit Ab production in one or more species. Individual antisera were screened in a selected assay methodology and their specific assay characteristics were explored. The development of various IA procedures for both ACV and AZT is described, with emphasis on some of the problems encountered and strategies developed to solve them.*

ACV (Zovirax®) and AZT (Retrovir®) are two recently developed antiviral agents which have lately received regulatory approval in the U.S.A. and numerous other countries. For both compounds, development was substantially aided by the early establishment and use of IA technologies for determining their concentrations in specimens from pre-clinical studies and in samples from treated patients [1, 2].

GENERAL CONSIDERATIONS

Various different technologies have been applied to the analysis of the numerous antiviral agents which have been used clinically over the last 20 years (Table 1). At Burroughs Wellcome Co. in the U.S.A., complementing HPLC, priority has been given to the development of IA methods with an early emphasis on RIA's and ELISA's. More recently, TR-FIA's have been examined. PC-FIA was investigated for use with ACV, but no displacement was seen and a standard curve

†now at: Toxicology Research, Bowman Gray Technical Center, R.J. Reynolds Tobacco Co., Winston-Salem, NC 27102

**Abbreviations.*- Letters preceding IA (immunoassay): R = radio, E = enzyme, TR- (or PC-)F = time-resolved (or particle concentration) fluorescence. ELISA, enzyme-linked immunosorbent assay. (M)Ab, (monoclonal) antibody; ACV, acyclovir (Zovirax®); AZT, zidovudine (Retrovir®, ZDV, azidothymidine); GAZT, AZT 5'-glucuronide. See text for BSA, PABA and SPA.

Table 1. Analytical methods for selected antiviral drugs. Pre-1985 refs. are in [1]. For AZT consult preceding art. (Good).

Drug	Method	Author and year	
Acyclovir (ACV)	Biological	Moore	1981
	HPLC	de Miranda, Land	1979, 1981
	ELISA	Tadepalli	1986
	RIA	Quinn, Skubits	1979, 1982
Adenine arabinoside	HPLC	Glazko	1975
	TLC	Glazko	1975
	HPLC	Kinkel & Buchanan	1975
	HPLC	Pavan-Langston	1973
Cytosine arabinoside	Biological	Borsa, Buthala, Tally	1964, 1967, 1968
	Microbiological	Hanka	1970
	Biochemical	Baguley & Falkenhaug	1971
	HPLC	Pallavicini	1980
	RIA	Piall, Okabayashi	1979, 1982
Ribavirin	Biological	Smee	1981
	RIA	Austin	1983
	TLC	Miller	1977
	Mass spectromy.	Roboz & Suzuki	1978
Trifluoro-thymidine	Biological	Sugar	1973
	Electrometric	Rogers & Wilson	1969
Zidovudine (AZT)	RIA	Quinn, Puckett	1989
	FIA	Tadepalli	1989
	Fluorescence polarization	Granich	1987
	ELISA	Tadepalli	1989
	HPLC	Good and others	1988

could not be generated with the antisera chosen for testing. A decision tree is used to help decide which type of IA method is to be attempted (Scheme 1). The choice depends somewhat on available Ab's. This is exemplified by the chronology of IA development for ACV:

1974 U.S. Patent to H. Schaeffer
1977 ACV's antiviral activity reported
- ACV-succinate as immunogen
- Rabbits injected with immunogen
- First bleedings from rabbits
- Antisera testing with 'ring test', Ouchterlony, tanned sRBC titrations
- ACV-succinate tyr Me ester made - iodinated antigen prepared

1978 HPLC method (Good & de Miranda) - problems analyzing urines
- More rabbits immunized
- Binding studies with iodo materials
- ^{3}H-ACV advent; RIA developed
- RIA validation & field testing
- Phase I trials - RIA used for urines

1979 More rabbits immunized
- Mice immunized; MAB's pursued
- Assay dispersed to 2 Universities & to Wellcome Res. Labs., U.K.
- Phase II studies in U.S.A.

1980 MAb's for ACV developed

1981 Submission for Topical and IV Formulations

1982 Approval for these
- Submission for Oral Formulations
- ELISA for ACV developed
- MAb WAC04 for Ganciclovir ESISA

1984 Attempted development of FIA

1985 Approval for Oral Formulation

1989 Quick RIA developed
- SPA Reagent (Amersham) tested

Scheme 1. 'Decision tree' for the choice of assay methodologies, usually with an HPLC method already available.

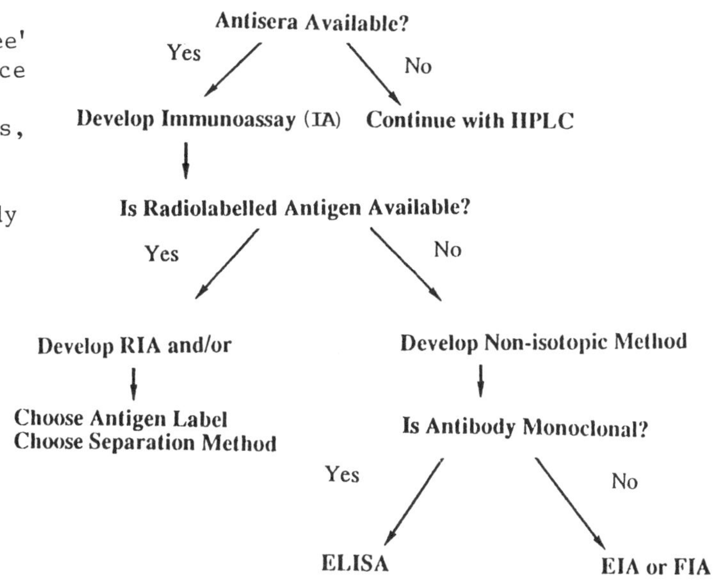

The length of time between the first attempt to synthesize ACV-succinate for use as an immunogen and the synthesis of high specific activity ^{3}H-ACV for use as an antigen was ~1 year (the first antigen used was an iodinated derivative but it had no advantage over the subsequent ^{3}H-derivative). Within a year a single-tube method had been devised using the ^{3}H-antigen, validated in-house and at two U.S. medical schools, and was being used for the early clinical trials of ACV. The Ab supply during these early studies was very limited until the advent of homologous carrier proteins for preparing immunogens, and still only low-titre antisera were available until immunization schemes were improved.

MAb's to ACV were finally obtained in 1980 by Frank Shand and Alan Bye at Wellcome in Beckenham, U.K. With these MAb's the development was achieved, by 1982 [3], of an ELISA which in turn allowed the development of an ELISA for ganciclovir (recently approved for U.S. use in CMV disease) which Wellcome and Syntex were then developing; the ELISA was used extensively for ganciclovir in clinical trial samples.

IMMUNOGEN PREPARATION

Standard methods were used to prepare ACV immunogens [4] (Fig. 1a). In addition, a number of 8-substituted conjugates were made and found to be non-immunogenic. Besides conjugates made using serum albumin, bovine (BSA) or rabbit, and thyroglobulin, conjugates were made with several other carrier proteins and tested. The BSA conjugate gave few responders (2/12), the thyroglobulin conjugate was better

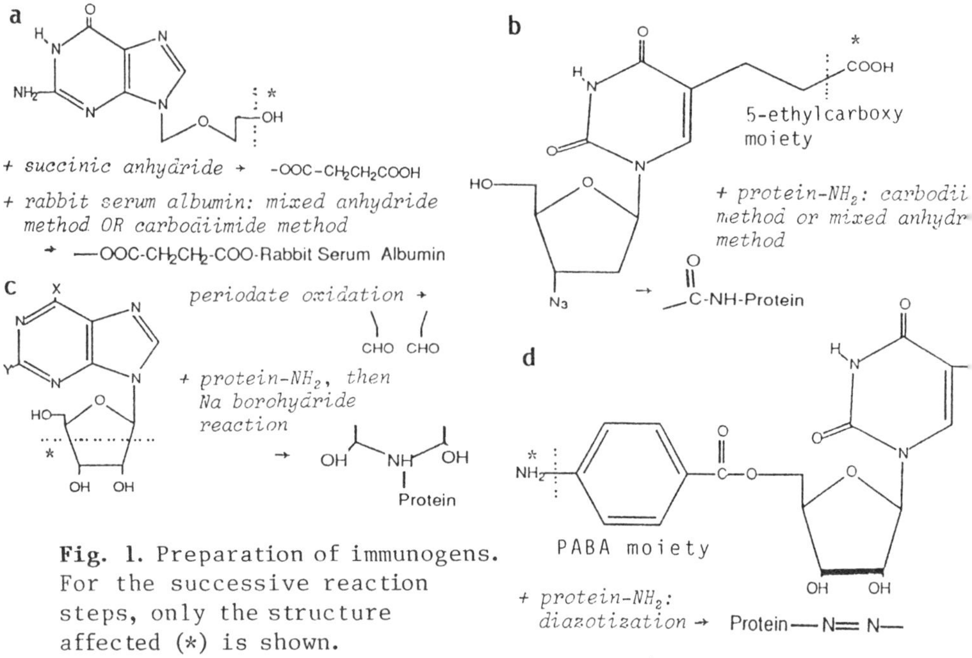

a

+ *succinic anhydride* → -OOC-CH_2CH_2COOH

+ *rabbit serum albumin: mixed anhydride method OR carbodiimide method*

→ —OOC-CH_2CH_2-COO-Rabbit Serum Albumin

b

5-ethylcarboxy moiety

+ *protein-*NH_2*: carbodii method or mixed anhydr method*

→ C-NH-Protein

c

periodate oxidation →

CHO CHO

+ *protein-*NH_2*, then Na borohydride reaction*

→ OH NH OH
Protein

d

PABA moiety

+ *protein-*NH_2*: diazotization* → Protein—N=N—

Fig. 1. Preparation of immunogens. For the successive reaction steps, only the structure affected (*) is shown. **a**, ACV (succinate prepared). **b**, 5-ethylcarboxy-AZT. **c**, Sugar oxidation/reduction approach. **d**, PABA-thymidine.

(6/16), but the most efficient was the homologous carrier protein rabbit serum albumin where 100% of the rabbits responded. In mice, the highest responsiveness (~50%) was found with mouse serum albumin as judged by binding of radioactive antigen. By ELISA the proportion of responders was much higher for the non-homologus proteins.

Immunogens for AZT were prepared similarly (Fig. 1b): the 5-substituted analogue (5-ethylcarboxy-AZT) was conjugated to proteins using the same chemistry [5]. The homologous serum albumin was again preferable as carrier protein, and the mixed anhydride reaction was favoured although two other chemical approaches are worth noting: coupling through a ribose moiety (Fig. 1c) and a diazotization approach as illustrated for the *p*-aminobenzoic (PABA) derivative of thymidine (Fig. 1d).

When conjugates were made for use as coating antigens in solid-phase non-radioactive IA methods, similar chemistry was employed, but a different coupling method than that used to prepare the immunogen and a different carrier protein were used [3, 6]. The substitution of hapten molecules on the carrier protein to produce an acceptable coating antigen was ~1-2 moles of hapten per mole of protein; for use as an immunogen, much higher hapten-to-carrier ratios are desirable.

Table 2. RIA protocol for determining antiviral agents in human plasma. PBS denotes phosphate-buffered saline, as also used to dilute the antigen and Ab. Additions (ml) were as tabulated; the first two columns represent normal-serum blanks and reference (B_0) tubes respectively; 'total' = radiolabel as added.

Solution added	*no Ab*	*+ Ab*	Standard	Unknown	Total
Buffer - PBS, pH 7.2	0.3	0.2	0.1	0.2	-
Plasma, *control* OR unknown	*0.1*	*0.1*	*0.1*	0.1	-
Drug standard	-	-	0.1	-	-
Antigen, ^{3}H or ^{125}I	0.1	0.1	0.1	0.1	0.1
Ab, mouse MAb or rabbit polyclonal	-	0.1	0.1	0.1	-

Contents mixed and incubated for at least 2 h; bound then separated from free by the chosen precipitation method:
A). *Ammonium sulphate precipitation.*- Contents mixed and incubated at least 4 h at room temp. or overnight at 4°. After 50 µl carrier protein (bovine γ-globulin, 10 mg/ml), add 0.5 ml satd. ammon. sulphate at pH 7.2 to all tubes except totals. Mix, incubate 30 min, centrifuge, and wash precipitate once. Decant, dissolve pellet in water + acid, add 2.5 ml scintillation fluid, cap, vortex to mix, and count.
B). *Second Ab precipitation.*- Contents mixed and incubated overnight at 4°. Add anti-rabbit or anti-mouse IgG. Mix; 4 h at room temp.; centrifuge. Decant; then as in **A**).
C). *SPA reagent.*- Contents mixed; 2 h at room temp. Add 100 µl SPA reagent (anti-mouse, anti-rabbit, or Protein-A). Shake tube for 2 h at room temp., then count. *(For 'SPA' see p. 192).*
D). *Pre-precipitated second Ab.*- Contents mixed; 2 h at room temp. Add Pre-precipitated anti-rabbit or anti-mouse IgG (INCStar Corp.), mix; 30 min at room temp. with shaking, then centrifuge. Decant; then as in **A**).

When RIA's have been developed, our preferred method of separation of bound from free has been the use of ammonium sulphate (Table 2, A) whereas that in the U.K. has been the use of second-Ab precipitation (B). Other precipitation schemes are available. At Wellcome we have occasionally used charcoal separation or polyethylene glycol. Newer types of separations are described below.

ACYCLOVIR IMMUNOASSAYS

The poor bioavailability of ACV by the oral route has limited some of its potential clinical use. Analogues developed over the last few years, in an attempt to achieve higher plasma levels after oral administration, included A134U, 2,6-diamino-9-[(2-hydroxyethoxy)methyl]-9H-purine, and also desciclovir, 2-[(2-amino-9H-purin-9-yl)methoxy]ethanol: each is converted to ACV after intestinal absorption. Although specific

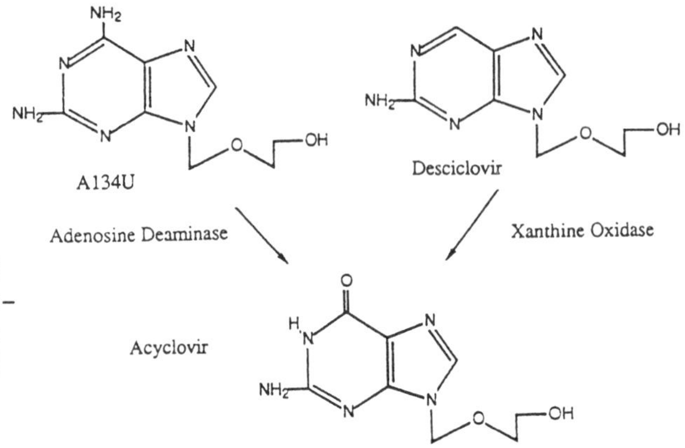

Fig. 2. Use of enzymatic interconversions for the analysis of ACV prodrugs.

IA systems for both compounds (an RIA for A134U, and both an RIA and an ELISA for desciclovir) were eventually developed [7, 8], our first approach in each case was to effect the conversion enzymatically *in vitro* (Fig. 2); this enabled both ACV and its prodrugs to be determined easily using parallel assays.

As mentioned above, MAb's to ACV were developed and used to devise an ELISA system. As judged by inhibition of binding to mouse MAb WAC-03, cross-reactivities were as follows, the first pair being the known metabolites of ACV: 9-carboxymethoxymethylguanine (CMMG), 1.0%; 8-hydroxy-ACV [8-hydroxy-9-(2-hydroxyethoxy)methylguanine], 27%; adenosine, nil; guanosine, 0.18%. Evidently, since 8-hydroxy-ACV is not present in ACV-treated patients, metabolite cross-reactivity is not a problem. With ganciclovir, the cross-reactivity was 9.0%, such that an ELISA, with good sensitivity, could be developed for its assay in the absence of ACV [3]: the concentrations were read directly from the ACV standard curve which had been calibrated with ganciclovir standards.

The sample types analyzed to provide data supporting the preclinical development and clinical use of ACV included the following, needing no pre-extraction: plasma, serum, urine, tissue culture fluid, amniotic fluid, CSF, aqueous humour, vitreous humour, vaginal secretions and washings, tears and saliva. Tissue and faecal extracts have also been analyzed.

ZIDOVUDINE (AZT) IMMUNOASSAYS

The HPLC approach used at the outset [9], as for ACV, had the notably useful capability of determining both AZT and the only known metabolite, GAZT, in man [10]. RIA's for AZT were developed before the other IA technologies

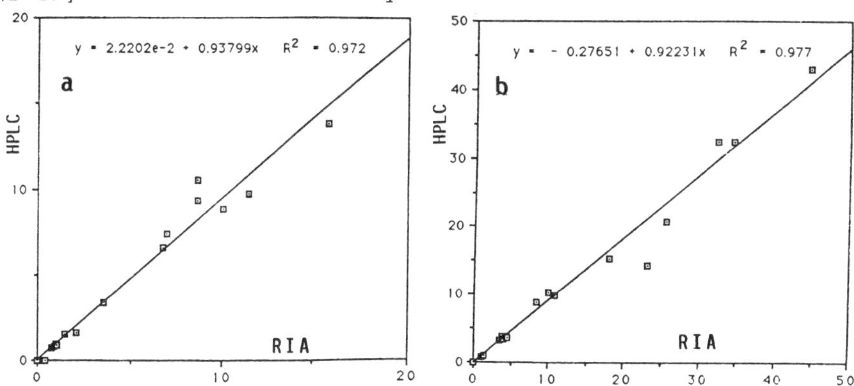

Fig. 3. Levels (μM) in human plasma of (**a**) AZT, (**b**): GAZT: HPLC *vs.* ZDV-TRAC Kit. GAZT was split by *E. coli* β-glucuronidase.

were investigated. The availability of ^{3}H-antigen [11] concurrently with the demonstration of the anti-HIV activity of this compound was responsible for the swift development of the original RIA system [5] (Table 2, A). This method was used for the direct assay of paired semen/serum samples from AIDS and ARC patients: the AZT concentration ratio was typically ~10 [12].

An agreement was subsequently made with INCStar Corp. for the development, to determine AZT levels, of a commercial kit which would be made available world-wide. Our company initially provided this firm with anti-AZT sera. A kit based on a different type of separation scheme (D, below Table 2) was prepared and validated (Fig. 3a) *vs.* the AZT/GAZT HPLC system of Good *et al.* [10]. Tadepalli [13] was able to develop an enzymatic method using β-glucuronidase for converting GAZT to AZT, whereby the kit could give values for GAZT: these were likewise validated *vs.* the HPLC method (Fig. 3b). The kit enabled GAZT to be determined in a Phase I trial of AZT in neonates, where sample size would have been insufficient for HPLC.

Recently, coating antigens for AZT have been produced, allowing the development of both an ELISA and a TR-FIA [6]. One difficulty encountered with the MAb used in the ELISA was significant (34.5%) cross-reactivity with GAZT. Since the FIA employs a polyclonal Ab without such interference, it will probably be adapted where non-radiometric assays are preferable to the ELISA.

Sometimes development of advanced assay technology can lead to unexpected applications. Thus, the ELISA for AZT was used to investigate the nature of the febrile reactions of several patients receiving this drug. Hapten-carrier-coated plates prepared for the ELISA were used instead to search for drug-induced Ab's [14].

NEW SYSTEMS

In the realm of RIA development, we have recently initiated the evaluation of the Scintillation Proximity Assay (SPA) Reagent from Amersham Corp., which has been available in the U.S.A. since July 1989. This reagent is appealing since it allows the development of homogeneous assay systems instead of the normal heterogeneous systems with their obligatory separation of bound from free antigen. Second-Ab is attached to an inert fluoromicrosphere which gives a signal only when in close proximity to a radioligand. No separation is necessary and no liquid scintillation fluid is required, thereby eliminating the problems encountered with the complications of disposal of such reagents.

We have tested this reagent in our ACV system using an MAb (Table 2, C). The influence of volume of SPA reagent has been studied (Fig. 4a). Relatively low volumes of reagent suffice for most uses. More notable was the influence of temperature of incubation on the efficiency of the reagent (Fig. 4b): a negative temperature coefficient was observed. We are currently investigating the use of the anti-rabbit and the Protein-A reagent in a rabbit polyclonal first-Ab system.

We will continue to develop these and any other technologies which fulfil our needs of speed coupled with sufficient sensitivity and selectivity.

Acknowledgements

We are indebted to a great number of people at the Wellcome Research Laboratories, both in the U.S.A. and in the U.K., and at the various medical schools where we have collaborations*. Special thanks are due to Paulo de Miranda, Steve Jeal and Steve Good. Thanks are also expressed to Larry Puckett at INCStar Corp. and Dr. Henry Balfour and Barbara Chinnock at the University of Minnesota.

*including Johns Hopkins Univ. and Univ. of California San Diego

References

1. Quinn, R.P., Scharver, J. & Hill, J.A. (1985) *J. Pharmacol. Exp. Ther. 30*, 43-65.
2. Blum, M.R., Liao, S.H.T., Good, S.S. & de Miranda, P. (1988) *Am. Med. J. 85 (Suppl. 2A)*, 189-194.
3. Tadepalli, S.. Quinn, R.P. & Averett, D.R. (1986) *Antimicrob. Agents & Chemother. 29*, 93-98.
4. Quinn, R.P., de Miranda, P., Gerald, L. & Good, S.S. (1979) *Anal. Biochem. 98*, 319-328.

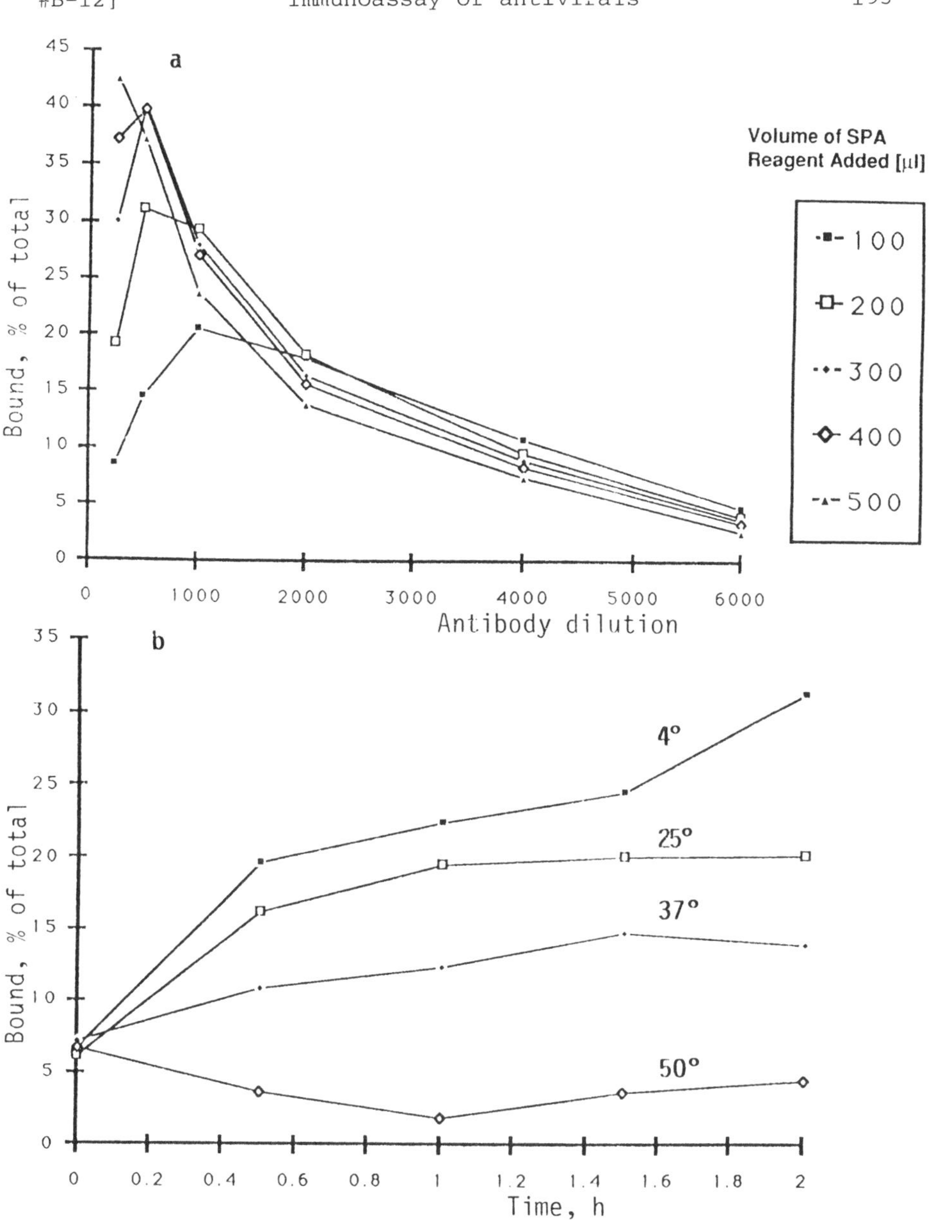

Fig. 4. Effects of (**a**) vol. of SPA reagent added (incubation at 25°), and (**b**) of incubation temperature, on the efficiency of precipitation of anti-ACV Ab in the RIA for ACV. In (**a**) note that above a 1:1000 dilution there is little effect of vol. of reagent, while below this concentration the % bound is proportional to the vol. of reagent used. In (**b**) note the inverse temperature relationship (vol. of SPA reagent added = 100 µl).

5. Quinn, R.P., Orban,P. & Tadepalli, S. (1989) *J. Immunoassay 10*, 177-189.
6. Tadepalli, S. & Quinn, R.P. (1989) *FASEB J. 3*, A885.
7. Quinn, R.P., Good, S.S., Gerald, L. & Sabatka, J.J. (1983) *Anal. Biochem. 134*, 16-25.
8. Quinn, R.P., Gerald, L. & Tadepalli, S. (1987) *J. Immunoassay 8*, 247-265.
9. Good, S.S., Durack, K.T. & de Miranda, P. (1986) *Fed. Proc. 45*, 444.
10. Good, S.S., Reynolds, D.J. & de Miranda, P. (1988) *J. Chromatog. 431*, 123-133.
11. Hill, J.A. & Freeman, G.A. (1988) *J. Labelled Compounds & Radiopharmaceuticals 25*, 277-280.
12. Henry, K., Chinnock, B.J., Quinn, R.P., Fletcher, C.V., de Miranda, P. & Balfour, H.H. (1988) *J. Am. Med. Ass. 259*, 3023-3026.
13. Tadepalli, S., Quinn, R.P., Puckett, L., Orf, J.W., Goldensop, C.R. & Donner, J.E. (1989) *Clin. Chem. 36*, 1187.
14. Jacobsen, M.A., McGrath, M.S., Joseph, P., Molaghan, J.B., Tadepalli, S. & Antimicrob. (1989) *J. Acquired Immune Deficiency Syndromes 2*, 382-388.

#ncB

NOTES and COMMENTS relating to

ANTI-INFECTIVE DRUGS AND THEIR METABOLITES

also immunosuppressives and anti-cancer drugs

Forum comments relating to the preceding main arts. and to the 'Notes' that follow appear on pp. 227-230

Reinforcing citations furnished by Senior Editor *(KEY: pp. 230-231) grouped similarly to the* #B & #ncB *arts.:-*

#ncB-1

A Note on

HPLC DETERMINATION OF CEFMETAZOLE AND ITS DEGRADATION PRODUCT, 1-METHYL-1H-TETRAZOLE-5-THIOL, IN BIOLOGICAL SAMPLES

P.A. Bombardt, W.M. Bothwell, K.S. Cathcart, M. Courtney, H. Ko and ⊗G.W. Peng

Drug Metabolism Research Laboratories,
The Upjohn Company, Kalamazoo, MI 49001, U.S.A.

CFMZ-Na* is a semi-synthetic derivative of cephamycin C with a broad spectrum of activity against Gram-positive and -negative bacteria (MIC's 0.5-4 μg/ml) [1]. It also appears to be resistant to several β-lactamases which deactivate many cephalosporin and cephamycin antibiotics. Phase III clinical trials in the U.S.A. have confirmed the effectiveness of CFMZ for the treatment of infections - urinary tract, skin and soft tissue, lower respiratory tract, abdominal, gynaecological - and for the prevention of post-operative wound infections [2]. This evaluation of the efficacy of CFMZ-Na has been supported by pharmacokinetic studies [3, 4]. Because renal tubular secretion is the major route of elimination of CFMZ in man, pre-administration of probenecid has been utilized to prolong the half-life of CFMZ [3].

A RP-HPLC assay method was developed to monitor CFMZ concentrations in plasma, serum and urine samples to provide data for clinical pharmacokinetic evaluation of CFMZ in the presence of probenecid [3, 5]. A simplified method was also developed to assay CFMZ in biological samples to support pharmacokinetic studies with CFMZ-Na alone. NMTT, a degradation product of CFMZ and other cephalosporin and cephamycin antibiotics containing the NMTT group, is thought to be linked by a still undefined mechanism to side-effects - hypoprothrombinaemia and platelet dysfunction - associated with clinical applications of these antibiotics [6]. For example, some patients treated with CFMZ-Na had a prolonged prothrombin time [7]. To monitor clinical safety, an HPLC method was developed to assay NMTT concentrations in plasma samples [8]. This assay method for NMTT and the methods for CFMZ are summarized here.

⊗addressee for any correspondence

**Abbreviations (formulae overleaf).*- CFMZ(-Na), cefmetazole (sodium); NMTT, a thiol (see title); MIC, minimum inhibitory concentration; i.s., internal standard; r, correlation coefficient.

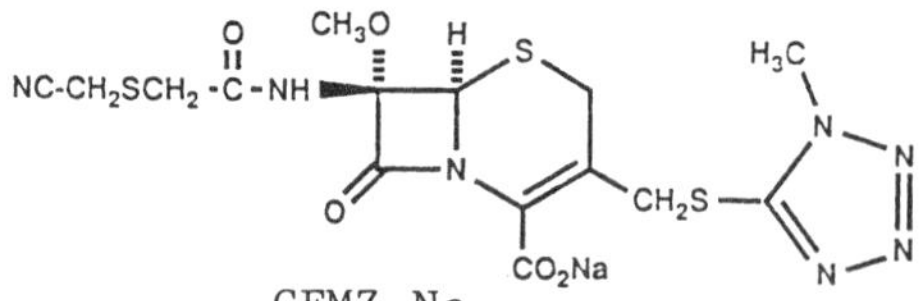

CFMZ-Na NMTT

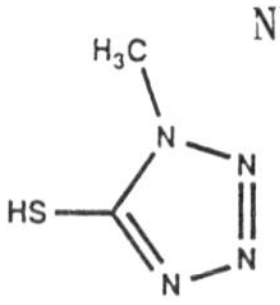

HPLC ASSAY OF CFMZ

Sample preparation.- To 1 ml aliquots of plasma or serum, 0.1 ml water and 2 ml methanol containing trichloroacetic acid (99.5:0.5 by vol.) and barbital (1.5 mg/ml) as i.s. were added. After vortex-mixing for 1 min and centrifugation for 10 min at 600 **g**, the supernatant was removed and mixed with 1 vol. of 0.2 M pH 5.4 sodium citrate/citric acid buffer; 50 µl aliquots were taken for HPLC. For urine the procedures differed only in that the 0.1 ml addition was of citrate buffer, not water, and the concentration of i.s. was doubled.

Chromatography.- The analytical column (250 × 4.6 mm) was 7 µm Zorbax C18 or 5 µm Supelcosil LC18. Elution was with acetonitrile-0.01 M [otherwise as above] citrate buffer (12:88 by vol.) at 2 ml/min. Detection was at 254 nm, and quantitation by peak-height ratios and calibration curves of concurrently analyzed standards.

A column switching system with a time-actuated pneumatic valve (Valco CV-6-HPax) was used for samples from subjects pre-dosed with probenecid. The analyte in the prepared sample was chromatographed on a clean-up column (Brownlee OD GU RP18, 5 µm, 30 × 4.6 mm) with the mobile phase delivered from pump I at 2 ml/min for 1.5 min, the eluent being directed into an analytical column. Then the switching valve was actuated to reverse the flow of the mobile phase, delivered from pump II, through the clean-up column for 10 min to elute probenecid and other retained substances. At the same time, the mobile phase from pump I, bypassing the clean-up column, completed the separation of CFMZ and i.s. on the analytical column. The clean-up column was re-equilibrated with mobile phase in the forward direction before the next sample was injected (re-equilibration time 3.5 min).

Results

As illustrated for plasma in Fig. 1, CFMZ ran at 6-7 min and i.s. at 10-11 min, and other peaks did not interfere. Although a small peak was observed at the retention window of CFMZ in some samples, this small interference did not cause any difficulty for the assay at the sensitivity required to adequately evaluate the pharmacokinetics of CFMZ at clinically effective dose levels. Serum samples behaved identically: for serum *vs.* plasma, linear regression analysis of peak height ratios gave a slope of 0.9981 with an intercept not significantly different from zero ($P > 0.05$) and with $r = 0.9999$.

The calibration curves had good r values with linearity over the range 2-200 µg/ml for plasma or serum and 15-2000 µg/ml for urine. The limits of quantitation were 2 µg/ml in serum and 15 µg/ml in urine. Comparison of serum with and without probenecid, using CFMZ calibration samples, showed essentially identical CFMZ concentration values utilizing the two chromatographic separation conditions. Linear regression of the two sets of results gave a slope of 0.9915 with a good r value, 0.9995.

For control samples containing 4.86, 24.37 and 121.55 µg/ml CFMZ in serum (n = 17), the concentrations found were 4.83 ±0.41, 24.37 ±1.03 and 11.87 ±3.37 µg/ml respectively. For 60, 243.1 and 600.2 µg/ml in urine (n = 25), the values were 61.8 ±1.4, 243.1 ±5.6 and 586.2 ±14.5 µg/ml. Therefore the accuracy (recovery) for the serum and urine samples was within ±4% of the seeded concentrations with relative S.D. <10%. For serum or plasma the within-day slopes had relative S.D.'s of 0.2-2.6%, and the between-day relative S.D.'s of 11.2%. For urine the corresponding values were 2.8-8.4% and 6%.

HPLC ASSAY OF NMTT

Sample preparation.- Aliquots (~0.5 ml) of plasma were centrifuged at 4° in Centrifree filters (Amicon Corp., Danvers, MA) at 1255 **g** for 90 min. **Chromatography,** using 40 µl aliquots of the filtrate, was performed with a Hypersil C18 column (3 µm; 30 × 4.6 mm; Shandon Southern Insts. Sewickley, PA) with a mobile phase of buffer-acetonitrile (95:5 by vol.) at 1 ml/min flow-rate; the buffer (pH 5.75) contained 5 mM tetrabutylammonium phosphate and 58 mM NaH_2PO_4. The effluent was monitored at 240 nm, and NMTT was quantitated from peak-height measurements and calibration curves. Finally, 10 min after sample injection, the column was back-flushed for 8 min with buffer-acetonitrile (75:25 by vol.), delivered by a second pump at 2 ml/min, to remove retained non-polar compounds. The column was then re-equilibrated with mobile phase for 3 min with the normal flow direction and rate, and thus made ready for the next sample analysis.

Results

NMTT eluted from the column, separated from endogenous substances in the filtrate, at the retention window of 5-7 min (Fig. 2). The total cycle time for each sample was ~21 min, which included the back-wash to clean up the column. Although no i.s. was used, the NMTT peak heights were linear with its concentration over the range 70 ng/ml to 70 µg/ml, with calibration curves having a slope range 5.210-5.870 and r 0.9981-unity; the slope C.V.'s were 3.5% between-day and 0.28-1.4% within-day, and the quantitation limit was 70 ng/ml. The C.V.'s of the calibration standards from different sets of calibration curves (usually n = 11), with spike levels as stated, were as follows:-

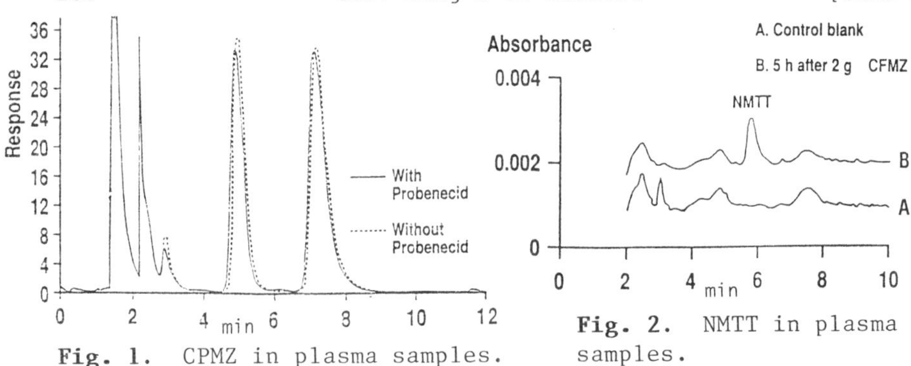

Fig. 1. CPMZ in plasma samples.

Fig. 2. NMTT in plasma samples.

0.071 µg/ml, 26; 0.141, 21; 0.282, 13; 0.565 (n = 10), 13; 1.13, 6; 2.26, 7; 4.52, 5; 9.03, 5; 18.1, 4; 36.1 (n = 8), 5; 72.3 (n = 7), 3%.

Assay accuracy (recovery) and precision were measured by determining the concentrations in control plasma samples containing 0.63, 1.58 and 6.3 µg/ml NMTT. The concentrations determined were 92, 96 and 99% of the seeded concentrations respectively, with C.V.'s <8%.

CONCLUDING COMMENTS

HPLC methods have been developed for the assay of CFMZ in plasma, serum and urine samples in the presence and absence of probenecid and for the assay of NMTT in plasma. These methods feature simplicity in sample preparation and chromatographic separation, and were capable of supporting large-scale clinical pharmacokinetic and drug-safety studies in phase III development of CFMZ-Na in the U.S.A.

References

1. Stapley, E.O. & Birnbaum, J. (1981) in *Lactam Antibiotics, Mode of Action, New Development and Future Prospects* (Salton, M. & Shockman, G.D., eds.), Academic Press, New York, pp. 327-351.
2. Griffith, D.L., Novak, E., Greenwald, C.A., Metzler, C.M. & Paxton, L.M. (1989) *J. Antimicrob. Chemother. 23 (Suppl. D)*, 21-33.
3. Ko, H., Cathcart, K.S., Griffith, D.L., Peters, G.R. & Adams, W.J. (1989) *Antimicrob. Agents Chemother, 33*, 356-361.
4. Ko, H., Novak, E., Peters, G.R., Bothwell, W.M., Hosley, J.D., Closson, S.K. & Adams, W.J. (1989) *Antimicrob. Agents Chemother. 33*, 508-512.
5. Bothwell, W.M., Cathcart, K.S. & Bombardt, P.A. (1989) *J. Pharm. Biomed. Anal. 7*, 987-995.
6. Bang, N.U., Tessler, S.S., Heidenreich, R.O., Marks, C.A. & Mattler, L.E. (1982) *Rev. Infect. Dis. 4 (Suppl.)*, S546-554.
7. Holloway, W.J., Winslow, D.L. & Reinhardt, J. (1989) *J. Antimicrob. Chemother. 23 (Suppl. D)*, 47-54.
8. Bombardt, P.A., Courtney, M. & Ko, H. (1990) *J. Pharm. Biomed. Anal.*, in press.

#ncB-2

A Note on

ANALYTICAL PROBLEMS WITH CICLOSPORIN AND ITS METABOLITES

K-Fr. Sewing, U. Christians, J. Bleck & S. Strohmeyer

Abteilung Allgemeine Pharmakologie,
Medizinisches Hochschule,
Konstanzy-Gutschow-Str. 8, D-3000 Hannover, F.R.G.

Ciclosporin (cyclosporin A) is a cyclic undecapeptide with a structure open to attack by drug-metabolizing enzymes and with physicochemical properties that complicate its therapeutic handling and its monitoring. Temperature-dependent distribution between blood cells and serum makes it advisable to monitor ciclosporin in blood rather than in serum or plasma. Another analytical problem is its strong binding to lipoproteins, necessitating mild procedures to remove the drug from its binding sites. The RIA's so far available measure either only the parent compound or an unknown mixture of the parent compound and metabolites.

Under these circumstances HPLC seems to be the method of choice to overcome at least some of the difficulties. We therefore embarked on developing an HPLC method based upon extraction of ciclosporin and its metabolites from biological material (blood, bile and urine), using disposable solid-phase extraction columns and a RP-HPLC system with elution by a concave water-acetonitrile gradient. Thereby it is possible to obtain a reasonable separation of ciclosporin and its metabolites mainly according to their lipophilicity. The use of an internal standard allows quantitative analysis of ciclosporin and its metabolites in biological material.

The advantage of therapeutic monitoring of ciclosporin by HPLC is that it is easy to detect changes in ciclosporin metabolism in diseases associated with liver and/or kidney dysfunction or produced by drugs influencing drug metabolism or elimination. Analytical problems can arise from inadequately coagulated blood, from hydrolysis of Phase II metabolites (conjugates) during extraction and chromatography, and from components of laboratory material (plasticizers) with a UV-absorption similar to that of ciclosporin and its metabolites.

Reference *noted by Senior Editor (précis overleaf)*
- supplementing the foregoing adaption of K-Fr. Sewing's Forum abstract (no publication text received)

Christians, U., Zimmer, K-O.,Wonigeit , K. & Sewing, K-Fr. (1987) *J. Chromatog.* *413*, 121-129.

Points from the cited ref.

The extraction of cyclosporin A and 4 metabolites from blood was performed with glass columns containing 100 mg of LiChroprep RP-8, with dichloromethane as eluent. The extract was injected onto a 5 µm LiChrosorb RP-8 column (250 × 4 mm). With flow-rate 1.4 ml/min for the mobile phase, the acetonitrile was increased from 55% to 63% during 14 min; the eluate was monitored at 210 nm. The detection limit for cyclosporin A was 20 ng.

#ncB-3

A Note on
ASSAY OF FLEROXACIN

Herwig Eggers*, Peter Heizmann and Dennis Dell

Department of Drug Metabolism and Pharmacokinetics,
F. Hoffmann-La Roche Ltd., 4002 Basel, Switzerland

Requirement — *An analytical method for the quantification of fleroxacin and its two metabolites in biological materials suitable for evaluating pharmacokinetics in man and experimental animals.*

End-step — *RP-HPLC with fluorescence detection (ex.: 290 nm; em.: 450 nm).*

Sample handlng — *Protein precipitation of plasma with acetonitrile, evaporation of supernatant and reconstitution with mobile phase; method modified if interferences encountered. Internal standard (i.s.) added at outset.*

Comments — *Chromatographic conditions have been carefully optimized, and a variant procedure devised for clean-up. Quantification limit 10 ng/ml for plasma and - with a different i.s. - 1 μg/ml for urine.*

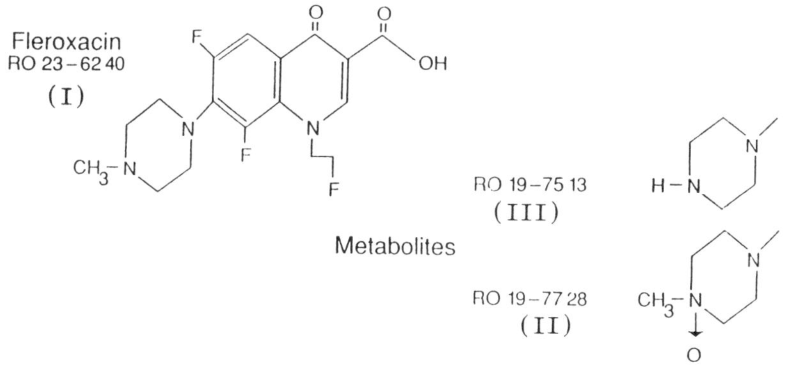

Fleroxacin
RO 23-6240
(I)

Metabolites
RO 19-7513
(III)
RO 19-7728
(II)

Fleroxacin (Ro 23-6240) is a new fluoroquinolone drug under clinical development [1]. The main metabolic pathways have been shown to be demethylation and *N*-oxidation [2].

Sample preparation is as follows, method (c) being used if interferences are encountered with (a) or (b).

(a) To 0.2 ml plasma, 1 ml acetonitrile containing 60 ng of the i.s. (AM 735; a difluoro analogue of the drug) is added. After vortex-mixing and centrifugation, the supernatant is removed and dried down (N_2). The residue is reconstituted with 400 μl of the eluent, and 20 μl injected.

*now at 7880 Grenzach-Wyhlen, F.R.G. (same company)

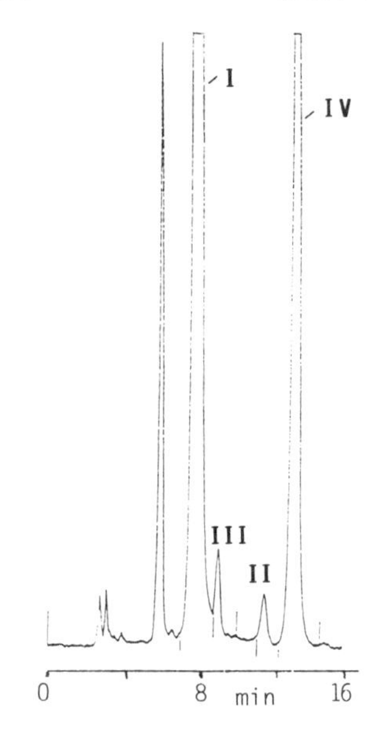

Fig. 1. Chromatogram of a plasma sample following a 400 mg oral dose to a healthy volunteer.
I, fleroxacin (1130 ng/ml);
II, fleroxacin *N*-oxide (24 ng/ml);
III, *N*-desmethylfleroxacin (24 ng/ml);
IV, i.s. specified for plasma*.
Column: 250 × 4.6 mm column packed with Toyo Soda 5 μm TSK ODS-120T.
Eluent (1 ml/min): methanol added (to 28% v/v) to aqueous 10 mM tetra-butylammonium hydrogen sulphate-50 mM potassium phosphate adjusted to pH 2.6.
Detection was by a F 1000 Merck-Hitachi fluorescence detector.

**not* the i.s. used for urine (shorter run time); interferences preclude its use for plasma.

(b) To 0.1 ml urine are added i.s. (pipemidic acid, 10 μg in 0.1 ml) and 5 ml of the eluent; 20 μl is injected.
(c) To 0.1 ml of plasma or urine, buffered to pH 7.5, is added 4 ml dichloromethane-hexane (9:1 by vol.). After mixing, the organic phase is separated and washed with pH 5 buffer. Finally the organic phase is dried down and the residue reconstituted with 500 μl of the eluent; 30 μl are injected.

HPLC and quality criteria.- For conditions see legend to Fig. 1, which shows a typical pattern for plasma. Quantification limits for fleroxacin and its two metabolites were 10 ng/ml in plasma and 0.5-1 μg/ml in urine. The mean inter-assay precision was 5.5% for plasma and 4% for urine.

Comments.- Chromatographic retention and separation was found to be strongly dependent on the type of RP material, the pH, and the type and concentration of the ionic additives in the eluent. Secondary ionic interactions obviously played an important role. For sample preparation under routine conditions, protein precipitation was preferred to liquid-liquid extraction because of its simplicity and high sample throughput. However, for some patient samples, serious interferences were observed which called for the latter approach, based on the sample preparation procedure of Dell & co-workers [3], to give more selectivity.

References

1. Manck, N., Andrews, J.M. & Wise, R. (1986) *Antimicrob. Agents Chemother. 30*, 330-332.
2. Weidekamm, E., Stöckel, K. & Dell, D. (1987) *Drugs Exp. Clin. Res. 13*, [85-90.
3. Dell, D., Partos, C. & Portman, R. (1988) *J. Liq. Chromatog. 11*, 1299-1312.

#ncB-4

*A Note** on

CHIRAL STATIONARY PHASES *VERSUS* CHIRAL DERIVATIZATION FOR THE QUANTITATON OF OFLOXACIN ENANTIOMERS

K-H. Lehr and P. Damm

Hoechst AG, Radiochemisches Laboratorium,
Postfach 80 03 20, D-6230, Frankfurt 80, F.R.G.

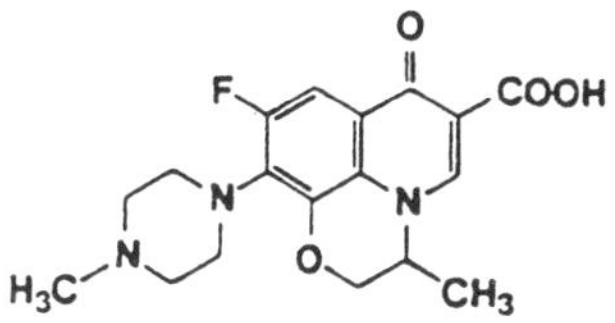

Ofloxacin - a fluoroquinoline - exists as a pair of enantiomers. Two different methods for separating the enantiomers are available [1].- (1) Direct resolution on a chiral stationary phase, immobilized bovine serum albumin (BSA), is feasible. Alternatively (2), derivatization may be performed with the chiral reagent L-leucinamide followed by separation of the diastereoisomers formed. Both methods have strengths and weaknesses. Examples are given (Figs. 1 & 2) of patterns from HPLC runs with these methods on plasma from a dosed volunteer or with spiked plasma.

Amplification *(see also legends to Figs.)*

Stock solutions of the racemate and individual enantiomers (from Daiichi Seiyaku, Tokyo) were prepared by adding 9 ml pH 7 phosphate buffer to a solution of 10 mg in 1 ml 0.2 M NaOH. Stability was satisfactory at 4° for at least 2 months or, for spiked plasma or urine, at -20°. To the samples, viz. 0.5 ml [1 ml in method (1)] of plasma/serum or 0.2 ml urine, 2 [or 5] ml dichloromethane was added. After shaking, and centrifuging for 5 min at 2500 **g**, the aqueous layer was discarded.

The organic phase, for method (1), was dried down under N_2 and the residue redissolved in 200 µl of mobile phase. For method (2) it was treated with phenylphosphinyl chloride /triethylamine (in 1 ml) and then ~150 mmol (in 0.5 ml) L-leucinamide (pre-prepared from its hydrochloride); the product was extracted into HCl and chromatographed (100 µl). Analyte detection was by fluorimetry (ex. 298 nm, em. 458 nm); peak areas were measured.

**Editor's précis* from [1]; author contributed Figs. and start.

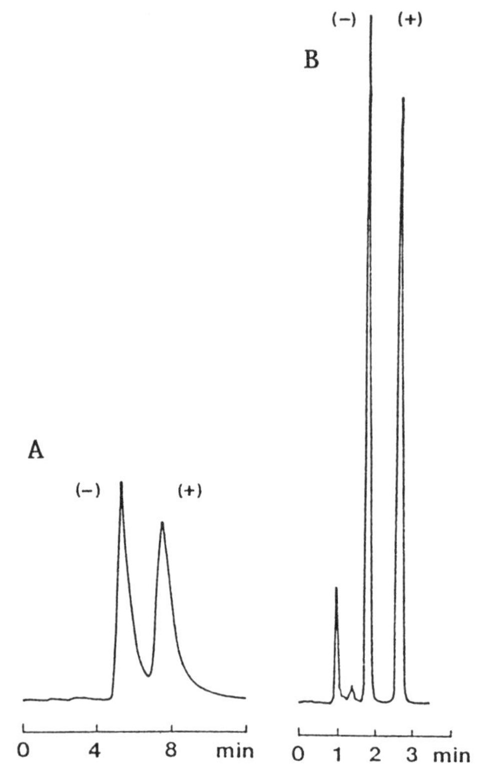

Fig. 1. HPLC patterns for plasma from a volunteer after 400 mg of the racemate orally. Method (1), **A**; (2), **B**. For (1): column, ET 150/8/4 Resolvosil-BSA-7 (150 × 4 mm i.d.) at ambient temp.; CT 30/6/4 guard column; pH 8 0.2 M phosphate buffer/propan-2-ol (97:3 by vol.), 1.0 ml/min.

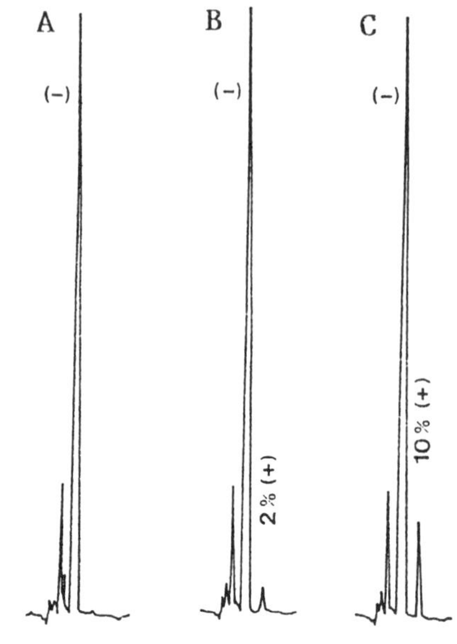

Fig. 2. HPLC patterns, by method (2), for control serum samples spiked with 1 µg/ml (-)-ofloxacin; additionally 2% (**B**) or 10% (**C**) of (+)-ofloxacin. *Applicable also to* **B**, *Fig. 1*:- column: Nucleosil 120-5 C-18 (125 × 4.6 mm; 5 µm); acetonitrile/0.2 M phosphoric acid adjusted to pH 1.85 with tetraethylammonium hydroxide (20:80); 1.5 ml/min; run at 40°.

Whilst method (1), with HPLC conditions guided by literature (leaflet from the column manufacturer, Macherey-Nagel) entails simpler sample preparation, (2) is preferred because it uses ordinary, long-lived HPLC columns and, moreover, reveals traces of one isomer even if the other dominates (Fig. 2). The derivatization procedure [2] is rapid, and gives a pure product, as verified by ^{1}H-NMR. Validation has been published [1]; detection limits were 3 ng/ml for plasma and 80 ng/ml for urine.

References

1. Lehr, K-H. & Damm, P.(1988) *J. Chromatog.* *425*, 153-161.
2. Bernasconi, S., Comini, A., Corbella, P., Gariboldi, P. & Sisti, M. (1980) *Synthesis*, 385.

#ncB-5

A Note on

AN IMIDAZOLE DERIVATIVE WITH ANTIBACTERIAL ACTIVITY *IN VITRO* BUT NOT *IN VIVO*

G. Dean and ⊗C.W. Vose

G.D. Searle & Co. Ltd., High Wycombe, HP12 4HL, U.K.

The compound, 'IPP' (see Fig. 1 for name and structure) was very active *in vitro* against anaerobic bacteria, especially *B. fragilis*. However, it was essentially inactive *in vivo* after i.v. administration to rat and mouse. This article describes the investigation, using ^{14}C-labelled compound, of the origin of these differences. Interpretation hinged on the finding that an α,β-unsaturated ketone metabolite, 'IPP-M' (*Editor's term*; Fig. 1) is rapidly formed from IPP both *in vivo* and in the antimicrobial test system.

⊗addressee for any correspondence, at Hoechst UK, Milton Keynes, Bucks.; GD is now at Huntingdon Research Centre, Huntingdon, Cambs.

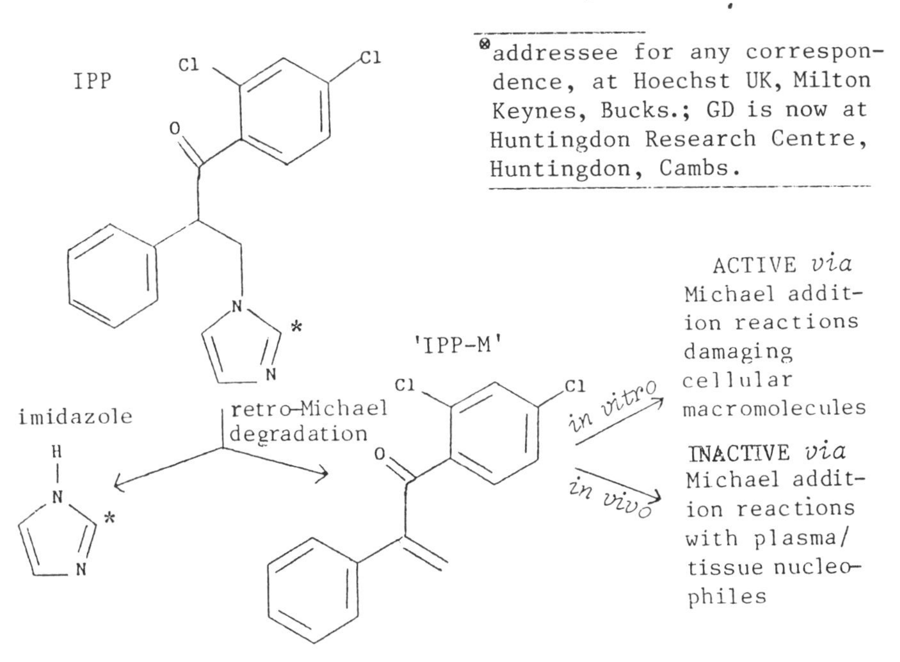

Fig. 1. Structures of IPP (* denotes position of ^{14}C label) and its degradation products, and mechanisms leading to activity or inactivity. IPP is 3-(imidazol-1-yl)-2-phenyl-1-(2,4-dichlorophenyl)propan-1-one; IPP-M is 3-(2,4-dichlorophenyl)-3-keto-2-phenyl-prop-1-ene.

Materials including solvents were mostly from BDH Chemicals. 2-[^{14}C]-Imidazole IPP was synthesized by Dr. R.L. Dyer of the company's Chemical Development Dept. **Extraction** was by 'differential' SPE[®]: samples of plasma or the supernatant (up to 5 ml) from tissue homogenates or culture media were applied to C-18 BondElut cartridges (200 mg; Jones Chromatography, Clwyd, U.K.) pre-conditioned with 2 ml methanol and 2 ml water under vacuum. The 'column' was eluted with 2 ml water, then 2 ml methanol. Aliquots of each eluent (0.05-0.2 ml) were scintillation-counted ('Mk. III', Denley Insts., Crawley) in PCS (10 ml; Amersham Intl.) and analyzed by HPLC.

Analysis was on a μBondapak C18 column (300 × 3.9 mm i.d.; 10 μm) with a Corasil C18 pre-column (20 × 3.9 mm; 37 μm). The system comprised two M6000 pumps, a WISP 710B autosampler, gradient controller (Waters) and a Pye 4020 UV detector set at 280 nm. The mobile phase gradient was aqueous methanol, 15% rising to 85% (v/v) over 30 min at 1.5 ml/min total flow-rate. Fractions (0.5 ml) were collected into scintillation vials and counted after adding 10 ml PCS. The HPLC method resolved IPP and imidazole (Fig. 2).

METABOLISM STUDIES

***In vivo* studies.**- Liver, kidney and plasma were taken from mice dosed i.v. with 40 mg/kg ^{14}C-IPP in PEG 400. Tissues were homogenized in water (1 vol.), and 50-200 mg portions of homogenate and plasma were combusted (Sample Oxidiser, Packard 306) and counted; other portions were taken for analysis (SPE, HPLC). Extraction and analysis as above were performed on weighed portions (up to ~5 ml) of homogenate supernatants (2000 **g**, 10 min) and plasma.

Total ^{14}C concentrations (Table 1) were ~5-15 times the *in vitro* MIC (~0.8 μg/ml) of IPP at 5 min. The extraction and HPLC showed that within 10-20 min the total ^{14}C was mainly imidazole with IPP concentrations <0.8 μg/ml, its *in vitro* MIC (Fig. 2).

***In vitro* experiments.**- ^{14}C-IPP (15 μg/ml, 20,000 dpm/ml) was incubated in mouse plasma at 37° or in water at 22° for 2 h. Samples were removed at selected times, extracted and analyzed by HPLC as above. A similar experiment was carried out with ^{14}C-IPP in the agar broth growth medium at 37° in N_2. Broth samples (~1.5 ml) were collected at selected times, diluted with water (2 vol.), centrifuged (2000 **g**, 10 min) and extracted and analyzed as above.

The results (Fig. 3) showed that ^{14}C-IPP was stable in water but was rapidly degraded to [^{14}C]imidazole in plasma ($t_{\frac{1}{2}}$ ~10 min). The degradation was unaffected by NaF (5 mM),

[®]*Abbreviations.*- SPE, solid-phase extraction; MIC, minimum inhibitory concentration. PCS is a scintillant (Amersham International).

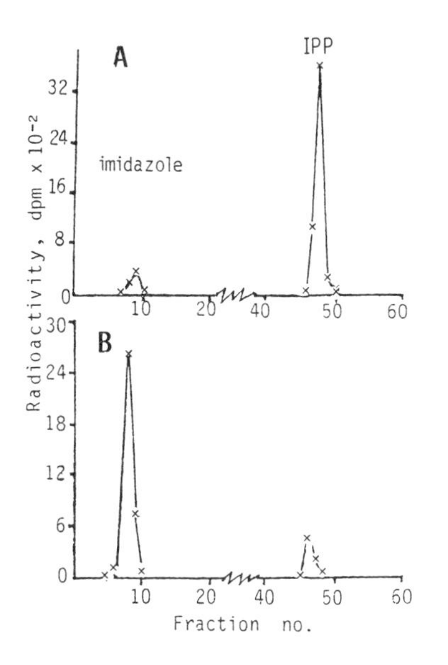

Fig. 2. Radiochromatograms of methanolic extracts of (**A**) ^{14}C-IPP in control mouse plasma, (**B**) mouse plasma 10 min after a 40 mg/kg i.v. dose of ^{14}C-IPP.

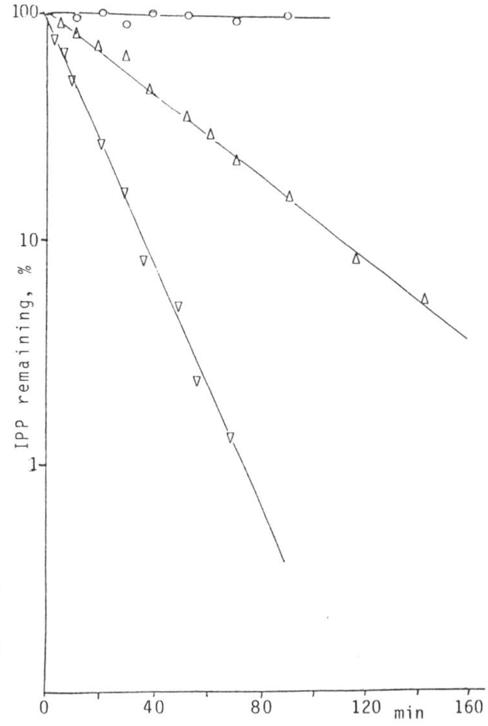

Fig. 3. *In vitro* degradation of ^{14}C-IPP in: o, water at 22°; Δ, agar broth at 37° under N_2; ∇, mouse plasma at 37°.

Table 1. Tissue distribution of total radioactivity and IPP after a 40 mg/kg i.v. dose of ^{14}C-IPP. IPP is expressed as µg/g, and radioactivity as µg-equivalent/g.

Tissue	5 min		10 min		20 min	
	^{14}C	IPP	^{14}C	IPP	^{14}C	IPP
Plasma	3.9	1.3	2.8	0.8	3.3	0.37
Liver	12.6	3.0	10.6	2.0	8.9	0.40
Kidneys	6.8	2.1	6.4	1.2	5.7	0.32

(Text, continued)

suggesting it was not metabolic. A similar degradation occurred in agar broth but more slowly ($t_{\frac{1}{2}}$ ~40 min). If plasma were added to the agar broth, no ^{14}C-IPP was left by 2 h.

CONCLUSIONS

The SPE procedure gave efficient recoveries (>90%) of added ^{14}C-IPP and ^{14}C-imidazole from plasma, tissue samples and agar broth. Although ^{14}C-IPP was recovered exclusively in the methanol eluate from the C-18 cartridge, the ^{14}C-imidazole eluted partly (~20%) in the water wash, presumably because of its greater polarity. The HPLC analysis showed some degradation (~10%) of ^{14}C-IPP to ^{14}C-imidazole after addition to the control biological matrices although added IPP was recovered quantitatively from aqueous solution.

Application of the assay showed that although after i.v. administration drug-related material rapidly reached high concentrations in target tissues (liver, kidney and plasma), the parent drug represented a minor component. Most was converted to ^{14}C-imidazole; this resulted in IPP levels below the MIC within 20 min, thereby preventing any anti-microbial effects.

The *in vitro* experiments showed that ^{14}C-IPP underwent rapid non-metabolic degradation to ^{14}C-imidazole in plasma and in the bacterial culture medium. The most likely mechanism for this degradation was a retro-Michael elimination of the α,β-unsaturated ketone 'IPP-M' (Fig. 1) to leave ^{14}C-imidazole as the only radioactive component. The ketone was synthesized and shown to be a potent inhibitor of *B. fragilis* in the bacterial test system.

Thus it appears that IPP was active *in vitro* through degradation to the α,β-unsaturated ketone which presumably reacted with bacterial macromolecules, thereby inhibiting growth. However, *in vivo* this reactive intermediate was probably 'mopped up' by tissue nucleophiles and thus failed to reach the target organism.

#ncB-6

A Note on

METABOLISM OF THE IMIDAZOLE GROUP IN AN ANTI-INFECTIVE TRICHLOROPHENYLHYDRAZONE DERIVATIVE

J.W. Firth, P.M. Stevens, R.D. Brownsill, N. Lopez, C.M. Walls and †C.W. Vose

G.D. Searle & Co. Ltd., High Wycombe HP12 4HL, U.K.

2-(Imidazol-1-yl)-1-phenylbutan-1-one-2,4,6-trichlorophenylhydrazone (IPTH; Fig. 1), as its hydrochloride salt, was an anti-infective imidazole active *in vitro* and *in vivo* against anaerobic bacteria, particularly *B. fragilis*. Although the compound was well absorbed (>90%) orally in rodents, it was significantly less active than after parenteral administration.

Preliminary pharmacokinetic studies entailed assay by HPLC and showed an oral bioavailability of 20-30% in rat and dog. During these studies two metabolite peaks, M1 and M2, were observed in extracts of rat and dog plasma, similar to - and now shown to be identical with - peaks detected in mouse plasma during pharmacology studies. This article concerns these plasma metabolites - their detection, isolation (by adapting the HPLC assay procedure; ether extraction initially) and identification (using MS⊗). They turned out to have the identities indicated in Fig. 1.

†addressee for any correspondence, now at Hoechst UK, Walton Milton Keynes, Bucks MK7 7AJ; other authors at: Finnigan-MAT, Hemel Hempstead (JWF); Sigma Ltd., Poole (PMS); Servier R & D, Fulmer, Slough (RDB); Napp Labs., Cambridge (NL); Rhone-Poulenc, Dagenham (CMW).

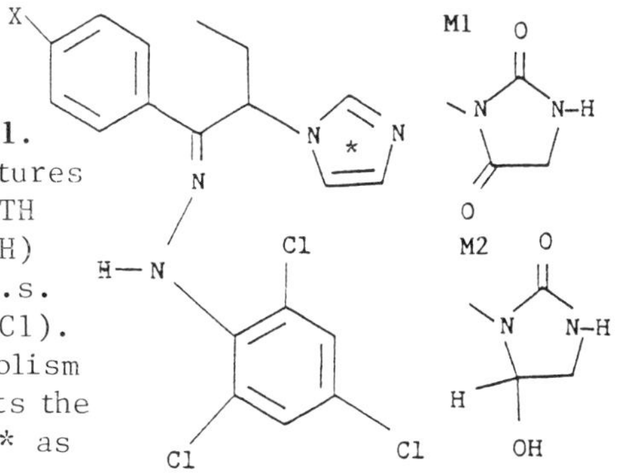

X N N * N H—N Cl Cl Cl M1 O N N-H O M2 O N N-H H OH

Fig. 1. Structures of IPTH (X = H) and i.s. (X = Cl). Metabolism affects the ring * as shown.

⊗*Abbreviations*.- MS, mass spectrometer/spectrometry (CI, chemical ionization; EI, electron impact); i.s., internal standard. See text for IPTH (drug); M signifies metabolite; ether connotes diethyl ether.

Materials were mostly from BDH Chemicals. IPTH.HCl and i.s. (Fig. 1) were from the company's Chemical Development Dept.; methanolic solutions were diluted to give stock solutions. **Extraction** of plasma (0.25 ml), after adding 1 µg i.s. (in 0.75 ml water) and 0.5 M NaOH (0.25 ml) and mixing vigorously, was with redistilled ether (2 × 2.5 ml, 10 min; centrifugation at 2500 **g**, 5 min). The combined ether layers were dried down under N_2 at 30°, and the residues dissolved in 0.1 ml of HPLC mobile phase. **Analysis** was as stated in the legend to Fig. 2, using a 50 µl sample.

HPLC profiles for mouse, rat and dog plasma (Fig. 2) after oral administration of IPTH showed peaks for the drug and drug-derived components (M1, M2), well resolved from endogenous components. IPTH detectability was ~20 ng/ml; accuracy was 95-102%, and the overall C.V. (25-2000 ng/ml) 8-15%.

Metabolite isolation and purification.- IPTH was given orally to fasted rats (200 mg/kg) in 5% aqueous PEG400, and to beagle dogs (600 mg/kg) by capsule. Plasma samples were collected between 2 and 6 h, and 10 ml of plasma was extracted as above, with scaling-up. The residue was dissolved in methanol-10 mM pH 7.5 phosphate buffer (70:30). HPLC analysis (Fig. 3) gave good resolution of M1 and M2. Each was collected, ether-extracted at pH 9.5, and dried down (N_2, 30°) for MS examination.

Identification.- The conditions for MS (Finnigan 4000; 6110 data system) were: solid-probe EI - source at 220°, emission current 0.48 mA, electron energy 49 eV; similarly for CI, with isobutane reagent gas at 22.6 Pa source pressure. M1 showed molecular and $(M+H)^+$ ions with the characteristic isotope pattern for 3 Cl's (Table 1), 32 amu higher than for IPTH and hence containing two additional oxygen atoms. The equivalent ions for M2 showed a 34 amu mass shift relative to IPTH. The metabolism of IPTH to M1 and M2 could occur in the phenyl, imidazole or alkyl groups. The key fragment ions to define the metabolic site(s) were those involving loss of phenyl or imidazole moieties, and cleavage at the N-N bond (Table 1).

The fragment ions formed with loss of the imidazole moiety from M1 and M2 (a, c, d & e) showed no mass shift relative to those ions in the IPTH spectra, whereas those ions retaining this group (b, f & g) showed the 32 and 34 amu mass shifts observed in the molecular ions of M1 and M2. Thus metabolism had occurred only in the imidazole ring - in particular (judged by the absence of M-16 fragment ions typical of N-oxides [1, 2] in their EI spectra) at the carbon atoms, forming the hydantoin (M1) and hydroxyimidazolidinone (M2) shown in Fig. 1. The spectra did not allow

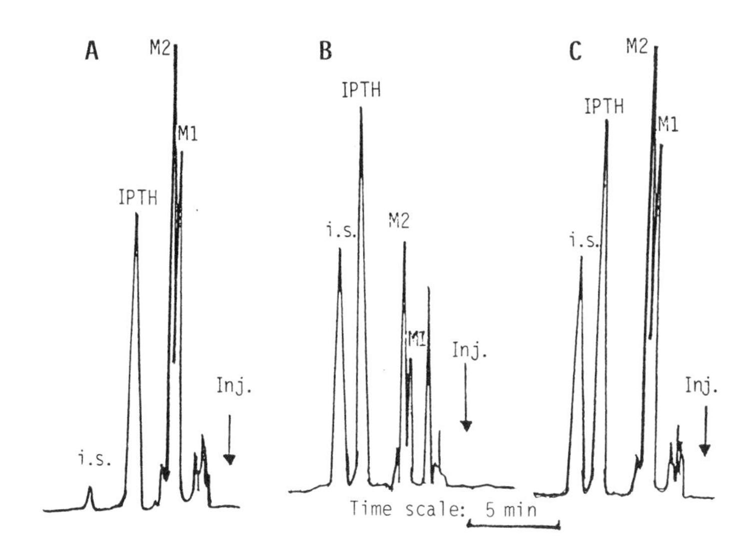

Fig. 2, *above (direction ←)*. Patterns for plasma (**A**, mouse; **B**, rat; **C**, dog) 2 h after an oral IPTH dose (100 mg/kg). **Column:** μBondapak C18 (300 × 3.9 mm i.d.; 10 μm). Pre-column: Corasil C18 (20 × 3.9 mm; 37 μm). **Eluent:** methanol-10 mM phosphate buffer pH 7.5 (80:20 by vol.; 2 ml/min). Pump: 750/05, ACS (Luton). Valve: Rheodyne 7125. Detector: PU 4020 (Pye, Cambridge); set at 280 nm. Integrator: HP3390, Hewlett Packard.

Fig. 3, *right*. Preparative HPLC: dog plasma 3 h after 600 mg/kg IPTH orally. Conditions as in Fig. 1, but methanol:buffer ratio 30:70.

definitive assignment of structures, but 2',5'-substitution is evident, by analogy with that reported previously for econazole [3] and imidazole and histidine [4]. The mouse plasma metabolites were shown by co-chromatography to be identical with M1 and M2 from rat and dog.

CONCLUSIONS

Modifications to a bioanalytical method allowed isolation, purification and subsequent identification of hydantoin (M1)

Table 1. Ions observed (m/z values) in mass spectra of IPTH, M1 and M2.

EI: *key fragment ions*

Cpd.	M^+	a	b	c	d
IPTH	406/408/410	339/341/343	212	144	77
M1	438/440/442	339/341/343	244	144	77
M2	440/442/444	329/341/343	246	144	77

CI (isobutane): *key fragment ions*

Cpd.	$(M+H)^+$	e	f	g
IPTH	407/409/411	339/341/343	109	67
M1	439/441/443	339/341/343	141	99
M2	441/443/445	339/341/343	143	not seen

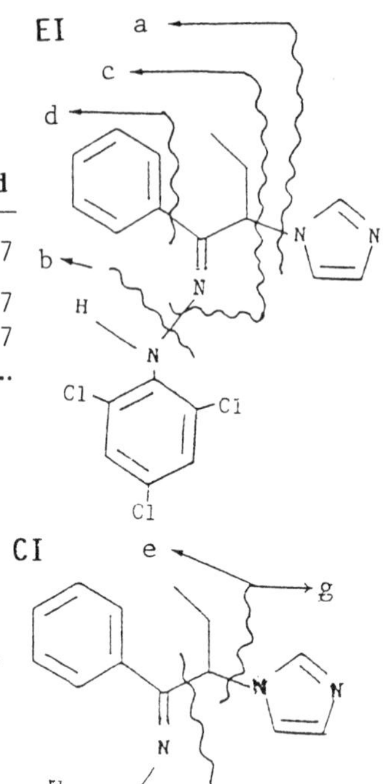

and hydroxyimidazolidinone (M2) metabolites of IPTH detected in plasma during kinetic studies. The rapid metabolism of the drug by these pathways appears to account for the low bioavailability and decreased activity of orally administered IPTH. Modifications to block this metabolic route may yield more potent orally active compounds in this series. Imidazole ring metabolism has been reported previously ([3-5]; see above). Methimazole is similarly converted to its thiohydantoin [6]. The hydantoins may ring-open to give substituted ureas [4, 6]. This may be a further pathway for the metabolism of M1 formed from IPTH.

References

1. Grigg, R. & Odell, B.G. (1966) *J. Chem. Soc. (B)*, 218-219.
2. Becker, H.G.O., Beyer, D. & Timpe, H.J. (1970) *Prakt. J. Chem. 312*, 869-876.
3. Girkin, R., Brodie, R.R, Midgley, I., Hawkins, D.R. & Chasseaud, L.F. (1983) *Eur. J. Drug Metab. Pharmacokin. 8*, 109-116.
4. Marriaggi, N., Cornu, A. & Rinaldi, R. (1973) *Analysis 2*, 485-488.
5. Midgley, I., Biggs, S.R., Hawkins, D.R., Chasseaud, L.F., Darragh, A., Brodie, R.R. & Walmsley, L.M. (1981) *Xenobiotica 11*, 595-608.
6. Skellern, G.G. & Steer, S.T. (1981) *Xenobiotica 11*, 627-634.

#ncB-7

A Note on

PROBLEMS WITH THE MEASUREMENT OF ANTIVIRAL DRUGS IN BODY FLUIDS

R.J. Simmonds

Clinical Research and Development, Upjohn Ltd., Fleming Way, Crawley, W. Sussex RH10 2NJ, U.K.

Whilst this article revolves on problems encountered in assaying antivirals, there are lessons for the analysis of any drug which is related to bioconstituents and their metabolites. This category includes, besides antivirals, peptides and the like [which featured in Vol. 16, this series: *Ed.*]. Antiviral drugs do represent a notable example of the situation, since many of them are close analogues of ubiquitous nucleotides, nucleosides and their bases. However, the problems and the ameliorative measures discussed are applicable to many other HPLC assays where ng/ml sensitivity is required.

The essential problem in assay development is to attain adequate selectivity in the presence of numerous and potentially interfering substances present in blood, plasma, serum, urine and the like. For some nucleotides and major breakdown products the concentrations in human biofluids (much higher in small-animal biofluids) are of the following order [1] (see also [2-4]; some particular values for urine*, with the same units, *added by Editor*):

- ~0.5 µmol/L plasma: ATP, ADP, AMP;
- ~0.5-5 µmol/L plasma, 5-50+ µmol/L urine: uric acid ($\sim 30 \times 10^3$), hypoxanthine (~700), xanthine (~400), uridine, inosine, guanosine, xanthosine;
- 1-100+ nmol/L urine: tRNA breakdown products - pseudouridine (19×10^5), 7-methylinosine (1-methyl: $\sim 7 \times 10^4$), 1-methylguanosine, 6-methyladenine, N^2,N^2-dimethylguanosine (14×10^4);
- also caffeine, theobromine.

These concentrations are such that interference in the assays for drugs of a similar structure is more or less inevitable. Fig. 1 shows breakdown products of tRNA in human plasma using a procedure that minimizes manifestation of metabolites such as those listed above. Fig. 2 shows rat plasma spiked with 100 ng/ml acyclovir, a representative antiviral drug: it is quite polar, and elutes close to many endogenous compounds in RP-HPLC. Such drugs clearly present problems.

*Source: *Geigy Scientific Tables* (1981), 8th edn., Vol. 1 (Lentner, C., ed.), Ciba-Geigy, Basle [refs. therein].

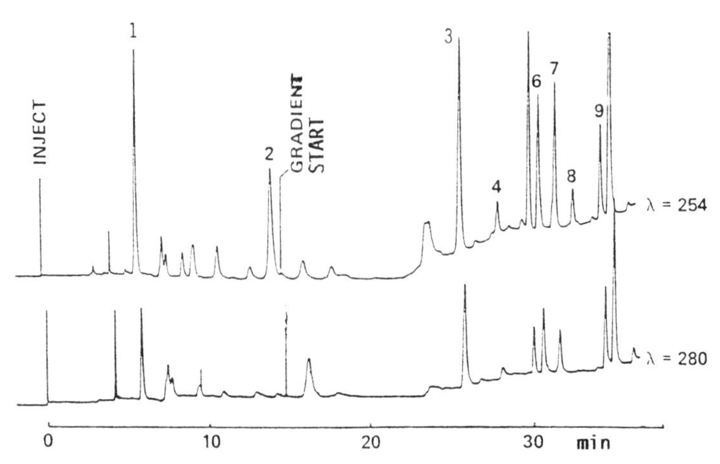

Fig. 1 *(above)*. HPLC chromatogram of nucleoside extract from human plasma, showing: **1**, pseudouridine; **2**, uridine; **3**, allopurinol (intl. std.); **4**, adenosine. Smaller peaks are unidentified, but are likely to interfere in a sensitive (ng/ml) drug assay. **Column:** Hypersil ODS (250 × 4.6 mm). **Eluent:** 4 mM KH_2PO_4 with initially 1% methanol but 20% in gradient inflow. Run at 30°, 1 ml/min. **Extraction:** TCA deproteinization; supernatant 'washed' with ether.

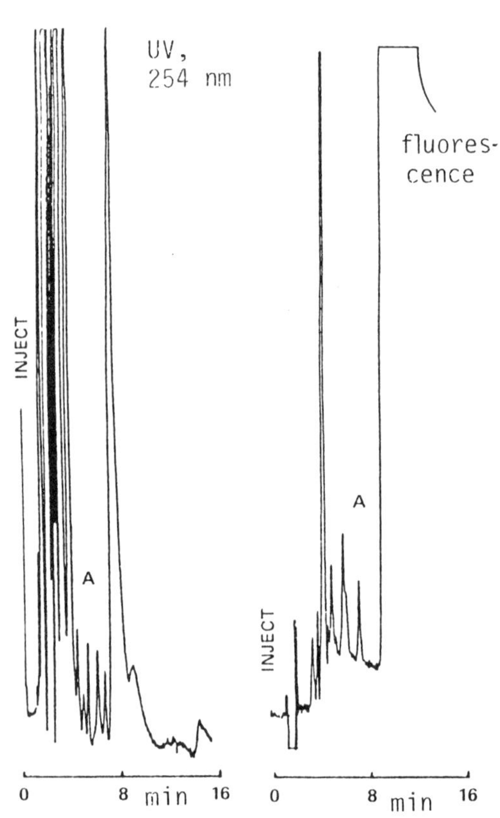

Fig. 2 *(right)*. HPLC chromatogram of rat plasma containing 100 ng/ml acyclovir (peak **A**). Conditions essentially as for the isocratic stage in Fig. 1, but fluorescence detection also performed, and 0.2 ml/min inflow of pH 2.5 0.5 M phosphate buffer to acidify. *Both Figs. from [4], by permission of CRC Press.*

The number of endogenous compounds and their chemical similarity to the analytes of interest make selective extraction from the biological matrix problematical. Nucleosides can be extracted with phenylboronic acid solid phases, or there may be other specific interactions that can be exploited, e.g. with silver-loaded thiol phases [5]; but generally, without access to an antibody, it is difficult to do better than a simple deproteinization.

Ultrafiltration or an enzymic method could be considered, but precipitation is likely to be the choice. Whilst any standard method, e.g. an organic solvent or salt addition, is effective, the preferred reagent is perchloric or trichloroacetic acid, which give the lowest chromatographic background for base and nucleoside analysis and are most probably optimal for analysis of analogous drugs. Conceivably, however, improved results with, for the particular drug, a better chromatographic 'window' might be obtainable with a different technique (compatible with the drug's polar nature) such as SPE or affinity cartridge extraction. Most probably, though, selectivity has to be achieved not during extraction but in chromatography or detection.

Chromatographic selectivity may be sought through exploration of the following variables:
column efficiency: length, particle diameter and size-homogeneity (gradient operation might have to be considered);
choice of bonded phase - RP: hydrocarbon chain length, carbon loading, degree of capping; - newer types: e.g. nitrile, phenyl, diol, carboxylic acid (CBA);
choice of mobile phase: manipulation in respect of buffer salt/pH and of organic solvent (concentration and nature -'solvent triangle'), counter-ion, ion-pair;
optimization of total system: temperature, injector, choice of pre-column, etc.

By careful choice of mobile and stationary phases, ion-pair etc., and with experience, it is generally possible to achieve some kind of assay. However, the assay will be delicate, need expert application, and will not be altogether suitable for large batches of samples.

Selective detection is hardly attainable with UV in this context, but often these compounds fluoresce at acid pH. A dilemma here is that an acidic mobile phase may not be optimal for analytical resolution. Post-column acidification is feasible, and allows for fluorescence detection which is more selective. This is exemplified (Fig. 2) for acyclovir in rat plasma, where some reduction in background has been achieved.

The sample often gets no attention or, worse, is taken for granted - particularly in respect of the actual sampling of the biofluid and the preparation and storage of a reasonably stable matrix before actual analysis. The numbers and concentrations of interfering substances seen in the assays can be profoundly affected by many factors:

- size of needle for blood sampling;
- nature of plastic of syringe or sample vial;
- nature of anticoagulant, if plasma chosen rather than serum: e.g. citrate/glucose, EDTA, NaF, heparin;
- state of volunteer/patient in respect of stress/disease /exercise;
- standing time of blood before deriving plasma/removing serum;
- sample storage time/conditions; freezing/thawing.

Some points such as the effect of plasticizer are obvious, but others are less so. If erythrocytes are damaged during sampling by use of too small a needle/too high a syringe vacuum, they will leak nucleotides which will quickly be degraded, as can occur too if blood is kept on the bench before centrifuging.

EDTA is generally preferred for analysis of bases and nucleosides. It probably inhibits leakage from erythrocytes and enzymic degradation. Drastic strategies to minimize such effects have been published for assays of clinical markers, such as adenosine [6], where meticulous procedures have to be followed to achieve meaningful values. Perusal of clinical biochemistry literature can be rewarding in respect of possibly, for antivirals, reducing interferences and so rendering an assay feasible or more robust.

The concentrations of such metabolites are also greatly increased after vigorous exercise [7] and in some not uncommon disease states: samples from patients may be very different from control samples! Optimizing such factors is time-consuming, and does put much responsibility upon clinical staff who may not grasp the procedural needs. What takes place in the clinic has been rather neglected by analysts, who can gain much by paying attention to minor details. Problems of analytical selectivity or specificity are best minimized by avoiding them.

References

1. Hartwick, R.A., Assenza, S.B. & Brown, P.R. (1979) *J. Chromatog. 186*, 647-658.
2. Hartwick, R.A., Krstulovic, A.M. & Brown, P.R. (1979) *J. Chromatog. 186*, 659-676.
3. Assenza, S.B. & Brown, P.R. (1984) *J. Chromatog. 289*, 355-365.
4. Harkness, R.A. & Simmonds, R.J. (1987) in *CRC Handbook of Chromatography: Nucleic Acids and Related Compounds* (Krstulovič, A. ed.), CRC Press, Boca Raton, FL, pp. 193-208.
5. Irth, I. Tocklu, R., Welten, G.J., Brinkman, U.A.Th., *et al.* (1989) *J. Chromatog. 491*, 321-330.
6. Ontyd, J. & Schrader, J. (1984) *J. Chromatog. 307*, 404-409.
7. Harkness, R.A., Simmonds, R.J. & Coade, S.B. (1983) *Clin. Sci. 64*, 333-340.

#ncB-8

A Note on

IN VITRO AND *IN VIVO* METABOLISM OF THE ANTI-HERPES AGENT 5-(2-CHLOROETHYL)-2'-DEOXYURIDINE

I. Szinai, Zs. Veres, K. Ganzler, J. Hegedus-Vajda and ⊗E. De Clercq

Central Research Institute for Chemistry of the Hungarian Academy of Sciences, P.O. Box 17, H-1525 Budapest, Hungary

and ⊗Rega Institute for Medical Research, Katholieke Universiteit Leuven, Minderbroedersstraat 10, B-3000 Leuven, Belgium

CEDU* has recently been synthesized and evaluated for its antiviral properties [1-4]. We aimed to study the enzymatic cleavage and metabolic elimination of CEDU in mice and rats. Both showed a novel stereoselective hydroxylation.

General methods.- 2-^{14}C-CEDU (5.2 MBq/mmol) was obtained from Sandoz Forschungsinstitut (Wien, Austria). It was given orally (50 mg/kg; 4.51 MBq/animal) to male mice (NMRI; 20-25 g) and male Wistar rats (180-200 g). Urine and faeces were collected daily for 3 days. Radioactivity in the samples was measured by liquid scintillation counting, direct (urine) or as a sequel to combustion (faeces) [5], with a Packard 2660 or LKB-Wallac Rackbeta 1210 instrument. **Enzymes.**- For phosphorolysis of ^{14}C-CEDU as previously described [6], uridine phosphorylase (UrdPase) was prepared from rat intestinal mucosa, and thymidine phosphorylase (dThdPase) from mouse liver. **Microwave extraction** of faeces samples was performed as for RGH 2981 [#ncC-4, this vol.] but using methanol-water (1:1); the supernatant was subjected to SPE.

Metabolite isolation entailed SPE with Extrelut (Merck) [5, 7], applied to urine (0-24 h) and faecal extracts (0-48 h). For **TLC** the SPE fractions were spotted onto 0.5 mm semi-preparative F_{254} plates (Merck), developed with benzene-acetone-methanol (**A**; 5:3:2 by vol.) and re-chromatographed with chloroform-acetone-methanol (**B**; 6.5:2:1.5). **HPLC** (on 10 µl) was performed as in #ncC-4 with variants: a Waters 204 W pump and a Pye Unicam rather than Isco detector (263 nm) were sometimes used; the NP column was 2-4 µm Supersil (Chromlab,

**Abbreviations.*- CEDU, the title compound; MS, mass spectrometry (CI, chemical ionization; EI, electron impact; FAB: fast atom bombardment); SPE, solid-phase extraction; NP/RP/ChP, normal/reversed/chiral phase.

Hungary; 150 × 4 mm) using chloroform-dioxan-methanol (7.5:2:0.5) at 0.5 ml/min. The eluent for the C-18 column was isopropanol-water (0.25:9.75) at 1 ml/min. The ChP column was Chiralcel OC (Dicell, Baker; 250×4.6 mm), eluted with ethanol-hexane (7.5: 2.5) at 0.5 ml/min. **MS**, generally as in #ncC-4, included CI with NH_3 or ethanol by 70 eV electrons, and FAB (AEI 902 with a saddle field fast atom gun) using argon at 10 kV, 1 mA (glycerol matrix). *TLC role:* 'screening', complementing HPLC.

RESULTS

Compared with analogues studied earlier [6], CEDU showed rapid phosphorolysis *in vitro* with higher Km's (mM): dThdPase, 41.0 ±0.5 (S.D.; n = 5); UrdPase, 10.0 ±1.5. In faeces and (predominantly) urine, a range of techniques (EI-, CI- and FAB-MS, NMR, IR and UV) disclosed 3 metabolites: the base derived from CEDU (**I**; CEU), 1-hydroxy-CEU (**II**) and one unidentified (**III**).-

% of total ^{14}C *applied to TLC plate: urine from mice or, shown ( ), rats.-* CEDU, 25.1 (24.1); **I**, 38.7 (56.1); **II**, 29.6 (17.5); **III**, 3.4.

TLC R_f's, **A** *and* **B** *respectively.-* CEDU, 0.44 & 0.67; **I**, 0.48 & 0.72; **II**, 0.38 & 0.67; **III**, 0.29 & 0.40.

HPLC R_t's, min.- **RP**: CEDU, 4.6; **I**, 5.7; **II**, 4.1; **III**, 3.2. **NP**: CEDU, 6.6; **I**, 4.6; **II**, 7.5; **III**, 11.5. **ChP**: **II**: 8.0, 9.5.

MS characteristic ions, e/z *values:* $[B+H]^+$, CEDU: 175; $[M+H-HCl]^+$, **I**, 139; **II**, 155; $[M+H-H_2O]^+$, **II**: 173; $[M+H]^+$, CEDU: 291; **I**, 175; **II**, 192; $[M+NH_4]^+$, CEDU: 308; **I**, 192; **II**, 208.

Besides cleavage of the ***N***-glycosidic bond as the main metabolic pathway of CEDU, we found a new pathway in both species investigated: **I** (CEU) underwent stereoselective hydroxylation at an α-carbon atom in the 2-chloroethyl side-chain, manifested by ChP-HPLC. This bioconversion could lead to some serious toxicological consequences that may help understand the carcinogenic [E. De Clerq & H. Sobis, unpublished data] side-effects of CEDU.

References

1. Griengel, H., Bodenteich, M., Hayden, W., Wanek, E., Streicher, W., Stutz, P., Bochmayer, H., Ghazzouli, I. & Rosenwirth, B. (1985) *J. Med. Chem. 28*, 1679-1684.
2. De Clercq, E. & Rosenwirth, B. (1985) *Antimicrob. Agents Chemother. 28*, 246-251.
3. Rosenwirth, B., Griengel, H., Wanek, E. & De Clercq, E. (1985) *Antiviral Res., Suppl. 1*, 21-28.
4. Griengel, H., Schwarz, W., Schatz, F., Skrobal, M., Werner, G. & Rosenwirth, B. (1986) *Chemica Scripta 26*, 67-71.
5. Veres, Zs., Szinai, I., Szabolcs, A., Ujszaszy, K. & Denes, G. (1978) *Drugs Exptl. Clin. Res. 13*, 7068-7074.
6. Veres, Zs., Szabolcs, A., Szinai, I., Denes, G. & Jeney, A. (1986) *Biochem. Pharmacol. 35*, 1057-1059.
7. Szinai, I. & De Clercq, E. (1989) *Drug Metab. Disp. 17*, in press.

#ncB-9

A Note on

DRUG SURVIVAL IN CLINICAL SAMPLES IRRADIATED AS AN ANTI-HIV PRECAUTION

[1]H. de Bree and [2]M.P. van Berkel

[1]Duphar BV, P.O. Box 900, 1380 DA Weesp, The Netherlands and [2]DSM Chemicals BV, P.O. Box 27, 6160 MB Geleen, The Netherlands

The analysis of drugs in biological samples involves the risk of infection by bacteria, parasites and viruses - notably HIV (AIDS) infection with its lethal outcome. In a bioanalytical laboratory processing thousands of human plasma samples per year (from volunteer and clinical studies) it is virtually impossible to discriminate between contaminated and safe samples. Moreover, it was recently reported [1, 2] that control, reference and pooled sera could be contaminated with HIV. Hence we listed the safety procedures applicable to high-risk laboratories and concluded that all biomedical and bioanalytical laboratories should be converted into bio-hazard laboratories, still with some risks remaining. A better approach would be to prevent contaminated samples entering the laboratory. This demands a sterilization procedure.

Thermal and chemical sterilization have been suggested in this connection [3, 4] but are generally inapplicable to samples containing low concentrations of relatively unstable pharmaceuticals and endogenous compounds. Moreover, chemical sterilization involves opening of sample vessels with the inherent risk of aerosol formation.

A third alternative, which could be efficient and elegant, is sterilization by gamma-irradiation, as routinely used for injection fluids, surgical equipment and certain food products [5, 6]. Its main advantage is that the container in which the samples are shipped can be kept closed and deep-frozen before and during sterilization. Unfortunately the literature only claims or assumes complete sterilization and lacks proof of efficacy. Together with colleagues in the Dutch Red Cross Blood Transfusion Service we set up a study to ascertain what irradiation dose was needed to inactivate HIV completely at 194°K (-78.5° C). In parallel we investigated whether an efficacious irradiation dose affected the levels of a number of drugs and endogenous components in plasma. As shown in this article, doses of 5 Mrad or higher inactivated the virus, without detriment to most test drugs or, except for blood enzymes, to constituents of interest to clinical chemists.

MATERIALS AND METHODS

Origin of the samples.- Drug-containing plasma samples were either from plasma pools originating from kinetic studies or were prepared by spiking at clinically relevant levels. The samples (each ~5 ml) were kept in polypropylene tubes (#12160, Greiner, The Netherlands). The drugs studied are listed below ('Analysis'). The clinical control sera were Precinorm U standards (Boehringer, Mannheim; BMC lot 1-504). HIV was inoculated into fresh human plasma by the Central Laboratory (Amsterdam) of the above-mentioned Service.

Sample packing and handling.- The 5-ml tubes with the frozen samples were placed into 25-ml Greiner polypropylene tubes containing sufficient cellulose tissue to absorb the sample in case of inner-tube leakage or damage. The 25-ml tubes were packed in polyethylene bags which were sealed. Subsequently they were placed in a cool-box (Souple-Curver, France; #504260: 18.5 × 29.6 × 40 cm). The box was topped-up with ~10 kg of solid CO_2 pellets (Hoek-Loos, The Netherlands) and sealed with cellotape. For each irradiation dose a separate container with the same composition was prepared. The containers were shipped to the irradiation facilities in compliance with the safety regulations for transport of extremely bio-hazardous substances.

Gamma-irradiation.- The samples were irradiated, whilst frozen at 194°K, by Gammaster BV (Ede, The Netherlands) using a Cobalt 60 Source (#JS 6500, AECL, Canada). The doses other than zero (see RESULTS) were controlled by Gammaster BV; the U.K. National Physical Laboratory calibrated the dosimeters.

Analysis.- Drug levels in the treated and control samples were measured by standard GC or HPLC procedures: clovoxamine [7], fluvoxamine [7], flesinoxan [8], idaverine [9], *N*-desmethyl-idaverine [9] and eltoprazine [10]. Standard biochemical analyses as adapted for the Cobas Bio automatic analyzer (Hoffman-La Roche, Basle) were used for endogenous constituents, including the first 4 enzymes in Table 1 below [11], GGT [12], bilirubin [13] and cholesterol [14].

HIV-deactivation was measured by determining the TCID50 (tissue infectious dose) before and after irradiation; the titres were determined by multiplicate titration using an H9 macroculture cell line. Viral replication was detected by examining the culture and measuring reverse-transcriptase activity in the supernatants [3].

RESULTS

HIV titre reduction (Fig. 1) was assessed from logTCID50 values, which were as follows.-

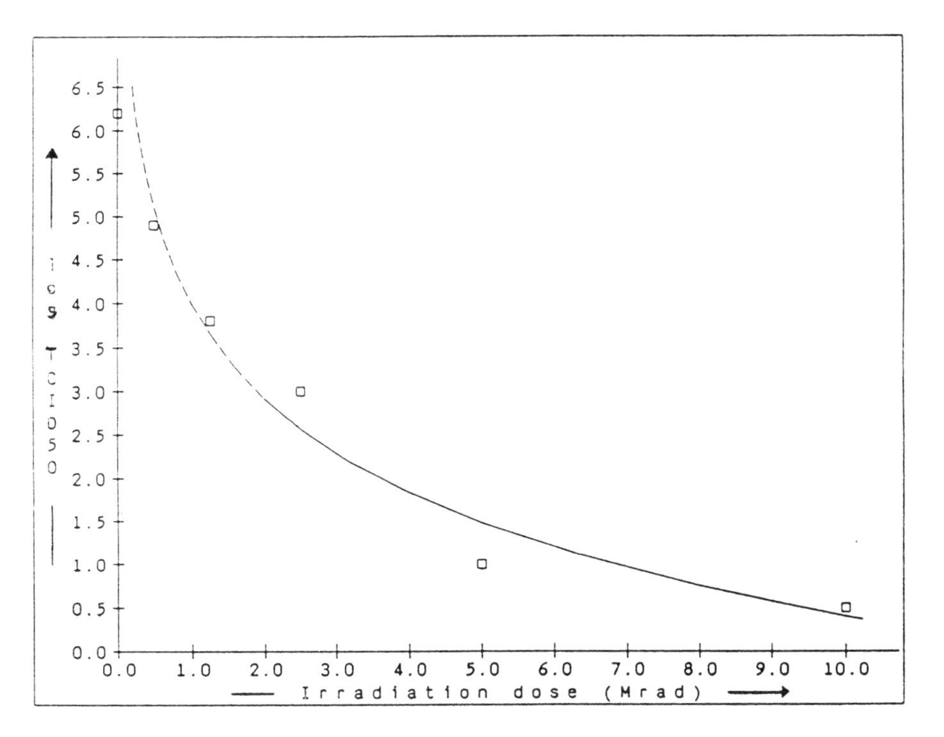

Fig. 1. Deactivation of Human Immunodeficiency Virus (HIV) by irradiation with gamma-rays at 194°K (solid CO_2 temp.).

nil Mrad (control), 6.2; 0.5 Mrad, 4.9; 1.25, Mrad, 3.8; 2.5 Mrad, 3.0; 5.0 Mrad, 1.0; 10 Mrad, 0.5.

Evidently irradiation with 5 Mrad results in a HIV-titre reduction of 5.2 logs, in good accord with a recent report [15] of 6 logs at a dose of 4 Mrad. A reduction of >5 logs is considered sufficient for plasma products [3].

Endogenous plasma constituents in control sera gave the analytical results (5 replicates averaged for each irradiation dose) listed in Table 1. Activity decreased significantly for all the enzymes studied, which is not surprising since this reflects enzyme denaturation. Amongst other constituents, only bilirubin and cholesterol showed a small but significant decrease in concentration, ~10% with 5 Mrad.

Drug levels are shown in Table 2 (at least 5 analyses averaged per irradiation dose). Evidently for no drug studied was the level significantly affected by doses up to 10 Mrad.

Table 1. Effect of different irradiation doses (Mrad) on the levels of endogenous components in control serum.

Component Dose:	nil	1.25	2.5	5.0	10.0
Aspartate aminotransferase (AST)[1]	112	104	88	71	57
Alanine aminotransferase (ALT)[1]	98	91	74	56	46
Lactate dehydrogenase (LDH)[1]	678	627	531	429	317
Alkaline phosphatase (ALP)[1]	311	291	253	221	173
Gamma-glutamyl transferase (GGT)[1]	75.8	70.9	58.9	49.1	33.4
Creatine phosphokinase (CPK)[1]	329	288	203	143	116
Bilirubin[2]	29.6	26.9	243	22.7	22.8
Glucose[3]	12.4	12.4	12.3	12.2	12.3
Total protein[4]	60	59	60	60	59
Albumin[4]	39.2	38.8	38.8	39.1	39.7
Urea[3]	13.9	13.6	13.7	13.9	13.6
Creatine[2]	246	249	250	247	249
Cholesterol[3]	3.13	2.92	2.92	2.80	2.63
Triglyceride[3]	0.67	0.68	0.66	0.64	0.67
Phospholipid[3]	1.77	1.69	1.75	1.60	1.75
Ca^{2+} [3]	2.87	2.92	2.90	2.86	2.86
Na^{+} [3]	150	150	151	151	149
K^{+} [3]	5.18	5.17	5.21	5.16	5.12

[1]U./L [2]µmol/L (µM) [3]mmol/L (mM) [4]g/L

Table 2. Effect of gamma-irradiation on plasma drug levels (means for at least 5 analyses per dose, as ng/ml; doses in Mrad; n.d., not determined).

Drug Dose:	nil	2.5	5.0	10
fluvoxamine	62	58	58	n.d.
clovoxamine	71	73	68	n.d.
flesinoxan	1.90	1.87	1.94	n.d.
idaverine	2.0	2.1	1.8	1.9
N-desmethylidaverine	2.0	2.0	1.8	2.1
eltoprazine	25.8	26.5	27.1	28.7

CONCLUSIONS

Gamma-irradiation of plasma or serum at a dose of 5 Mrad results in virtually complete inactivation of HIV: the virus titre is reduced by a factor of >5 logs. While a number of endogenous plasma components are stable, there is a considerable loss of the activity of the investigated blood enzymes.

Nevertheless, irradiation treatment prior to standardization yields suitable standard sera that carry no risk of infection. Doses up to 10 Mrad have no significant impact on the levels of a number of drugs from quite different structure classes. Hence this procedure can be used with those drugs, to render potentially hazardous samples harmless.

Acknowledgements

We gratefully acknowledge the help of Martin de Haan, who skilfully performed all assays for endogenous components. We also express gratitude to our colleagues in the Dutch Red Cross Blood Transfusion Service, Jan Over and Thijs Tersmette, who inoculated the samples with HIV and determined the viral activities.

References

1. Howanitz, P.J., McBride, J.H., Kilewer, K.E. & Rodgerson, D.O. (1986) *Clin. Chem. 32*, 773-777.
2. Bove, J.B., DePalma, L. & Weirich, F. (1987) *Clin. Chem. 33*, 308.
3. Tersette, M., de Goede, R.E.Y., Over, J., de Jonge, E., Radema, H., Lucas, C.J., Huisman, H.G. & Miedema, F. (1986) *Vox Sang. 51*, 239-243.
4. Horowitz, B., Wiebe, M.E., Lippin, A. & Stryker, M.H. (1985) *Transfusion 25*, 516-522.
5. Spire, B., Dormont, D., Barré-Sinoussi, F., Montagner, L. & Chermann, J.C. (1985) *Lancet, Jan. 26*, 188-189.
6. Van Duzer, J.P. (1984) *U.S. Patent #*4,620,908.
7. Hurst, H.E., Jones, D.R., Jarboe, C.H. & de Bree, H. (1981) *Clin. Chem. 27*, 1210-1212.
8. Van Berkel, M.P., de Bree, H. & Sierat, K. (1988) in *Bioanalysis of Drugs and Metabolites* (Reid, E., Robinson, J.D. & Wilson, I.D., eds.) [Vol. 18, this series], Plenum, New York, 209-212.
9. Van der Stel, D.J.K., Brockhoff, O.A.M., & de Bree, H. (1988) *as for* 8., 251-255.
10. Raghoebar, M. & Van Harten, J. (1989) *Pharm. Weekbl. Sci. Ed. 11, Suppl. H*, H5.
11. Recommendations of the German Society for Clinical Chemistry (1972) *Z. Klin. Chem. Biochem. 10*, 281-291.
12. Persyn, J.P. & Van der Slik, W. (1976) *J. Clin. Chem. Clin. Biochem. 14*, 421.
13. Rand, R.N. & Di Pasqua, A. (1962) *Clin. Chem. 8*, 570-578.
14. Harders, H.D. & Helger, H. (1977) *Med. Lab. 1*, 141-144.
15. Kitchen, A.D., Mann, G.F., Harrison, J.F. & Zuckerman, A.J. (1989) *Vox Sang. 56*, 223-229.

Comments on #**B-1**, R. Horton - BIOASSAY *vs.* HPLC FOR ANTIBIOTICS
#**B-3**, C. Town - CEFTETRAME ASSAY & DATA HANDLING

R. Horton, replying to R. Woestenborghs.- The bioassay calibration equation is of a generally accepted nature, despite the shortness of the linearity range, which one must extend by sample dilution. **L.A. Meyer asked** whether bioassay and HPLC can be compared where, as with neomycin, there is more than one active component. **Reply.-** The two isomers, B and C, in 'neomycin' have similar antibacterial activity, but that of C is more susceptible to impairment by the growth-medium salt, variations in which make bioassay results unpredictable. I believe there are several chromatographic methods which can resolve A and B.

P. Logue asked C. Town whether he had considered use of a bioassay, given that the isomer is inactive. **Reply.-** There is a published bioassay, but in general we are happier with an HPLC system since we can get better precision and accuracy; we fall back on a bioassay if we can't find a suitable HPLC assay. **Comment from Woestenborghs:** for inter- and intra-assay C.V.'s of 10-15%, r should be ≥0.98. **Tanner commented** on the risk that ester pro-drugs in blood may be enzymically converted to the drug even at the initial centrifugation stage: their remedy (published by J. Ayrton) is to transfer the blood from the syringe into acetonitrile (1 vol.) as an enzyme-quencher, then centrifuge. (See also a comment below on #B-10.- *Ed.*)

Comments on #**B-5,** O. Varoquaux - TRIMETHOPRIM, SULPHAMETHOXAZOLE
#**B-6,** K. Borner - QUINOLONE ANTIMICROBIALS

O. Varoquaux, answering V. NcNally.- We have not validated our assay near the sensitivity limit, where the TMP peak is only ~5 mm high. **Reply to P. Sallabank.-** The reduced retention times for TMP and SMZ in Method **II** compared with **I** (on the same column) are related not to the extraction-solvent difference but to the use of dichloromethane rather than chloroform in the mobile phase. A **further reply** on mobile phase influences, e.g. shortening of retention times by water, referred the questioner to the text (#B-5). **Answer to G. Land:** only now are we trying out SPE.

D. Dell remarked to K. Borner (response: Not tried!) that use of blood-bank plasma suits well for spiked standards, and would have obviated the poor recoveries and precisions that led him to spike into water instead of reconstituted commercial serum. **I.D. Wilson remarked** on the aptness of the analytes for a ^{19}F-NMR assay approach (which features in art. #ncC-5).

Comments on **#B-7**, P.V. Macrae - FLUCONAZOLE ASSAY
#B-8, R. Woestenborghs - ANTI-FUNGALS, -HELMINTHICS
#B-9, M.V. Doig - HYDROXYNAPHTHOQUINONES

Woestenborghs, to Macrae.- Since fluconazole has 4 N atoms in the relatively small molecule, the GC-NPD approach would seem appropriate. **Reply.**- GC-ITD was more selective, and enabled us to prove that the drug is not metabolized. **Response by Woestenborghs** to two questions: (1) the benzimidazoles can't be analyzed by GC because of their thermal instability and low volatility; (2) as confirmed with radiolabelled material, levamisole in tissue homogenates can be extracted (using heptane-IAA) with >80% recovery.

Woestenborghs commented to M.V. Doig that silanization of glassware was hardly appropriate since the hydroxynaphthoquinone structures have no basic functions (**reply:** the glassware is merely subjected to thorough cleaning), and queried whether the recovery losses were time-related.- **Doig replied** that the variability reflected not time-dependence but rather variation between plasticware batches: once the methylated metabolites were in the autosampler vials, no further losses occurred. **Comment by Tanner.**- We have had a problem of a lipophilic compound being adsorbed onto glass when making up dilute stock solutions. Silylation only made matters worse. Our remedy was to make the dilute solutions up in 20% aqueous plasma; the compound then stuck to plasma components rather than to the glass.

Doig, replying to Dell.- It was because of a change of batch of SPE cartridges that SPE problems came to attention when we started to assay larger numbers of samples. **S. Wood asked**, in connection with the abortive attempts to extract an analyte with RP cartridges, whether initial protein precipitation prior to SPE had been tried. **Doig's reply.**- TCA caused analyte loss with the protein; acetone was also tried, but SPE still failed. **Suggestion by Wood.**- The use of acetonitrile (4 vol.), optionally followed by dilution, might have led to success with SPE.

Comments on **#B-10**, D.M. Pierce - FAMCICLOVIR (A PRO-DRUG)
#B-11, S.S. Good - ZIDOVUDINE & ITS GLUCURONIDE

C.W. Vose asked Pierce about the esterase-inhibition efficacy of NaF, which in his experience could be poor. **Reply.**- NaF indeed substantially reduced ester hydrolysis, although not completely; but samples were centrifuged within minutes and the plasma stored on dry ice - the compounds then being stable for many months; (**in reply to S. Wood**) it is better that the NaF be coated onto the collection tube than added as a solid. Our tubes are made, with EDTA/NaF (5 mg/ml), by Teklab, County Durham. **S.S. Good, answering G.S. Muirhead/G. Land.**- In general we had no problems due to particulate matter when we loaded samples onto Bond Elut after heat-inactivation; but sometimes we have to centrifuge. Now we dilute with buffered saline before loading. Close control of temperature is important: if it reaches 60°, plasma coagulates. **Tanner asked Good** whether binding of AZT and/or GAZT to the ultrafiltration membranes was determined at concentrations expected in serum. **Reply:** indeed we checked binding to membranes at low concentrations, and found none for either analyte.

Comments on #**ncB-2**, K-Fr. Sewing - CICLOSPORIN MONITORING *made after a Forum contribution, here summarized (no publication text), by* J.M. Queraltó [cf. 1]:-

For reasons that include prevention of graft rejection, it is important to be able to monitor cyclosporin A (CyA; ciclosporin) in patients. HPLC approaches are handicapped by various difficulties, notably poor UV absorbance such that a large sample volume has to be taken and, with questionable efficiency, pre-extracted. For the immunoassays now tried, whole blood, diluted with isotonic NaCl, was used rather than plasma (cf. #ncB-2). The antibodies (Ab's) in the Sandoz RIA kits tested were polyclonal (kit #1) or monoclonal (#2, specific for CyA; #3, non-specific). Also investigated (#4), with diluted whole blood, was a kit (with analyzer; Abbott Labs.) for fluorescence polarization immunoassay (FPIA) of plasma. "Comparison of results [for patient samples] obtained by these techniques shows systematic, constant, proportional or mixed errors" [1]. Values with (4) were higher than with (1)-(3), as in a published study [2].

K-Fr. Sewing, replying to K-H. Lehr, said that none of the investigated assays correlated with ciclosporin toxicity; we don't know which ciclosporin-related compound is responsible for the toxicity. There is an especial problem concerning blood samples dispatched in the external quality assessment scheme: they have been frozen, and are not reliably extractable (column blockages occur). **R.J.N. Tanner.**- Whole blood that has been frozen and thawed is a very difficult matrix to pass through a SPE cartridge. Pre-dilution may help, although there may still be blockage, e.g. by cell debris.

1. Queraltó, J.M., Arcelus, R., Pedrono, X. & Amengual, M.J. (1989) in *Biologie Prospective [7th Colloq. Pont-à-Mousson]* (Galteau, M., Henny, J. & Siest, G, eds.), Libbey, Paris, pp. 723-726.
2. Vogt, W. & Welsch, I. (1988) *Clin. Chem. 34*, 1459-1461.
Other assay refs. noted by Ed. are on p. 235.

Comments on **#ncB-5**, C.W. Vose - AN ANTIMICROBIAL INACTIVE *in vivo*
#ncB-6, C.W. Vose - METABOLISM OF AN IMIDAZOLE GROUP
#ncB-8, I. Szinai - AN ANTI-HERPES AGENT

L.E. Martin asked Vose whether the instability of the TPP was due to base catalysis, accelerated by alkylation of plasma proteins. **Reply.-** Retro-Michael reactions are base-catalyzed, but the results do not merely reflect a change of pH. Other active compounds in the series do not show the same rate of degradation. Reaction of the α,β-unsaturated ketone with plasma constituents may drive the reaction by removing the product, and the reactivity of differently substituted α,β-unsaturated ketones may also have an effect on this phenomenon. **Reply** to a suggestion that the imidazoles might bind to cytochrome P-450: if the compounds had progressed we would have had to examine that possibility. **R. Woestenborghs asked** whether the two ketoimidazole metabolites underwent any further degradation, especially ring scission. **Reply.-** Indeed this probably happens; but the research on these compounds was stopped (for reasons unconnected with its merit).

Szinai, in reply to Dell, Logue and Tanner.- Trial of the microwave technique for tissue samples is at an early stage, but seems promising. Already we have extracted from brain as well as faeces. In the heating period of ~30 sec that we use, the oven (which has to be checked for homogeneity of the microwaves) gives a temperature of ~70-80°.

Comment by Ed.- From the brief description (in #**ncC-3**) it appears that the oven (no turntable?) is of higher wattage than most ovens on the U.K. household market. Those on the scientific market, made in the U.S.A. (CEM; U.K. agent: Processing Concepts, Wigston, Leicester) or France ('Prolabo', Rhone-Poulenc; U.K. source: May & Baker Lab. Products), are tailored to drastic operations such as Kjeldahl digestion, ashing for trace-metal analysis, or moisture determination.

***KEY to Editor's citations** that follow*

The listing order generally matches that of the #**B** and #**ncB** arts., reflecting therapeutic as well as chemical type. MATRIX: **plasma** (or serum) unless others indicated ('blood' = whole blood); inclusion of METABOLITE(S) in the assay denoted **metab(s)**.

Assay abbreviations.- LE, liquid-liquid extraction; SPE, solid-phase extraction (usually C-18); pptn., precipitation (of protein); →dry, dried down; deriv., derivatization. NP, normal-phase; RP, reversed-phase (usually C-18/ODS); col., column; fluor, fluorescence detection; sens. = detection limit. MeCN, acetonitrile; MeOH, methanol; AcOH, acetic acid (usually diluted); phos. = phosphate. *Other abbreviations* are well known: e.g. PCA, perchloric acid; TEA, triethylamine (aqueous); EC, electrochemical.

ANTI-INFECTIVES: *Citations contributed by Senior Editor*
*(**KEY opposite**) to widen the coverage (consult past vols. also)*

General *BACKGROUND for antibiotics: most have poor solvent-extractability & UV absorbance*

Arts. in an **antibiotic assay** book include: sample preparation approaches, and degradation problems such as pH- or air-instability (Kees, F., pp. 7-19); HPLC assay development for antibiotics (Borner, K., pp. 21-36; cf. #B-6); cation-exchange HPLC (Essers, L., pp. 37-44).- *High Performance Liquid Chromatography in Medical Microbiology* (Reeves, D.S. & Ullmann, U., eds.; 1986), G. Fischer, Stuttgart.

Antibacterial drug assay recommendations (Report of a Working Party on clinical evaluation): methodologies, and validation guidance and pitfalls such as co-elution of metabolites or endogenous interferants; don't rely on bioassay or HPLC only. AnMcAgCh* **23 Suppl. B**, 1-39 (1989).

Direct loading of serum is feasible with NP-HPLC (silica) with near-100% aqueous eluents, whereby protein does not precipitate and impair the column; retention data given for 16 agents. Adamouics, J.A. (1987) J. Pharm. Biomed. Anal. **5**, 267-274.

HPLC of **antibiotics** in clinical microbiology: dilution; RP-HPLC-UV (220, 230 or 278 nm) or -fluor, MeOH-phos. or acetonitrile-phos.; examples of sens.: **aspoxicillin**, 0.5 µg/ml; **ciprofloxacin**, 2.5 ng/ml. Knöller, J., **et al.** (1988) J Chr* **427**, 257-267. *Another monitoring review:* Wenk, M., **et al.** (1984) Clin. Pharmacokinet. **9**, 475-492.

β-lactams *- initially **General**, then **'cillins'** including β-lactamase inhibitors, notably* clavulanic acid

Survey of chemistry: *Recent Advances in the Chemistry of β-Lactam Antibiotics* (Brown, A.G. & Roberts, S.M., eds.), Royal Society of Chemistry, Cambridge (1985).

Penicillins, 3rd-generation **cephalosporins** & **aztreonam**, routine monitoring: LE (='wash'; MeCN-DCM); RP-HPLC-UV (rapid if 3 µm and short col.), MeCN-pH 5 acetate, sometimes + THAB or other ion-pair agent. Jehl, F., **et al.** (1987) J Chr **413**, 109-119.

Cephalosporins: review of recent methods. Toothaker, R., **et al.** (1987) AnMcAgCh **31**, 1157-63.

Detection modes for cephalosporins & decomposition products: fluor and EC compared with UV. Blanchin, M.D., Kok, W.Th. & Fabre, H. (1987) Chromatographia **24**, 625-627.

Amoxycillin (I) & **clavulanic acid** (II): dilution; for II, deriv. with imidazole; RP-HPLC-UV (I, 229 & II, 313 nm), MeOH-pH 3.2 phos. Evans, C.M., **et al.** (1988) AnMcAgCh **22**, 363-369.

Aspoxicillin & **metabs.**: urine; RP-HPLC (TSK-GEL; eluent unstated); post-col. deriv.; fluor; sens. 0.5 µg/ml. Suzuki, S. & Yanagida, T. (1987) J. Pharm. Sci. **76**, S11.

....

(KEY**, *continued)* ***'Streamlined' refs.: AnMcAgCh = Antimicrob. Agents & Chemother., J Chr = J. Chromatog., ClCh = Clin. Chem., IJClPh = Int. J. Clin. Pharmacol.

Azlocillin: pptn.; RP-HPLC-UV (220 nm), MeCN-pH 4.8 phos. Valenza, T. & Rosselli, P. (1987) Chromatographia **24**, 862-864.

Benzylcillin (& **probenecid**), body fluids: pptn. by MeCN; LE; RP-HPLC-UV (231 nm), MeOH-MeCN-buffer + TBAHB; sens. 0.5 μg/ml [probenecid: 0.25 μg/ml]. Van Gulpen, C., **et al.** (1986) J Chr **381**, 365-372.

Brobactam (6β-bromopenicillanic acid; serum & urine): RP-HPLC -UV (315 nm after post-col. degradative reaction), phos.-MeOH; sens. (serum) 50 ng/ml. Kissmeyer-Nielsen, A-M. (1988) J Chr **426**, 425-430.

Flucloxacillin (strongly protein-bound): HCl, then LE & →dry; RP-HPLC-UV (220 nm), MeOH-pH 7 phos. Bergan, T., **et al.** (1986) AnMcAgCh **30**, 729-732.

Isoazolyl penicillin (e.g. cloxacillin, dicloxacillin, flucloxacillin): pptn.; SPE; RP-HPLC-UV (220 nm), MeCN-pH 2 phos. (eluent optimization studied); sens. ~0.1 μg/ml. Hung, C.T., Lim, J.K.C., Zoest, A.R. & Lim, F.C. (1988) J Chr **425**, 331-341.

Mecillinam (an amidinopenicillin): bioassay (no details); sens. <2 μg/ml. Moukhtar, I., Nawishy, S. & Sabbour, M. (1987) IJClPh **7**, 59-62.

Mezlocillin (plasma & urine): pptn. by MeCN; LE to wash; RP-HPLC-UV (214 nm), MeCN-pH 7 phos. Jankke, D.M., **et al.** (1988) AnMcAgCh **32**, 777-779.

Mezlocillin & **piperacillin** (acylureidopenicillins): pptn. (MeCN); LE to wash; RP-HPLC-UV, MeCN-pH 7 phos.; sens. 0.5 & 1 μg/ml resp. Mortens, M.G., **et al.** (1987) AnMcAgCh **31**, 2015-2017.

Tiarcillin: extraction; RP-HPLC-UV (242 nm); pH 4 buffer-MeOH. Certain cephalosporins co-eluted. Schull, V.H. & Dick, J.D. (1985) AnMcAgCh **28**, 597-600.

Ticarcillin & clavulanic acid: (for latter) imidazole deriv.; RP-HPLC-UV (229 and 313 nm resp.), pH 2.1 phos.-MeOH in different ratios. Syrogiannopoulos, G.A., **et al.** (1987) AnMcAgCh **31**, 1296-1300.

Aztreonam: pptn. (MeCN); RP-HPLC-UV (290 nm), pH 3 buffer-TBHS. Lindner, K.R. & Pilkiewicz, F.G. (1986) pp. 75-82 in the book ref. at start of previous page.

Cefachlor, cefadroxil, cefixime, cephradine & **cephalexin:** pptn. (MeCN); RP-HPLC-UV (octyl; 240 nm), MeOH-pH 2.6 phos. McAteer, J.A., **et al.** (1987) ClCh **33**, 1788-1790.

Cefazolin (dog serum & tissues): extraction by aq. ethanol-MeCN and ultrafiltration; RP-HPLC-UV (phenyl, with ion-pairing; 230 nm), MeCN-MeOH-SDS-TEA-phosphoric acid; sens. ~10 ng/ml. Tyczkowska, K., **et al.** (1987) J. Liq. Chromatog. **10**, 2613-2624.

Ceftamet pivoxil (serum & urine): serum, pptn. (PCA); RP-HPLC -UV (265 nm), MeCN-pH 6.5 phos. (ester) or MeCN-aq. PCA (parent drug). Koup, J.C., **et al.** (1988) AnMcAgCh **32**, 573-579.

Cefixime (biological fluids): pptn. (TCA); RP-HPLC-UV, MeCN-phos. Falkowski, A.J., **et al.** (1987) J Chr **422**, 145-152.

Cefmenoxime (rat bile): RP-HPLC-UV (254 nm), MeCN-AcOH. Kita, Y., **et al.** (1986) AnMcAgCh **17**, 205-213.

Cefmetazole: pptn. (MeOH-TCA); RP-HPLC-UV (254 nm), MeCN-pH 5.4 buffer; sens. 0.1 μg/ml. Rodriguez-Barbero, J., **et al.** (1985) AnMcAgCh **28**, 544-547.

Cefonicid (plasma & urine): pptn. (plasma; MeCN & TCA, then LE to wash); RP-HPLC-UV (ion-pair; 254 nm), MeCN-buffer; sens. 0.1 μg/ml. Fillastre, J.P., **et al.** (1986) J. Antimicrob. Chemother. **18**, 203-211. Cf. Scaglione, F., IJClPh **9**, 313-317 (bioassay).

Cefotaxime & desacetyl (active **metab.**): pptn. (isoPrOH; LE; RP-HPLC-UV (240 nm), MeOH-pH 5.1 buffer-TBAH; sens. 40 (**metab.**) or 200 μg/ml. Trang, J.M., **et al.** (1987) J. Pharm. Sci. **76**, S16.

Same; also **ceftriaxone**: pptn. (MeOH); RP-HPLC-UV (254 nm), MeOH-pH 4.5 phos.; sens. 0.1 (**metab.**) or 0.25 μg/ml. Hakim, L., **et al.** (1988) J Chr **424**, 111-117.

Cefoperazone + **sulbactam**: LE (HCl, Et acetate; then into pH 7 phos.); RP-HPLC-UV (cyano; 215 nm), MeCN-phos.-TBAH. Reitberg, D.P., **et al.** (1988) AnMcAgCh **32**, 42-46 [& see 51-56: Johnson, C.A., **et al.** (same lab.): phenyl col.].

Cefoperazone (serum & urine): pptn. (serum; MeOH); RP-HPLC-UV (254 nm), MeCN-pH 4 buffer. Feldman, S., **et al.** (1986) AnMcAgCh **30**, 874-876. **Cefotiam** & isomer: Kees, F., **et al.** (1990) J Chr 484-489.

Cefotetan (isomers & tautomer separable): pptn. (MeCN); RP-HPLC-UV (254 nm), MeCN-phosphoric acid. Browning, M.J., **et al.** (1986) AnMcAgCh **18**, 103-106.

Cefpirome: pptn. (MeCN); RP-HPLC-UV (270 nm), MeOH-AcOH. **Bioassay** comparison: good accord. Kavi, J., **et al.** (1988) J. Antimicrob. Chemother. **22**, 911-016.

Cefpiramide (plasma & urine) - micro-method, col. switching: take to pH 5; 2-step RP-HPLC-UV (270 nm), pH 3.0 phos.-TEA & (final stage) MeCN; sens. 0.25 μg/ml. Demotes-Mainard, F., **et al.** (1987) J Chr **419**, 388-395. *Also* (1988) J. Pharm. Biomed. Anal. **6**, 407-413:- **ceftriaxone**: similar HPLC method (but not switching); sens. 0.5 μg/ml. *See also* J. Chr. **490**, 115-123 (1989).

Cefpiramide (biological samples): pptn. (serum; MeCN, & LE to wash); RP-HPLC-UV (254 nm), aq. ammon. acetate-MeCN. Brogord, J.M., **et al.** (1988) AnMcAgCh **32**, 1360-1364.

Cephalexin & **cadralazine**: microbore RP-HPLC-UV (254 nm), MeOH-phosphoric acid (ion-pair agent helpful), or (cadralazine) MeCN-phos. buffer. Rouan, M.C. (1988) J Chr **426**, 335-344.

Cephoridine: pptn. (MeOH); RP-HPLC-UV (254 nm), MeCN-buffer. Hayashi, T., **et al.** (1988) AnMcAgCh **32**, 912-918.

Penems (various) & **metabs.**: RP-HPLC-UV (254 or 320 nm): [**1**] MeCN-pH 4.1 buffer; [**2**], incl. urine: pptn. (plasma; ammon. sulphate); C-8, pH 6 phos.; sens. 0.4 μg/ml; [**3**] (urine, with radiolabel) C-8, MeCN-pH 2.5 phos. [**1**] Lin, C., **et al.** (1986) pp. 83-90 in book ref. at start of p. 231. [**2**] Godbillon, J., **et al.** (1988) J Chr **427**, 269-276; [**3**] Benedetti, M.S., **et al.** (1989) J. Antimicrob. Chemother. **23 Suppl. C**, 173-177.

Thienamycin & **cephamycin C**: RP-HPLC, 38°, phos. Williamson, J.M., **et al.** (1985) J. Biol. Chem. **260**, 4637-4647: *cited in* AnMcAgCh **28**, 478-484 (consult this too).

N-Formimidoyl-thienamycin (β-lactamase resistant): pptn. (by MeCN); RP-HPLC-UV (300 nm), pH 7.2 borate; sens. 0.8 μg/ml (suits paediatric samples). Bloh, A.M., **et al.** (1986) J Chr **375**, 444-450.

Glycosylated compounds

Clindamycin (a lincomycin): pptn. (MeCN); RP-HPLC-UV (198 nm), MeCN-pH 6.7 phos.-TMAC. Flaherty, J.F., **et al.** (1988) AnMcAgCh **32**, 1825-1829.

Gentamicin: pptn. (MeCN, & LE to wash); deriv. (with benzene sulphonyl chloride, to improve UV absorbance); ⇾dry; RP-HPLC-UV (230 nm), MeCN-DCM-water-MeOH; sens. 0.2 μg/ml. Good agreement with EMIT method. Larsen, N-E., **et al.** (1980) J Chr **221**, 182-187.

Gentamicin components: deriv. (OPA); silicic acid col., MeOH to elute; RP-HPLC-fluor, MeOH-TEA; sens. 0.08 μg/ml for each. Rumble, R.H. & Roberts, M.S. (1987) J Chr **419**, 408-413.

Kanamycin & **dibekacin:** pptn. (PCA & Na octanesulphonate); RP-HPLC-fluor (post-col. deriv. with OPA), MeCN-pH 3.5 aq. sulphonates. Kubo, H., **et al.** (1985) AnMcAgCh **28**, 521-523.

Nikkomycin antifungal components (from microbial culture media): cation-exchange chromatography, & eluate concentrated; RP-HPLC-UV (290 nm), pH 4.7 formate with heptane sulphonic acid. Engel, P., **et al.** (1989) Prep. Biochem. **19**, 321-328.

Sisomicin*& **gentamicin:** RP-HPLC-UV, and comparison with **bioassay.** Barends, D.K., **et al.**, (1981) JChr **222**, 316-323.

Teicoplanin glycopeptide components: pptn. (MeCN); deriv. (fluorescamine; obviates need for a gradient); RP-HPLC-fluor, pH 7.5 'Waters PicA'-MeOH-nBuOH; sens. 2.5 μg/ml. Joos, B. & Lüthy, R. (1987) AnMcAgCh **31**, 1222-1224 [also 1255-1262: Falcoz, C., **et al.**].

Vancomycin (urine) + **ceftazidime:** HPLC; sens. 0.8 μg/ml for latter. Boeck, M., **et al.** (1988) AnMcAgCh **32**, 92-95 (*& see* **24**, 333; *also*, for vancomycin, McClain **et al.**, J Chr **231**, 463).

--------- [**& see* J Chr **425**, 143-152

Macrolides

Erythromycin & its esters: LE (ether); ⇾dry; RP-HPLC-EC (amperometry), MeCN-MeOH-pH 7 buffer; sens. 10 ng. Croteau, D., **et al.** (1987) J Chr **419**, 205-212.

Erythromycin & ester: SPE; RP-HPLC-EC, MeCN-pH 6.3 phos.; sens. ~0.2 μg/ml. Stubbs, C. & Kanfer, I. (1988) J Chr **427**, 93-101.

Erythromycin & **roxithromycin:** LE; ⇾dry; RP-HPLC-EC (C-8), MeCN-pH 6.7 phos.; sens. 0.01 μg/ml. Grgurinovich, N. & Matthews, A. (1988) J Chr **433**, 298-304.

Josamycin (plasma & blood cells), automated method: SPE (water to elute); RP-HPLC-UV (230 nm) with a pre-col.; sens. 5 ng on-col. Rader, K., **et al.** (1985) J Chr **344**, 416-421.

Spiramycin: SPE (MeOH to elute); ⇾dry; RP-HPLC-UV (231 nm); sens. 50 ng/ml. Carlhant, D., **et al.** (1989) Biomed. Chromatog. **3**, 1-4.

Tylosin (muscle): pptn.; LE; RP-HPLC-UV (280 nm), MeCN-MeOH-phos.; sens. <0.1 µg/g, better than for **bioassay** (poor accord). Moats, W.A., **et al.** (1985) J. Assoc. Off. Anal. Chem. 413-416.

Cytotoxic / Immunosuppressive agents

For anti-cancer drugs consult Vol. 14 too*

Adriamycin: add a volatile alkaline buffer and iPrOH; centrifuge; LE (chloroform); >dry; RP-HPLC. Van Lancker, M.A., **et al.** (1986) J Chr **374**, 415-420. (*Synonym:* **Doxorubicin.**)

Adriamycin & **metabs.** (monitoring): RP-HPLC-fluor, MeCN-pH 2.3 formate; sens. ~1 ng/ml (for metabs. also). Mahdadi, R. **et al.** (1987) Biomed. Chromatog. **2**, 38-40.

Azathioprin & **6-mercaptopurine** (6MP; **metab.**), conjointly by a novel approach: deriv. (6MP; by NEM); LE (Et acetate); >dry; RP-HPLC-UV (280 nm), MeCN-phos.; sens. <10 ng/ml. The deriv. step renders 6MP extractable and measurable at 280 nm. Nabi, E.M. & Rosen, A. (1986) pp. 237-243 in *Development of Drugs and Modern Medicines* (Gorrod, J.W., **et al.**, eds.), Horwood/VCH.

Ciclosporin assay: Schran, H.F., Robinson, W.T., Abisch, E. & Niederberger, W. (1986) Prog. Allergy **38**, 73-92. (Review.)

Particular cyclosporin A (ciclosporin) refs. (& see p. 229):-

[1] **Serum:** pptn. (MeCN-phosphoric acid); col. switching RP-HPLC-UV (214 nm): C-8, then C-2 col., 70°, aq. MeCN. Gmur, D.J., **et al.** (1985) J Chr **344**, 422-427; also **425**, 343-352:- blood, 2 **metabs.**; sens. 20 ng/ml. [2] **Plasma/Blood** distribution (esp. **metabs.**), e.g. temp. influence: SPE; RP-HPLC. Lensmeyer, G., **et al.** (1989) ClCh **35**, 56-63, *also* **33**, 196-201 & (blood, **metabs.**) 1841-1850, and **34**, 1269). *See also* J. Chr. **413**, 131-150.

Blood as matrix.- [1] Pptn. (MeCN); charcoal adsorption, Et acetate elution, >dry; RP-HPLC-UV (206 nm), 75°, MeCN-MeOH-water; sens. 50 ng/ml. Aravind, M.K., **et al.** (1985) J Chr **344**, 428-432. [2] Acidify, then LE (Me-tBu ether - novel; then wash); >dry; RP-HPLC-UV (C-8, microbore; 214 nm); sens. allowed small sample vols. Annesley, T., **et al.** (1986) ClCh **32**, 1407-1409. [3] With standards isolated by LE/SPE/RP-HPLC from bile, 9 **metabs.** were assayed in blood. Lensmeyer, C.L., **et al.** (1987) ClCh **33**, 1841-1850. [4] Automated assay ('AASP') for monitoring: pptn. (MeCN); centrifuge; SPE (C-8); RP-HPLC-UV (210 nm; 53°), EtOH-hexane, re-cycled; sens. 12.5 ng/ml (0.2 ml sample). Lachno, D.R., **et al.** (1990) J Chr **525**, 123-132.

Idarubicin (anti-cancer) & **metab.**: LE (chloroform-1-heptanol); RP-HPLC-fluor (C-2 col.), MeCN-phosphoric acid; sens. 1 ng/ml. Eksborg, G. & Nilsson, B. (1989) J Chr **488**, 427-434.

Suramin (patients): pptn. (MeOH-TBAB); centrifuge; RP-HPLC-UV (313 nm), MeOH-pH 7.5 phos.-TBAB. Sens. ~0.1 µg/ml. Tjaden, U.R. **et al.** (1990) J Chr **525**, 123-132,

Vinca alkaloid: **navelbine** & **metab.** (desacetyl): RP-HPLC (cyano col.). Jehl, F., **et al.** (1990) J Chr **525**, 225-233.

Abbreviations *(incl. Journals): foot of pp. 230 & 231*

For* **5-FU *see p. 237*

***Quinolines / Various antimalarials** (& see art. #B-6)*

Review - **antimalarials** & **metabs.**: amodiaquine, chloroquine, mefloquine, primaquine, proguanil, pyrimethamine & sulphadioxime. Monitoring context; HPLC generally the preferred approach. Bergqvist, Y. & Churchill, F.C. (1988) J Chr **434**, 1-20.

Quinine: pptn. (MeCN); RP-HPLC-fluor (nitrile col.); sens. 50 ng/ml. Rauch, K., **et al.** (1988) J Chr **430**, 170-174.

Dabequin (blood & urine): 2-step LE (chloroform); GC-NPD; sens. (blood) 0.05 µg/ml. Tapanes, R.D. (1989) J Chr **493**, 202-206 [cf. 196-201:- analogous assay for **chloroquine**, not lucidly presented].

Mefloquine (biol. fluids): LE (pptn.; DCM); deriv. with phosgene for capillary GC-ECD, 265°; sens. 10 ng/ml. Bergqvist, Y., **et al.** (1988) J Chr **428**, 281-290.

Difloxacin & **metabs.** (body fluids): add SDS and ultrafilter (serum); RP-HPLC-fluor or -UV (280 nm), MeCN-phos.-SDS (serum) or pH 5.3 phos. (urine). Granneman, G.R., **et al.** (1986) AnMcAgCh **30**, 689-693. **Temafloxacin** & 2 other fluoroquinolones (biological fluids): 2-step LE (DCM); RP-HPLC-fluor, MeCN-phos.-TBAB; sens. (serum) 0.01 µg/ml. Koechlin, C., **et al.** (1989) J Chr **491**, 379-387.

Nitroxoline (8-hydroxy-5-nitroquinoline; plasma & urine): add HCl (and, for urine, hydrolyze), then LE (chloroform); →dry; RP-HPLC-absorbance (436 nm), MeOH-pH 7.4 phos., + analogue (anti-tailing); sens. (plasma) 80 ng/ml. Sorel, R.H.A., **et al.** (1981) J Chr **222**, 241-248. *A cogent study; amplification on p. 372.*

Arteether (antimalarial, related to Qinghaosu), rat plasma: LE (CH_2Cl_2); →dry; cleavage step (HCl-MeOH; 53°, 15 min); RP-HPLC-UV (254 nm), MeCN-water; sens. <50 ng/ml. Idowu, O.R., **et al.** (1989) J Chr **493**, 125-136.

Halofantine (antimalarial) & **metab.**: pptn.; SPE; RP-HPLC-fluor (C-8;), MeCN-AcOH-Na lauryl sulphate; sens. 1 ng/ml. Gawienowski, M., **et al.** (1988) J Chr **430**, 412-419.

Various antibacterials

Chloramphenicol & ester (biological samples): [ref. 1] LE (Et acetate), →dry; [ref. 2] directly onto guard-col.; RP-HPLC-UV ([1] 272, [2] 280 nm), [1] MeCN-pH 6.4 buffer, [2] MeCN-AcOH; [1] sens. 5 µg/ml (sample only 50 µl). [1] Aravind, M.K., **et al.** (1980) J Chr **221**, 176-181; [2] El-Yazigi, A., **et al.** (1987) ClCh **33**, 1814-1816. *Also* Weber, L. (1990) J Chr **525**, 454-458.

Chloramphenicol: EMIT assay accorded with RP-HPLC-UV (214 & 254 nm, disclosing any interference), MeOH-phos. White, L.O., **et al.** (1988) J. Antimicrob. Chemother. **21**, 673-686 (& accompanying 'letters').

....

***Abbreviations** (incl. Journals): at foot of pp. 230 & 231*

Cilastatin (antibacterial) & **metab.** (body fluids): SPE; RP-HPLC-UV (210 nm), MeCN-phos.-octane sulphonic acid; sens. 1 μg/ml. Hsieh, J.Y-K., **et al.** (1987) J. Pharm. Sci. **76**, S12.

Doxycycline (serum & urine): SPE (C-18); RP-HPLC-UV (340 nm), MeCN-AcOH-phos.; sens. ~25 ng/ml. Sheridan, M.E. & Clarke, G.S. (1988) J Chr **434**, 253-258.

Nalidixic acid & active **metab.**: LE ($CHCl_3$); HPLC-'dynamic IEC'; sens. copes with minimal sample size. *See p. 372.* Sorel, R.H.A. Hulshoff, A. & Snelleman, C. (1980) J Chr. **221**, 129-137.

Rifampicin (rifampin; antitubercular) & **metabs.**: LE (chloroform); ➾dry; RP-HPLC-UV (340 nm), MeCN-pH 4 phos.; sens. 0.1 μg/ml. Ishii, M. & Ogata, H. (1988) J Chr **426**, 412-416 (*& see* **525**, 495-497).

Sulphasalazine (antibacterial) & **metabs.**: HPLC, Hypersil-MOS col. (fluor & diode-array UV), MeOH-phos. stepwise gradient. Rona, K., **et al.** (1987) Chromatographia **24**, 720-724.

Various antifungals and antiparasitics (*see also p. 235*, **Suramin**)

Amphotericin B (a polyene antifungal): pptn. ([ref. 1] MeCN; [2], MeOH); [1] HPLC-UV with 'CLC'-TMS col. (405 nm), MeCN-pH 7.4 acetate buffer; [2] RP-HPLC-UV (405 nm), MeOH-MeCN-EDTA; sens. [1] adequate for monitoring; [2] 0.01 μg/ml. [1] Hosotsubo, H., **et al.** (1988) AnMcAgCh **32**, 1103-1105; [2] Brassinne, C., **et al.** (1987) J Chr **419**, 401-407.

Flucytosine (antifungal): an enzymic assay (rapid; NH_3 split off by creatine iminohydrolase) was compared with pptn. (TCA) followed by RP-HPLC-UV (276 nm). Washburn, R.G., **et al.** (1986) J. Antimicrob. Chemother. **17**, 673-677.

Named **S-containing compounds** (antifilarial) & **metabs.** (including an active isothiocyanate) whose moieties include (1) piperazine *or* (2) benzthiazole *or* (3) piperidinylbenzthiazole (biological fluids): pptn. (MeCN); RP-HPLC-UV *or* (2) NP-HPLC-UV, MeCN *alone* (3), *or* (1) + MeOH-ammonia, *or* (2) + water; sens. (1) 25 ng/ml; (3) 5 ng/ml. Bhatia, S., **et al.** (1988) J Chr **434**, (1) 288-295; (2) 296-302; (3) 303-307. *Study in labs. of Ciba-Geigy, India.*

Antivirals of nucleoside or related type

A **guanine derivative** (anti-herpes), (*R*,*S*)-9-[4-hydroxy-2-(hydroxymethyl)butyl]guanine: pptn.; RP-HPLC heart-cut (pH 2 phos.) re-run on coupled col.:- RP-HPLC-fluor, pH 2 phos.-MeOH-decyl sulphate. Lake-Bakaar, D.M., **et al.** (1988) AnMcAgCh **32**, 1807-1812.

Bromovinyldeoxyuridine (BVDU; anti-herpes; serum, urine & infected cells): dilution (urine) or pptn. by PCA; RP-HPLC-UV (292 nm; 48°), gradient: pH 5 phos.-TBAB and inflowing MeOH; sens. (serum) 0.64 μM. Ayisi, N.K., **et al.** (1986) J Chr **375**, 423-430.

5'-Deoxy-5-fluorouridine (DFUR; pro-drug for **5-fluorouracil**, 5-FU) and **BVDU** *as above* [latter studied as possible anti-cancer potentiator; all 3 agents assayed]: LE (Et acetate); ➾dry; RP-HPLC-UV (~260 nm), aq. MeOH, with gradient after 10 min. Iigo, M., **et al.** (1989) Biochem. Pharmacol. **38**, 1885-1889.

Azidothymidine (Zidovudine; cf. arts. #B-11 & -12; plasma & urine).- [1] LE (Et acetate-ether); →dry; RP-HPLC-UV (266 nm), MeCN-phos. Unadkat, J.D., **et al.** (1988) J Chr **430**, 420-423.
[2] (incl. **metab.**, GAZT): col.-switching variant of method of Good, art. #B-11. Lacroix, C., **et al.** (1990) J Chr **525**, 240-245.
[3]: pptn. (PCA) or on-line SPE (pH 5, polymer); on-line trace enrichment (pH 11.6, Ag-loaded thiol polymer); RP-HPLC-UV (267 nm); sens. 10 nM. Irth, H., ...& Frei, R.W. (1989) J Chr **491**, 321-330.

Carbovir: SPE; RP-HPLC-UV (252 nm), MeOH-pH 7 phos.-TEA; sens. 50 ng/ml. Remmel, R.P., **et al.** (1989) J Chr **489**, 323-331.

Ribavirin (a ribotriazole carboxamide; biol. fluids & lung homogenates): ultrafilter; affinity chromatography on phenyl boronate gel (elute with dil. formic acid); →dry; RP-HPLC-UV (207 nm), pH 5.1 phos.-MeOH (1%); sens. 0.1 ug/ml. Smith, R.H.A. & Gilbert, B.E. (1987) J Chr **414**, 202-210; for application to tissue extracts see AnMcAgCh **32**, 117-121 (1988). For **ribavirin** pharmacodynamics in AIDS patients see Roberts, R.B., **et al.** (1987) Clin. Pharmacol. **42**, 365-373.

Various antivirals

Amantadine (biol. samples): LE (Et acetate) at pH 10.5; deriv. to isothiocyanate (CS_2); wash with aq. acid; →dry; GC-MS (SIM); sens. presumably adequate for pharmacokinetics as well as monitoring. Narasimhachari, N., **et al.** (1979) Chromatographia **12**, 523-526.

Enviroxime (a benzimidazole; pulmonary route): RP-HPLC-UV (215 nm), MeCN-water. Wyde, P.R., **et al.** (1988) AnMcAgCh **32**, 890-895.

Foscarnet (Na_3 phosphonoformate hexahydrate; anti-herpes): GC-MS (few details); sens. 0.05 µg/ml. Ringden, O., **et al.** (1986) J. Antimicrob. Chemother. **17**, 373-387. *Also* (plasma & urine, with precautions against adsorption and any contaminating HIV): ultrafilter; pptn. (EtOH); RP-HPLC-EC, MeOH-pH 5.8 phos.-THAHS; sens. 33 µM. Hassanzadah, M., **et al.** (1990) J Chr **525**, 133-140.

Section #C

APPROACHES FOR VARIOUS DRUGS AND METABOLITES

#C-1

ON THE SELECTIVITY OF SOME RECENTLY DEVELOPED RIA's

R. Woestenborghs, I. Geuens, H. Lenoir, C. Janssen and J. Heykants

Department of Drug Metabolism and Pharmacokinetics, Janssen Research Foundation, B-2340 Beerse, Belgium

Chromatographic determination of a drug in plasma is sometimes unachievable if the drug has unfavourable physico-chemical or pharmacokinetic characteristics resulting in either an insensitive assay method or very low plasma concentrations. An immunoassay may then be the only alternative. Some of our recently developed RIA's are discussed, mainly with regard to selectivity towards optical isomers and metabolites.*

BIOANALYTICAL TECHNIQUES: SELECTION CRITERIA

The choice of assay approach is governed by the physico-chemical properties of the drug (UV-spectrum, mol. wt., molecular structure, volatility, extractability) and by its pharmaco-dynamics (potency, clearance, metabolism). Most assays in our laboratory, in order of preference, entail HPLC, GC or RIA. HPLC is versatile, can be automated, and can handle all non-volatile, large and unstable molecules; it is precluded only when high sensitivity is needed, especially for molecules with poor spectral properties. GC, although semi-eclipsed by HPLC during the last two decades, remains useful for thermostable and relatively volatile compounds, is very selective and sensitive and, advantageously, is linkable to MS.

RIA is often the only possible way out when chromatographic methods fail. Once developed - this being very time-consuming, and critical in comparison to a chromatographic procedure - an RIA offers many advantages. Detection limits are sometimes in the lower pg/ml range and the analysis time per sample is drastically reduced. The assay selectivity is determined by the nature of the Ab's but may be varied by the use of different sample extraction procedures and even by the assay conditions. Selectivity towards metabolites, active or otherwise, may be an advantage or a disadvantage, depending on the goal of the plasma level determination; at any rate the assay specificity of the overall procedure must be settled for both metabolites and optical or other isomers. Over 15 years we have developed some 20 RIA's,

---

**Abbreviations*.- RIA, radioimmunoassay; Ab, antibody.

and used them to elucidate the pharmacokinetic behaviour of our compounds, now listed together with the RIA-development year (and, parenthetically, any relevant analogues): neuroleptic: **haloperidol**, 1976 (bromperidol, moperone, trifluperidol); narcotic analgesics: **fentanyl**, 1976, and **bezitramide**, 1977; neuroleptic: **benperidol**, 1977 (droperidol, spiperone); major tranquillizer: **pimozide**, 1975 (fluspirilene, clopimozide); antihelminthic: **mebendazole**, 1978 (flubendazole); antidiarrhoeal: **loperamide**, 1977 (cf. its oxide); narcotic analgesics: **alfentanil** and **sulfentanil**, 1980; gastrokinetic: **domperidone**, 1977; anti-allergic: **oxatomide**, 1977; antifungal: **ketoconazole**, 1983 (terconazole); antihypertensive: **ketanserin**, 1979; antihistaminics: **astemizole**, 1981, and **levocabastine**, 1983; antidiarrhoeal: **loperamide oxide**, 1988; antipsychotic: **risperidone**, 1988; β_1-blocker: **nebivolol**, 1989.

Some of the more recently developed RIA's will be discussed further, mainly with regard to their selectivity towards optical isomers and metabolites.

RIA: GENERAL DEVELOPMENT STRATEGY⊗

Taking account of the drug's structure and known or predicted metabolism, a hapten is made (commonly by joining on a primary amine or carboxylic acid by a short alkyl chain, n = 2-4); after careful inspection for purity and complete dissolution, it is conjugated (our routine being the carbodiimide-BSA procedure). Rabbits are immunized by regular intradermal injections of the conjugate emulsified in Freund's adjuvant, with monitoring of Ab titre, sensitivity and selectivity using a suitable ^{3}H-tracer. The finally selected or pooled antisera are freeze-dried and stored at 4°.

Developing and validating a practical RIA with the antiserum involves careful selection of incubation (shaking, pH, temp., time) and separation method (charcoal) for bound/unbound radioactivity, determination of method accuracy, precision and sensitivity, determination of specificity towards stereoisomers and metabolites and, last but not least, the dilution test. With this strategy (more details in [1]) generally we have succeeded in raising Ab's usable in effective assay procedures. Insolubility of the hapten during conjugation prevented production of useful Ab's in only two cases, and in one was remedied by using cyclodextrins to enhance hapten solubility.

SOME RECENTLY DEVELOPED RIA's

Astemizole

Astemizole (see Fig. 1) is a non-sedative and potent, long-acting H_1-antagonist, undergoing in man first-pass metabolism

⊗*Note by Ed.*- There are pertinent arts. in earlier vols., e.g. #2, #16 & #18: helpful Index entries include Antisera, Conjug....., RIA. See also this vol., art. #B-12.

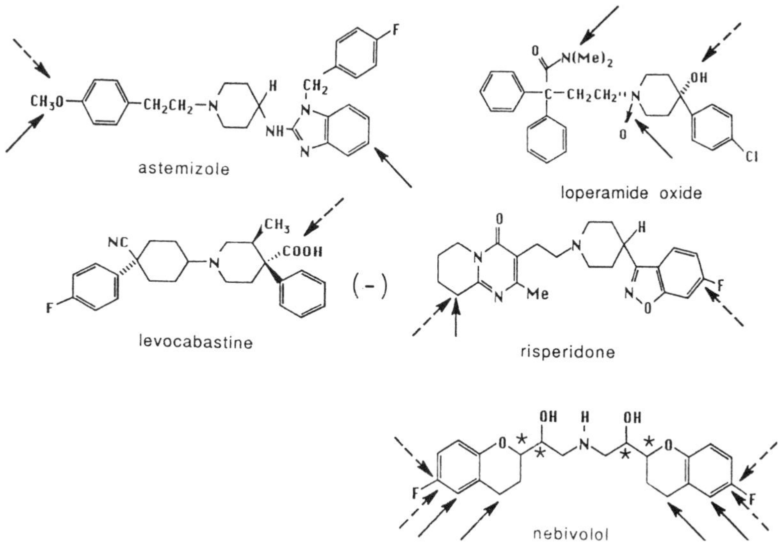

Fig. 1. Chemical structures, with *arrows* showing sites of metabolism (⟶) and hapten coupling(--->); *, a chiral centre.

to form desmethylastemizole which has a comparable pharmacological profile: the respective therapeutic plasma levels are 0.1-1 and 1-12 ng/ml. As the HPLC detection limit was 1-2 ng/ml [2], an RIA was developed [3]. The hapten-BSA linkage site was the 4-phenyl position (Fig. 1). Hence the cross-reactivity was the same whether the plasma was assayed direct or, as routinely done, was extracted with 1.5% (v/v) isoamyl alcohol in n-heptane at pH 7.8. The RIA measures the sum of astemizole and desmethylastemizole, which represents the total 'antihistaminic' plasma level.

For pharmacokinetic purposes, however, a discrimination between the unchanged drug and the metabolite was necessary. As desmethylastemizole is a phenolic compound, it was separable almost completely from the parent drug by extracting the latter at pH 12.5 as an additional step, leaving the phenolic metabolite in the aqueous phase; the difference in RIA values gave the metabolite concentration.

Levocabastine

Levocabastine (Fig. 1) is a potent and specific H_1-antagonist used for topical treatment of allergic rhinitis (nose spray; plasma level 2-10 ng/ml) or conjunctivitis (eye drops; plasma, 0.2-1 ng/ml); its half-life is in the range 30-45 h.

As chromatographic properties are very unfavourable, an RIA was established; this could be direct, since the laevorotatory (*cis*) isomer is dosed and it undergoes negligible metabolism. The Ab's nevertheless proved to be specific in tests with some postulated metabolites ($ID_{50} \geq 500$) and the stereochemical and optical isomers. The dextrorotatory *cis*-isomer still showed ~7% cross-reactivity, but the racemate of the *trans* isomers, with the fluorophenyl in the axial position relative to the piperidine ring, hardly bound to the Ab's (ID_{50} ~500).

Loperamide oxide

This pro-drug (as marketed; Fig. 1) of the antidiarrhoeal loperamide is reduced mainly in the gut contents and gut wall, giving longer exposure of the gut wall receptors to the active drug and diminished (50%) systemic absorption compared with loperamide. After a therapeutic dose of 4 mg, plasma levels of the pro-drug are <2 ng/ml and decline rapidly (half-life ~1 h). The liberated loperamide ranges from 0.05 to 0.4 ng/ml plasma and the half-life from 12 to 30 h. Loperamide is not amenable to GC or HPLC, and an RIA has been established [4].

The pharmacokinetics of loperamide and its oxide were studied by two approaches, one using the original loperamide Ab's. As cross-reactivity of loperamide oxide to these Ab's is negligible, a first RIA measures loperamide concentrations; then the sum of loperamide and its oxide is obtained by RIA after extraction and chemical reduction by $TiCl_3$: the difference gives the plasma levels of the oxide. The approach proved to be useful, but several precautions had to be taken to avoid inadvertent or insufficient reduction of the oxide.

In a second, more recent approach, specific Ab's to loperamide oxide were obtained, through conversion of the originally used hapten (the 4-oxobutanoic acid analogue) to the corresponding *N*-oxide by *m*-chlorobenzoic acid oxidation. The resulting hapten and BSA-conjugate proved to be stable towards reduction, and useful Ab's were obtained for the direct determination of the pro-drug; cross-reactivity of loperamide proved to be only ~1%.

Risperidone

This new neuroleptic (Fig. 1) with both S_2- and dopamine antagonism undergoes hydroxylation at the tetrahydro-pyrido-pyrimidine moiety yielding 9-OH-risperidone as the major metabolite with a pharmacological activity similar to the parent compound. Plasma levels of risperidone and its metabolite are in the lower ng/ml range, especially in pharmacokinetic studies in healthy volunteers, and the analysis by HPLC or GC failed because of insufficient sensitivity or low volatility. As human metabolism takes place only at one side of the molecule (Fig. 1), two different antisera were prepared. They were obtained from haptens linked to the parent molecule at the same site as the metabolite and at the opposite site, viz. at the 6-position of the benzisoxazolyl moiety.

Obviously the first antisera do not discriminate between the parent drug and the metabolite, and measure the total neuroleptic plasma level. The method is now being used routinely for monitoring psychotic patients. The second RIA was aimed at specific determination of the unchanged drug and used the Ab's obtained from the hapten at the opposite position. The cross-reactivity of the 9-OH metabolite was found to be $^1/_{15}$, which was considered as unacceptable, the more so as metabolite concentrations were found to be up to 10-fold the concentrations of the parent drug. A cross-reactivity of $^1/_{120}$ could be obtained by adapting the assay conditions (pH, incubation time) and applying an additional extraction step with a non-polar solvent prior to the RIA. Another approach to increase the specificity was the application of RIA to collected HPLC fractions. The latter method was found to be very susceptible to changes in chromatographic retention times, and as the metabolite (more polar) eluted prior to the unaltered compound, it still contaminated the latter fraction to some extent.

Nebivolol

This new antihypertensive drug (Fig. 1), under clinical investigation, is a racemate: *d*-nebivolol (*SRRR* configuration) is a β-blocker, and *l*-nebivolol (*RSSS*) has a positive inotropic effect. In man, nebivolol is metabolized by hydroxylation at different aromatic or alicyclic positions yielding numerous hydroxylated metabolites. Plasma levels of unchanged nebivolol are very low, and could only be measured by HPLC with fluorescence detection [5]. This approach, however, failed to measure the concentrations of the many polar metabolites and could not discriminate between the two enantiomers in the sub-ng/ml range. Furthermore, correlation between the plasma levels assayed by HPLC and pharmacological activity (blood pressure) was very poor, indicative of the presence of active metabolites. As a follow-up to the HPLC method for the unchanged drug we therefore had to develop an RIA capable of measuring fractions of active metabolites in patient samples and also discriminating, to a certain degree, between the different optical isomers and their metabolites.

As the RIA should be able to discriminate, for both enantiomers, between metabolites formed at the left- or right-hand side of the molecules, it was obvious that 4 different haptens had to be synthesized. This proved to be a critical and time-consuming phase, especially as, at different synthesis stages, the reaction products had to be separated into pure optical isomers. Finally, the 4 desired haptens, all 6-(2-aminoethyl)nebivolol analogues, were synthesized and proved to be optically pure. The resulting 4 Ab's could discriminate between the enantiomers (cross-reactivity 1-1.5%) and the opposite hydroxylated metabolites (cross-reactivity ~10%).

CONCLUDING REMARKS

Although often considered as 'black magic', RIA can be a powerful and flexible tool for the bioanalyst. Selection and synthesis of the hapten is crucial for assay development and performance. The preparation of useful Ab's is likely to succeed if the above guidance is followed. The final RIA development should be governed by the pharmacokinetic demands (measurement of unaltered drug ± active metabolites).

Empirically we found that the specificity of the Ab's was satisfactorily high ($ID_{50} \geq 100$) towards stereoisomers and for hydroxylated metabolites (often important) good to moderate (ID_{50} 10-20), increasable by additional extraction steps. Altogether cross-activity is a key concept in RIA methodology and, when used judiciously, adds to the flexibility of the method: it often provides the bioanalyst with the only possible breakthrough in the increasingly demanding fields of pharmacokinetics and drug monitoring.

References

1. Harlow, E. & Lane, D. (1988) in *Antibodies - a Laboratory Manual* (Cuddihy, J., ed.), Cold Spring Harbor Lab., New York, pp. 53-139.
2. Woestenborghs, R., Embrechts, L. & Heykants, J. (1983) *J. Chromatog. 278*, 359-366.
3. Woestenborghs, R., Geuens, I., Michiels, M., Hendriks, R. & Heykants, J. (1986) *Drug Development Res. 8*, 63-69.
4. Michiels, M., Hendriks, R. & Heykants, J. (1977) *Life Sci. 21*, 451-460.
5. Woestenborghs, R., Embrechts, L. & Heykants, J. (1988) in *Bioanalysis of Drugs and Metabolites, especially Anti-Inflammatory and Cardiovascular* [Vol. 18, this series] (Reid, E., Robinson, J.D. & Wilson, I.D., eds.), Plenum, New York, pp. 215-216.

#C-2

SOME RECENT DEVELOPMENTS IN HPLC AND KINDRED TECHNIQUES

R.J. Dolphin

Shandon Scientific Ltd.,
Chadwick Road, Runcorn, Cheshire WA7 1PR, U.K.

Advances that aid HPLC selectivity include a detector which distinguishes optically active components, and a particle beam interface whereby MS is more readily linked to HPLC. Separation selectivity, especially for isomer mixtures, may be improved by use of PGC as stationary phase.*

SFC offers a completely different range of selectivities, even when used with traditional HPLC packed stationary phases. Linking to MS or FTIR further increases the analytical power of SFC and is easier than with HPLC. Another 'chromatographic' but in fact electrophoretic technique, CE, that has come into vogue offers much promise, as is briefly reviewed.

Whilst chromatography may seem to be advancing less dramatically than 15 or 20 years ago, progress continues throughout a broad family of techniques. There have been advances that help attain selectivity, which hinges not only on between-phase partitioning but also on the detection mode. In the never-ending search for selectivity, one approach has been pre- or post-column derivatization of analytes. Such techniques, which can improve the selectivity both of extraction and of detection, have been surveyed by Brinkman in a previous volume [1; cf. #ncC-13 below]. For selective detection, fluorescence and electrochemical techniques are now widely used; but there is continuing interest in achieving even more specific detection modes and in gaining additional qualitative information from the components detected.

TRENDS IN DETECTION

Chiral detector.- With the increasing emphasis on drug enantiomers, there are interesting possibilities for a recently introduced chiral detector [2]. It has a collimated laser diode with a radiation source (820 nm), light from which

* *Abbreviations include:* CE, capillary electrophoresis; FT, Fourier-transform (IR); MS, mass spectro-metry/meter; PGC, porous graphitized carbon; RP-, reverse phase (HPLC); SFC, supercritical fluid chromatography.

is passed through a polarizing prism to produce plane-polarized light. Any optically active analyte in the flow-cell produces an amplitude modification of the carrier signal that is recovered by an amplifier/phase-sensitive detector combination. When the chiral detector is used in series with a conventional UV detector, it discloses the enantiomeric composition of a component without isomer separation. Thus, the enantiomer ratio of ibuprofen has thereby been established within ±2% without use of a chirally selective column. Sensitivity will depend on the enantiomer optical rotation values, and there could be interference by other optically active components. Yet the detector is a useful addition to the available range of selective detectors.

Coupling HPLC with NMR presents tremendous practical difficulties, but does offer the notably attractive possibility of extracting useful structural information on the solutes. In order to apply ^{1}H-NMR to column effluents, all protonated solvents must be eliminated. For an analytical study on amino acids and a group of antibiotics [3], a mobile phase consisting of a mixture of deuterium oxide and deuterated methanol was developed. Fig. 1a shows RP-HPLC-UV results for the *Streptomyces* antibiotic complex antimycin A, indicating a 2-component composition for each of the antimycins (A_1 to A_5). Figs. 1b and 1c show the on-line ^{1}H-NMR spectra for the two components of antimycin A_1. The spectra are clearly different and the chemical shifts indicated that antimycin A_{1a} has an isopentyl group while A_{1b} carries an isovaleryl substituent on the dilactone ring. Evidently HPLC-NMR can provide both quantitative information from the HPLC trace and qualitative confirmation of structural assignments from the NMR data.

HPLC-MS is another hyphenated technique that gives notable qualitative information besides quantitative data. While this technique has been available for some years, it has traditionally suffered from two significant disadvantages. Firstly, the flow rates typical in HPLC lead to large quantities of solvent vapour which must be removed from the interface to the MS. Secondly, as most HPLC-MS systems have operated in the chemical ionization mode, the resulting spectra have been rather uninformative. The recent development of the particle beam interface [4] allows the generation of classical electron-impact spectra that are library-searchable for positive compound identifications. This interface is compatible with NP or RP eluents, including gradients with typical HPLC flow-rates (0.1-1 ml/min). The interface is a simple sample transport device similar to a 2-stage jet separator, where the solvent vapour is pumped away while the analyte particles are concentrated into a beam, entering the MS source where

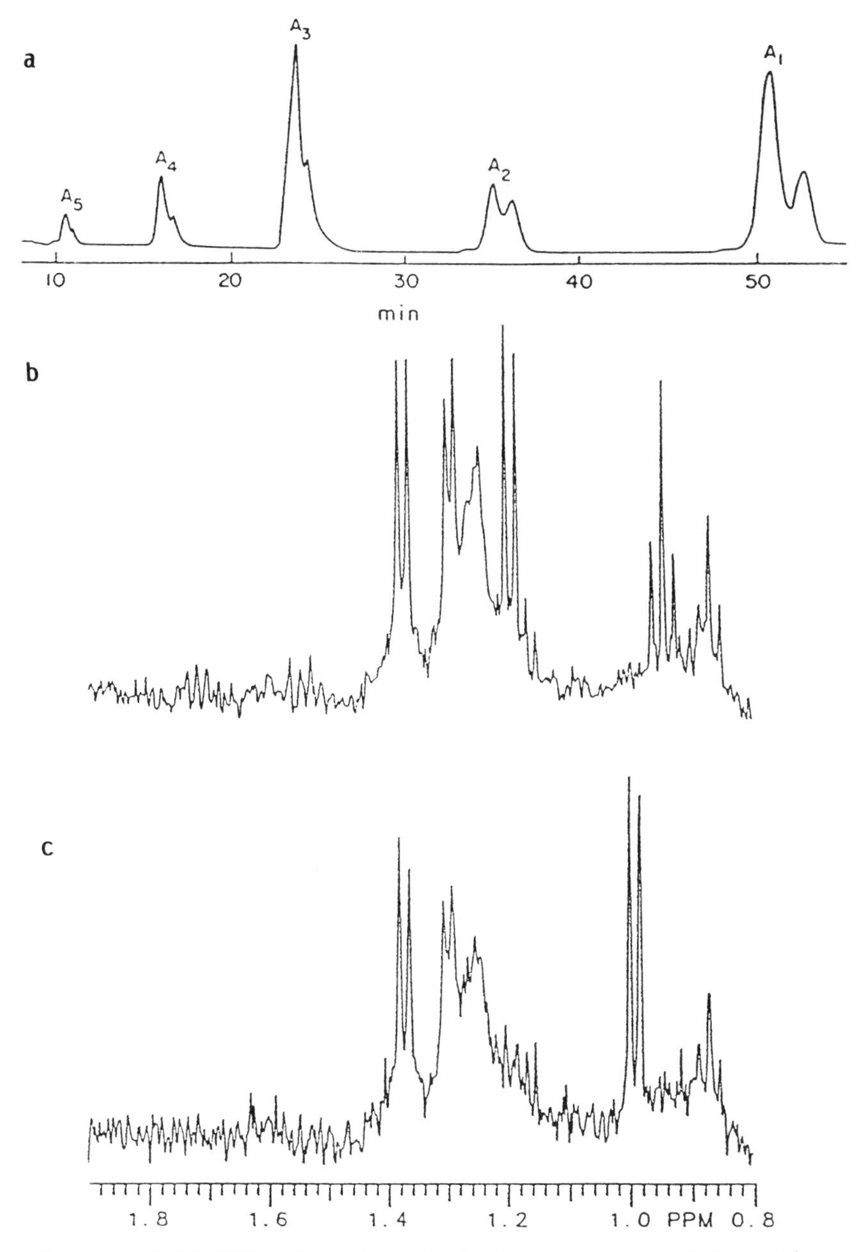

Fig. 1. HPLC-NMR of antimycin A: HPLC pattern (**a**) and ^{1}H-NMR spectra (**b**, antimycin A_{1a}; **c**, antimycin A_{1b}).
C-18 column (3 µm), eluted isocratically with methanol-^{2}H/ deuterium oxide, 75:25 by vol. at 1.5 ml/min; detection at 254 nm. On-column loads were of the order of 0.5 mg.
From [3], courtesy of the authors and the American Chemical Society.

they are vaporized and ionized by electron impact. One application has been to a mixture of corticosteroids which, in 100 ng amounts, were individually identified.

COLUMN TECHNOLOGY, PARTICULARLY POROUS GRAPHITIZED CARBON (PGC)

With HPLC, in contrast with GC, the mobile phase is involved in the separation process, whose selectivity can therefore be manipulated through the choice of mobile phase/ stationary phase system. Each year sees the advent of new stationary phases, many of which, such as chiral and affinity columns, have been developed for their selectivity towards particular types of compound. However, one recent introduction could have a more far-reaching effect. Knox and co-workers [5] have developed a PGC material in the form of spherical mesoporous particles offering a crystalline graphitic surface. Graphitized carbon provides a highly selective, homogeneous surface, utilizable across the entire range of aqueous (pH 0-14) and organic mobile phases. The material is produced by the graphitization of a phenol-formaldehyde resin-impregnated silica gel. Pore structure is imparted by dissolution of the silica gel template prior to graphitization.

While PGC has retention properties like classical RP type in nature, the graphitic structure lends the capability for substrate-solute interactions of classical electron donor-acceptor character. The graphite surface is sensitive to changes in the solute electron density caused by the electron-donating and -withdrawing ability of solute constituents and the number and position of electron-dense (Π) bonds in the solute. Furthermore, PGC is highly sensitive to steric changes that disturb the electron density of the solute molecule and the resulting interaction of the solute with the graphite surface. Hence PGC is uniquely selective to positional and stereo isomers.

Fig. 2 shows the stereoselectivity of PGC for antibiotic isomers. Diastereomeric pairs of the cephalosporin Axetil E47 (Δ-3-isomers) and synthesis by-products (Δ-2-isomers and anti-methoxy isomers) were easily resolved on PGC. Not only were the diastereomers separated for each of the three synthesis products, but both the Δ-2- and antimethoxy isomers were well separated from the desired active (Δ-3) isomers. In contrast, the Δ-2-isomers, differing from the active ones only by the location of the β-lactam double bond, were barely resolved from the active isomers by the bonded silica phase and are not resolved as a diastereomeric pair as on the PGC.

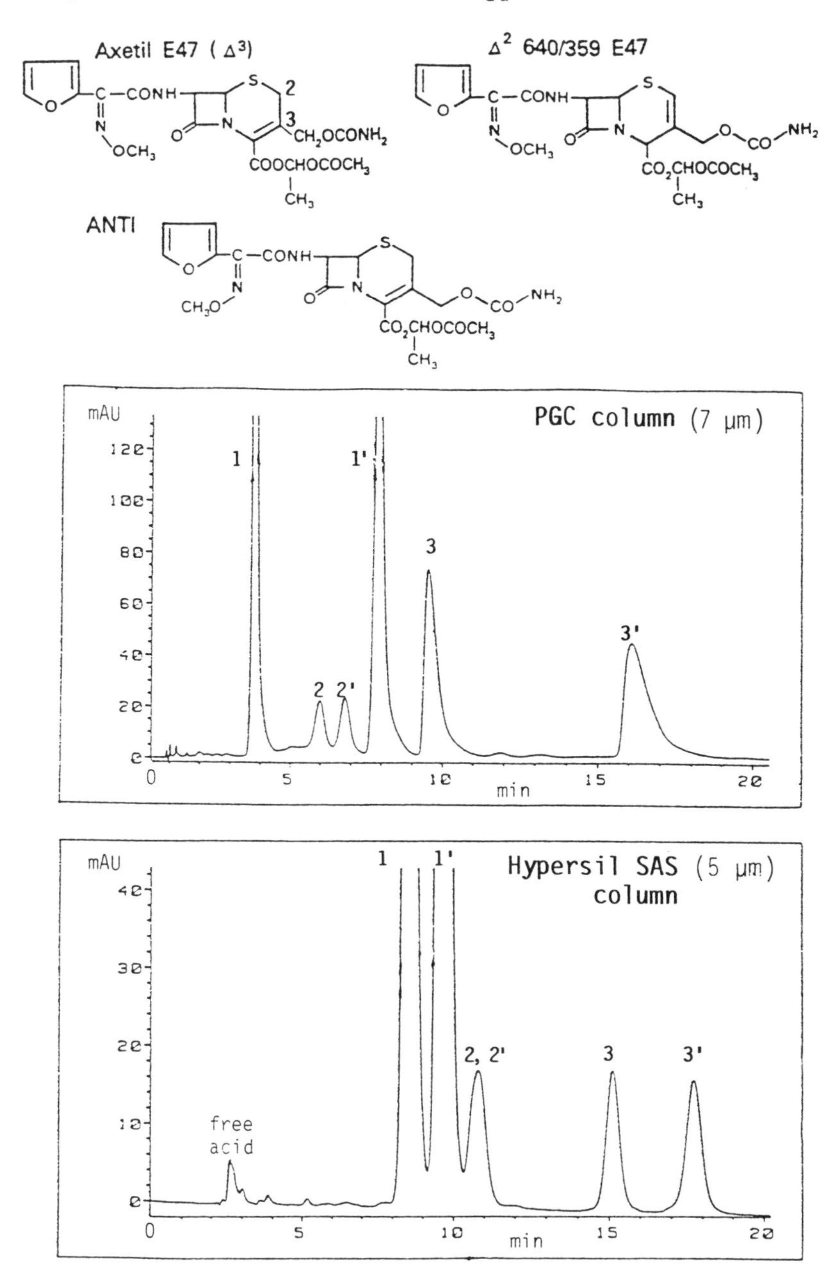

Fig. 2. HPLC patterns for cephalosporin isomers on 2 columns. Axetil E47, Δ-3-isomers (peaks **1** & **1'**); Δ-2-isomers (**2** & **2'**); antimethoxy isomers (**3** & **3'**). **PGC column**, 100 × 4.6 mm, at 30°: acetonitrile-water-methanol-dioxan, 35:20:35:10 by vol.; 1.3 ml/min. **Hypersil column**, 200 × 4.6 mm: methanol–0.2 M pH 4.2 ammonium phosphate buffer, 38:62; 1.0 ml/min. Detection at 276 nm.
Courtesy of N.W. Smith and D. Brennan ('Poster'; 1988, Washington DC).

Besides being unusually selective, even for chiral separations with an appropriate mobile phase additive, PGC columns are robust, and can be readily cleaned by washing overnight with 2 M NaOH.

SUPERCRITICAL FLUID CHROMATOGRAPHY

A major attraction of SFC is that it offers a completely different range of selectivities, even when used with traditional HPLC packed stationary phases. Supercritical CO_2 provides a combination of physical, chemical and spectroscopic properties which are highly useful for both extraction processes (SFE) and chromatographic separations (SFC). By controlling the density of the supercritical fluid, different fractions may be selectively extracted from a complex mixture or sample-matrix. On decompression the extracted solutes are precipitated and may be collected for injection into the GC or SFC instrument. Bartle & co-workers [6] have described a simple on-line system where the extracted solutes are deposited from the end of a restrictor into the loop of the micro-injection valve of a capillary SFC instrument. These authors have also illustrated, using coal-tar pitch, the usefulness of the SFE-SFC combination for difficult samples that are involatile and dissolve poorly in common HPLC solvents.

'Hyphenated' techniques now play an increasingly important analytical role, and those involving SFC, such as SFC-MS, seem particularly promising. Thus, there is a report [7] on coupling of capillary SFC to charge-exchange MS to produce library-searchable spectra of electron-impact character, exemplified in Fig. 3 for ibuprofen in equine urine 4 h post-administration: its mass spectrum (identical with that of pure compound) enabled it to be identified even though another extracted component markedly overlapped the peak.

FTIR allows specific identification of individual functional groups; hence the SFC-FTIR combination offers tremendous scope for identifying compounds that are not amenable to other analytical techniques. As supercritical CO_2 is transparent through much of the IR region of the spectrum, a flow-cell interface for SFC-FTIR is feasible. The use of a solvent elimination interface offers the additional possibility of working with mobile phase modifiers such as methanol [6]: the solutes may be aspirated from a tapered capillary restrictor onto a KBr window, while the mobile phase is lost by evaporation. The SFC-separated solutes are deposited as spots on the window which is continually moved beneath the restrictor to provide spatial separation of the components in preparation for FTIR analysis. This, for an SFC-separated acetylated steroid, gave evidence of O-H stretching, of aliphatic C-H stretching, and of C=O bending indicative of ester groups; identification was thereby achieved.

Fig. 3. Charge-exchange SFC-MS total ion current profiles of: **A**, 25 ng on-column injection of ibuprofen; **B**, solvent extract of equine urine 4 h after administration of ibuprofen: see text for comment on the drug peak, **1**, in **B** (2 is a carboxylated metabolite). SFC performed with an SB octyl capillary column, 10 m × 50 µm i.d.; supercritical CO_2 at 150°. *From ref. [7], courtesy of the authors and the American Chemical Society.*

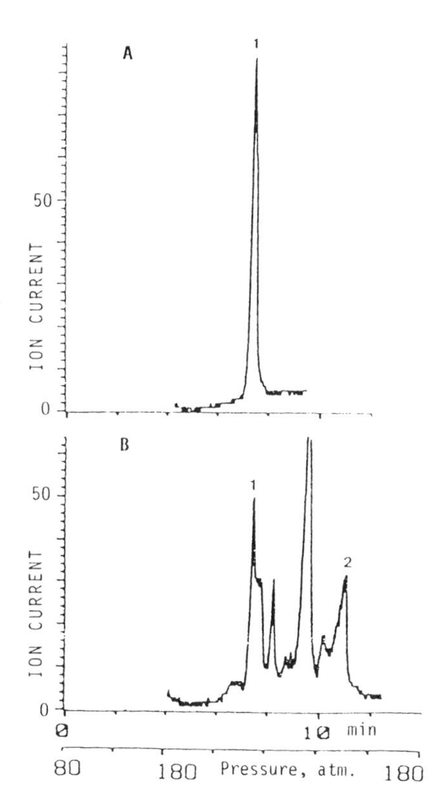

CAPILLARY ELECTROPHORESIS (CE)

CE may not, on first consideration, be regarded as a chromatographic technique. However, it is certain that in future CE will be considered alongside HPLC as a means of achieving fast, highly efficient separations of many biomolecules. CE need no longer be considered as applicable only to separating ionic species. The development of capillary electrokinetic chromatography, which uses an electrosmotically pumped micelle, has extended CE to the separation of neutral compounds. An ionic surfactant is added to the operating buffer at a concentration exceeding the critical micelle concentration. The surfactant monomers tend to form roughly spherical aggregates, or micelles, with the hydrophobic tail groups oriented towards the centre and the charged head groups along the outer surface. The system is thus composed of two phases, aqueous and micellar, and solutes can partition between them. The micellar phase is similar to a chromatographic stationary phase and the micelles have been termed a 'pseudo-stationary phase'. [For 'micellar LC' see J.G. Dorsey in Vol. 18.-*Ed.*]

The instrumental development of CE has also benefited from HPLC, there being close similarities in detector design and sample introduction.

Microcolumn separations offer particular promise for analysis of discrete biological systems. Ewing and co-workers [8] used CE for removing and separating cytoplasm samples from single nerve cells (Fig. 4a). For investigating the giant dopamine cell of the pond snail, 100-300 pl of cytoplasm was injected by electromigration 55 sec after exposure to ethanol, and the resulting capillary electropherogram compared with one for pure dopamine (Fig. 4b). The peak area in the sample corresponds to ~14 fmol of dopamine. This example illustrates well the capabilities of CE and perhaps indicates a future trend in separation science.

WHITHER NEXT?

It is almost mandatory to end such a review, even one as cursory as this, with a little speculation on what the future may hold. Separation science has developed very rapidly over the past years and undoubtedly there are many exciting developments still to come.

This review has focussed on means to improve the selectivity of a separation, and there is certainly a trend towards selective analyzers. The use of highly selective columns, reaction techniques and specific detectors opens the way to dedicated analyzers. Further impetus comes from the development of biosensors that can be selective to individual molecules. [See art. #ncC-7, this vol.- *Ed.*]

In another direction, the perennial request in HPLC has been to develop the sensitive universal detector. In some respects this is a clear terminological contradiction, in that the detector must respond to all solutes but not to the components of the mobile phase. Although there have been many attempts, no HPLC detector has yet matched the capabilities of the FID in GC. The closest approach to date has been the LC-MS, and there might be further developments in the interfacing of the two techniques that will solve the problem.

Lastly it is interesting to speculate whether the true personal chromatograph is on the horizon. Undoubtedly such a device would prove highly useful - for example, in a doctor's surgery to provide instant diagnosis of a disease from a body-fluid profile. The development of single-cell sampling in CE indicates that maybe this idea is not merely a pipe-dream.

References

1. Brinkman, U.A. Th., De Jong, G.J. & Gooijer, C. (1988) in *Bioanalysis of Drugs and Metabolites* [Vol. 18, this series], (Reid, E., Robinson, J.D. & Wilson, I.D., eds.), Plenum, New York, pp. 321-338. *[continued*

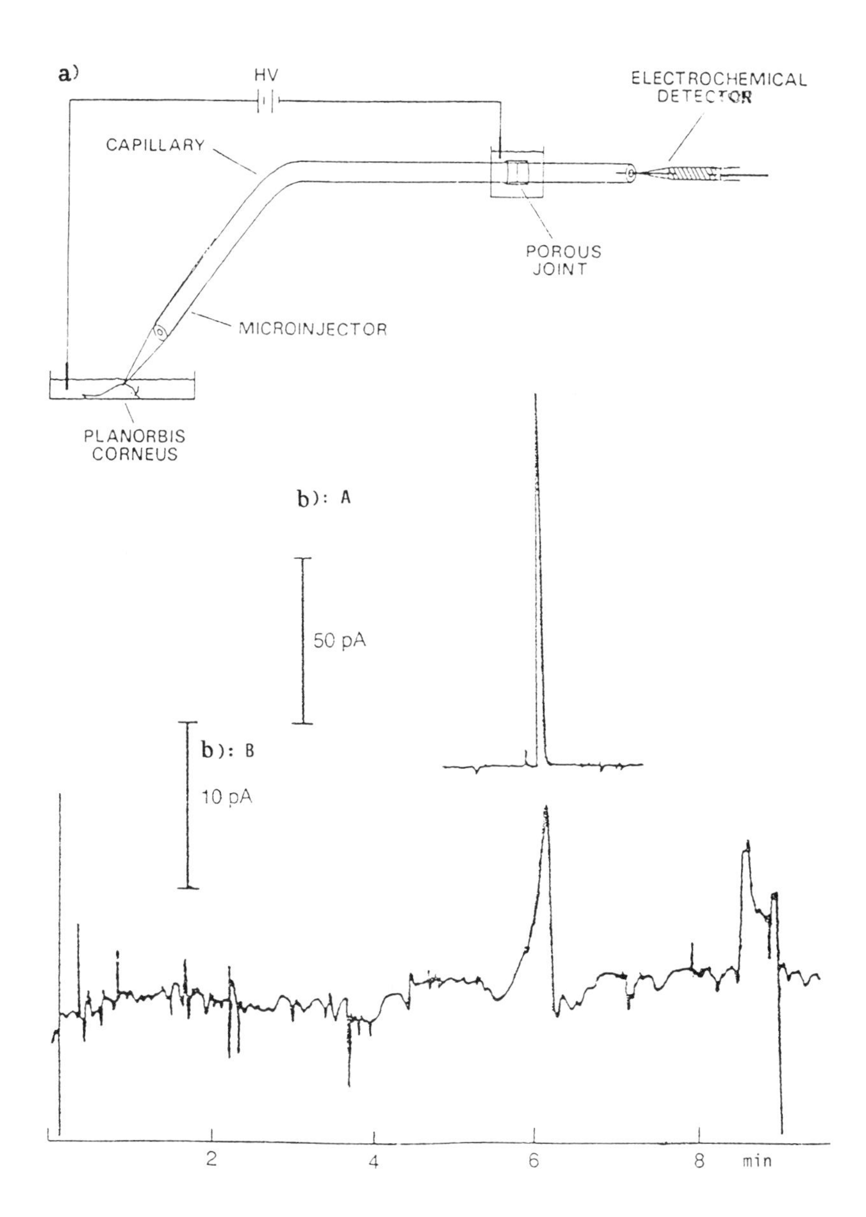

Fig. 4. Use of CE in studying single nerve cells.
a) System used for removal and separation of cytoplasmic samples from single nerve cells of *Planorbis corneus*.
b) Capillary electropherograms of dopamine: **A**, authentic compound; **B**, cytoplasm separated as in **a)**.
From ref. [8], courtesy of the authors and the American Chemical Society.

2. Lloyd, D.K., Goodall, D.M. & Scrivener, H. (1989), *Anal. Chem. 61*, 1238-1243.
3. Ha, S.T.K., Wilkins, C.L. & Abidi, S.L. (1989) *Anal. Chem. 61*, 404-408.
4. Winkler, P.C., Perkins, D.D., Williams, W.K. & Browner, R.F. (1988) *Anal. Chem. 60*, 489-493.
5. Knox, J.H., Kaur, B. & Millward, G.R. (1986) *J. Chromatog. 352*, 3-25.
6. Bartle, K.D., Raynor, M.W., Clifford, A.A., Davies, I.L., Kithinji, J.P., Shilstone, G.F., Chalmers, J.M. & Cook, B.W. (1989) *J. Chromatog. Sci. 27*, 283-292.
7. Lee, E.D., Hsu, S-H. & Henion, J.D. (1988) *Anal. Chem. 60*, 1990-1994.
8. Ewing, A.G., Wallingford, R.A. & Olefirowicz, T.M. (1989) *Anal. Chem. 61*, 292A-303A.

#C-3

BIOANALYTICAL SUPERCRITICAL FLUID CHROMATOGRAPHY

D.W. Roberts† **and I.D. Wilson**

Drug Kinetics Group, ICI Pharmaceuticals,
Alderley Park, Macclesfield, Cheshire SK10 4TG, U.K.

SFC has, after a long gestation period, started to be applied to drugs and to biological samples. The SFC of acidic and basic drugs on a variety of stationary phases using packed columns and converted HPLC equipment is described. The use of organic modifiers and substances such as TEA to improve chromatographic peak shape is described. Preliminary results are given, e.g. with urine extracts.*

SFC is now one of the fastest growing analytical techniques. The first-ever paper on SFC was published in 1962 by Klesper and co-authors [1]. Thereafter very few articles appeared till 1981 when Lee and co-authors [2] first described capillary SFC. Since then the literature has escalated, particularly in the areas of equipment design, interfacing with FTIR and MS, and application to different analytical areas. However, these rapid developments have not been inspired by any revolutionary new ideas; rather the progress has closely paralleled developments in capillary GC and in HPLC.

SFC applications are usually to compounds that are difficult to analyze by HPLC or GC due to thermal instability or lack of volatility or of a suitable chromophore. Here we present some of our initial results, obtained with HPLC equipment converted to enable it to function for SFC, on some polar acidic and basic compounds (non-steroidal anti-inflammatories and β-blockers).

PRINCIPLES OF SFC

Above the critical point, the density of a fluid and hence solute solubility can be modified by changing the pressure and temperature. If the pressure is increased at constant temperature, the fluid density approaches that of a liquid and so solute retention on the stationary phase decreases. The converse applies, but solute retention decreases

---

†now at Sterling Res. Group, Alnwick Res. Centre, Alnwick NE66 2JH

**Abbreviations.*- SFC, supercritical fluid chromatography; GC-type detectors: ECD = electron-capture, FID = flame-ionization, PI = photoionization; FTIR, Fourier-transform infra-red; MS, mass spectrometry/spectrometer; SPE, solid-phase extraction; TEA, triethylamine,

rather than increases when the rise in temperature exceeds a certain point such that the elevation of solute vapour pressure improves solubility. When pure fluids are used for the analysis of polar compounds by packed-column SFC, poor peak shapes and long retention times often result. To overcome these problems small amounts of organic modifiers such as methanol, 2-propanol and 2-methoxyethanol are added to the supercritical fluid. Such addition therefore allows us to manipulate retention, resolution and selectivity, although to stay above the critical point of the mixture often needs higher pressures and temperatures.

HARDWARE FOR SFC

Mobile phase delivery.- For capillary SFC, where low mobile phase velocities are required (typically 1-10 µl/min), syringe pumps are preferred, whereas for packed-column SFC (2-20 ml/min) normal reciprocating pumps with cooled pump heads to prevent cavitation are more appropriate.

Columns.- For separations which require high efficiencies - where numerous relatively non-polar components are present and detection calls for FID, FTIR or MS - capillary SFC is the method of choice. The capillaries are basically the same as for GC, although for optimal performance smaller internal diameters (<100 µm) are necessary.

For applications requiring high densities, as with polar, high mol. wt. and non-volatile solutes, SFC using standard HPLC columns is preferred. The analysis of basic compounds on silica-based bonded-phase columns does however suffer from residual silanol interactions. As with HPLC, these problems can be overcome by adding a competing organic base such as TEA or by using a polymer column - e.g. PLRPS (Polymer Labs.) or Rogel RP - or the carbon column (Hypercarb; Shandon) developed by Knox (see R. Dolphin, #C-2, this vol.).

Detection.- Most capillary SFC systems use FID, although more sophisticated systems such as ECD, FID, MS and FTIR are now gaining popularity, whereas most packed-column SFC applications have used UV or fluorescence detectors fitted with high-pressure flow cells. Several workers have explored the use of UV and fluorescence diode-array detectors, refractive index detectors and FID with stream splitting.

SFC SYSTEMS

Dedicated SFC systems are now available from a number of manufacturers. So far, however, the number of published applications using these systems is rather limited, due mainly

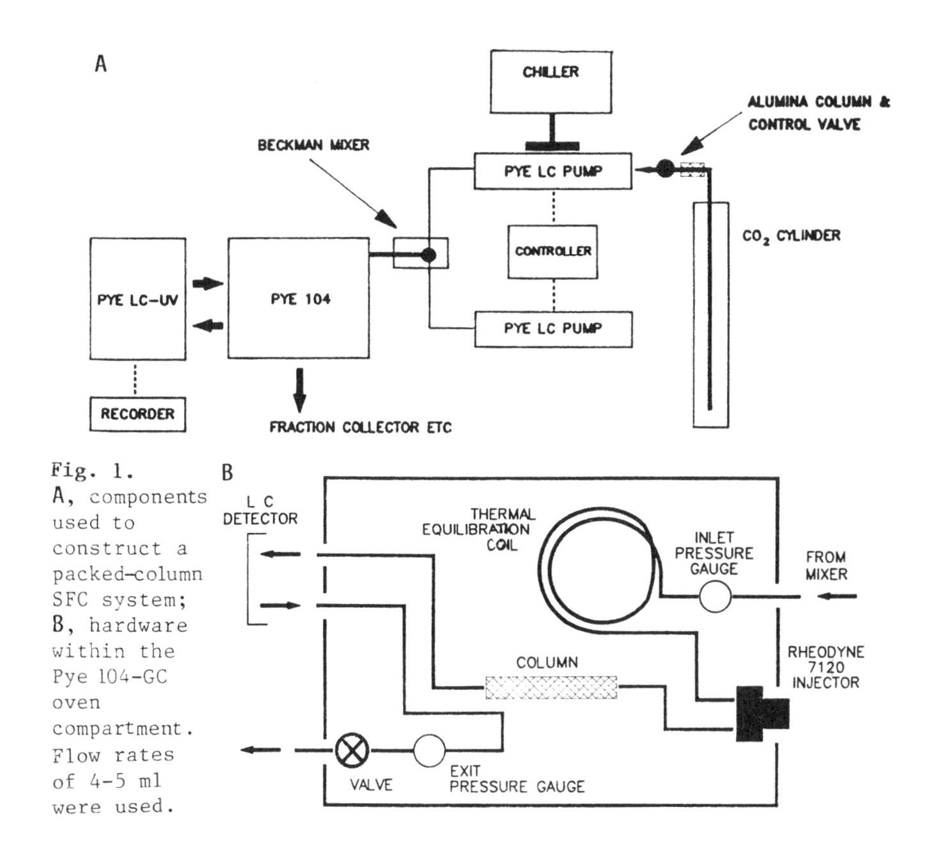

Fig. 1. A, components used to construct a packed-column SFC system; B, hardware within the Pye 104-GC oven compartment. Flow rates of 4-5 ml were used.

to their expense and the relative ease of constructing a home-made system from conventional GC and HPLC hardware.

In order to fully evaluate the technique for drug metabolism and bioanalytical applications we decided to build a packed-column SFC system from redundant equipment. Fig. 1A shows the layout. From a CO_2 cylinder, fitted with a siphon tube to deliver liquid from the bottom of the cylinder, CO_2 was passed through an alumina-filled column to a Pye Unicam LC-XP pump fitted with a clamp-on heat exchanger cooled by a modified refrigerator unit (Ash Insts., Macclesfield). Various modifiers were pumped using a second Pye Unicam LC-XP pump to a magnetically stirred chamber where mixing with the liquid CO_2 occurred. The proportions of modifier (e.g. methanol) and CO_2 were controlled by a Beckman 421A gradient controller. The modifier-containing fluid then passed via a pressure gauge and thermal equilibration coil to a Rheodyne 7120 valve fitted with an injection loop, 10 or 20 µl, which was mounted inside a modified Pye 104 GC oven (Ash Insts.; internal layout: Fig. 1B). The fluid, now in

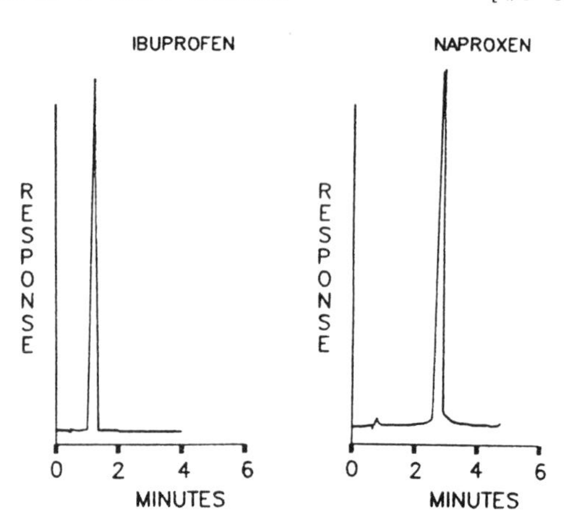

Fig. 2. Elution of two unrelated acidic compounds on an aminopropyl column (150 × 4.6 mm). Eluent: supercritical CO_2 containing methanol (20% v/v). Pressure: 3000 psi; temp.: 50° (likewise in SFC runs shown later). Amount loaded: ~50 µg. UV detection.

the supercritical state, passed from the injection loop to a cartridge column (Capital HPLC Specialists, Edinburgh), then out of the oven to a Pye Unicam LC-UV variable wavelength detector fitted with a high-pressure flow-cell (Ash Insts.); finally the fluid returned to the oven environment to a back-pressure regulator (Tescom Model 26 or Rheodyne 7037). From there the depressurized mixture of CO_2 and modifier were bubbled through water in a flask.

Packing materials.- Those examined were silica, silica coated with cyanopropyl, aminopropyl, nitro, octyl or octadecyl groups, porous graphitic carbon (Hypercarb; Shandon) and polymeric adsorbent (Rogel RP column). Particle size was 7 µm for the graphitic column and 5 µm for all the others.

RESULTS

Analysis of acidic compounds

Fig. 2 shows the chromatograms of two structurally unrelated compounds (ibuprofen and naproxen; for conditions see legend), without adding any acid to control analyte ionization. Evidently good retention and peak shape are achievable, and sample loads can be high even on small analytical columns.

Drug metabolism studies often involve isolating metabolites for structure determination. Traditional methodologies include liquid-liquid extraction and preparative TLC and HPLC. Each technique, however, requires the elimination of solvent after the extraction or separation has been performed. With SFC there is no such problem, since most of the mobile phase after depressurization is a gas and is therefore easily eliminated. In order to evaluate semi-preparative SFC for metabolite isolation, a sample of freshly voided human urine was spiked with 3 acidic drugs (A, B & C) at relatively high concentrations (typically >1 mg/ml). Following

Fig. 3. Evaluation of preparative SFC for isolating compounds from a biological matrix (human urine, subjected to SPE), using supercritical CO_2 containing methanol (20% v/v). Column, psi and temp. as in Fig. 2.

The three acidic compounds used for spiking were structurally unrelated.

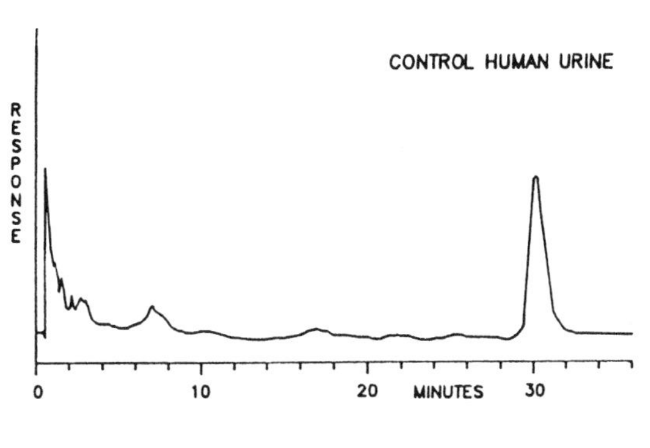

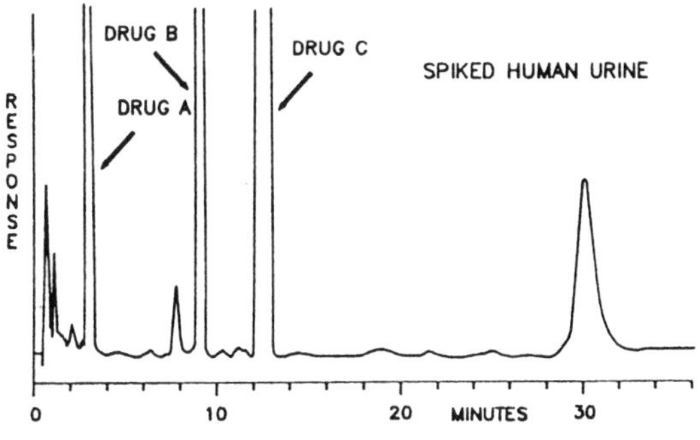

a concentration step using C-18 SPE, aliquots of the extracts from control and spiked urine were analyzed by SFC (Fig. 3; conditions in legend). The control chromatogram had very few peaks in the region of interest, but a very late-eluting peak was observed: it was subsequently identified as hippuric acid, a common constituent of human urine. The chromatogram for the spiked urine showed the 3 drugs, well resolved from endogenous components and having peak shapes which were excellent for the column loading (>500 µg). We are currently investigating the use of radio-flow cells for detecting ^{14}C-metabolites and the problems of efficient collection of eluting fractions.

A comparison between gradient-elution RP-HPLC and SFC, the solvent containing 1% formic acid in each technique, is shown in Fig. 4 for a partially purified faecal extract. The components A and B are both aromatic mono-hydroxylated metabolites of an acidic drug. The separation in the HPLC run was achieved by use of a series of isocratic and linear gradients and took some considerable time to develop. The SFC separation was achieved by isocratic elution with CO_2/methanol on the aminopropyl column as used earlier. SFC again gave adequate resolution with a notably short run time using very simple chromatographic conditions. Moreover, the components were eluted without large volumes of solvent, and samples could be applied in either methanol or dichloromethane.

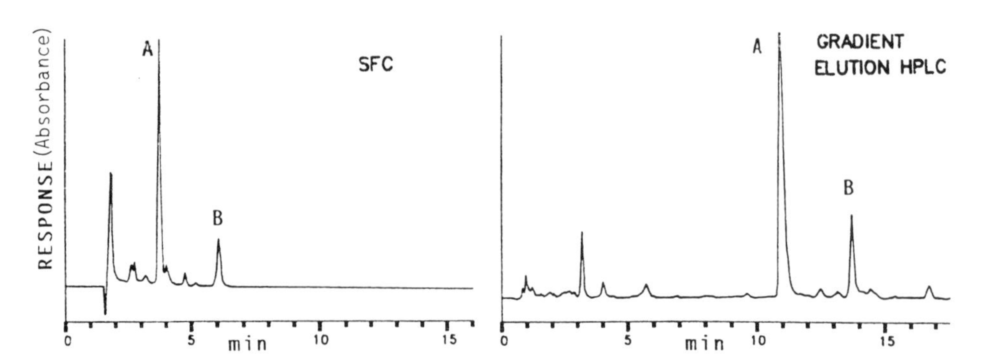

Fig. 4. Gradient elution HPLC *vs.* SFC for a partially purified faecal extract containing 2 monohydroxylated metabolites (**A**, **B**) of an acidic compound currently under development at ICI. **SFC.**- Column: aminopropyl-silica, 150 × 4.6 mm. Eluent: 80:20 (by vol.) CO_2/methanol containing formic acid, 1% w/v; 3000 psi, 50°. **HPLC.**- Column: Zorbax C8, 150 × 4.6 mm. Eluent: initially water in mixer; gradient generated by inflow of methanol containing 1% w/v formic acid.

Analysis of basic compounds

The analysis of basic compounds such as β-blockers by HPLC on RP-silica is often hindered by excessive peak tailing unless a competing base such as TEA is added to the mobile phase. Fig. 5A shows SFC of a β-blocker using CO_2/methanol with no basic modifier. Unlike HPLC only slight peak tailing occurred. With a cyanopropyl instead of an aminopropyl column even high methanol concentrations failed to elute the compound; but with added TEA elution was achieved (Fig. 5B). No elution was achieved on alumina, nitro or bare silica columns even after adding TEA. The polymer and graphitic columns gave poor peak shapes. The conditions of Fig. 5B were effective for various β-blockers *(list below)* but failed to elute labetalol. The resolution of 'Compound X' from its internal standard showed that the technique could be used for routine analyses.

SFC as in Fig. 5B: k' *values for various β-blockers:-*
timolol, 2; atenolol, 10; betaxolol, 3; pindolol, 1; bupranolol, 2; pronethalol, 4; oxprenolol, 4; practolol, 10; Compound 'X', 4, and its internal standard, 7; labetalol, -.

CONCLUSIONS

SFC has great potential for drug-metabolism and bioanalytical studies, using the capillary mode or, more appropriate for polar compounds, the packed-column mode. The conversion of redundant GC and HPLC equipment for use with supercritical

Fig. 5. Alternative SFC approaches tried for a β-blocker, using CO_2/methanol (9:1 by vol.). **A**, aminopropyl column. **B**, cyanopropyl column, and TEA (0.4%) present in the methanol.

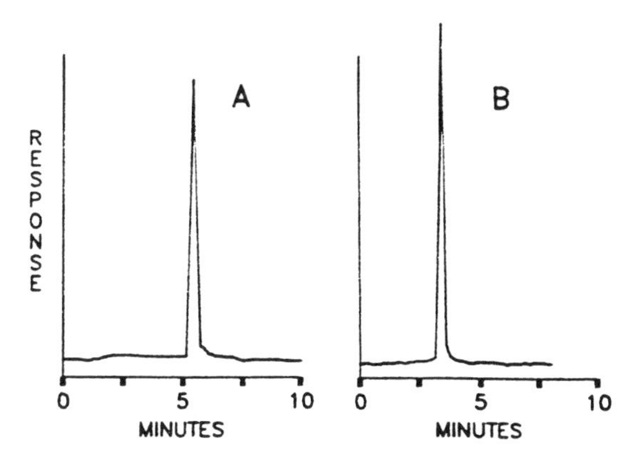

fluids is relatively easy and inexpensive, and the equipment can revert to its original purpose if required.

Both analytical and semi-preparative separations show excellent efficiencies, selectivities and resolution, with notably short run times. High loadings are possible even with an analytical column. Acidic and basic drugs and metabolites can be analyzed with no need for complex modifiers or buffers. So far we have used mainly UV detection, but already it appears that ^{14}C-compounds are measurable, using a radio-flow cell. Work is being pursued on the use of SFC for routine, automated bioanalyses and chiral separations.

References

1. Klesper, E., Corwin, A. & Turner, D. (1962) *J. Org. Chem.* *27*, 700-701.
2. Novotny, M., Springston, S.W., Peaden, P.A., Fjelsted, J.C. & Lee, M. (1981) *Anal. Chem.* *53*, 407A.

#C-4

A FULLY AUTOMATED ANALYTICAL SYSTEM USING SOLID-PHASE EXTRACTION: APPLICATION TO THE DETERMINATION OF CARBAMAZEPINE AND TWO METABOLITES IN PLASMA

J.B. Lecaillon, M.C. Rouan, J. Campestrini and J.P. Dubois

Laboratoires Ciba-Geigy,
Biopharmaceutical Research Center,
B.P. 308, 92506 Rueil-Malmaison Cedex, France

Several HPLC systems (multi-column and cartridge/column) are available for the automated determination of drugs in biological fluids. Automation is extendable, using a new system (ASPEC), to sample preparation on a sorbent cartridge followed by on-line injection directly into the HPLC system. The ASPEC system allows precisely defined operating conditions, with volume transfers at a controlled flow-rate, air-drying, air-mixing and injection onto the HPLC column. The system was used to develop an automated method for determining CBZ and its epoxide and dihydroxy metabolites in plasma. The mean relative S.D. for replicate determinations of the three compounds on spiked samples was 0.9% at 10 μmol/L and 3.4% at 0.1 μmol/L [CBZ mol. wt. = 236]. Besides having good reproducibility and accuracy, the technique is promising in respect of speed: it allows the determination of one sample within ~10 min.*

More and more options are now available for determining drugs and metabolites in biological fluids, to meet the need of pharmacokinetic units for improvements in sample throughput, accuracy and precision, sensitivity, reliability and ease of use, and ease of transfer of methods from one laboratory to another.

These targets should be largely fulfillable by analytical automation. Dealing with HPLC systems and full automation, several approaches can be considered [see also Vols. 16 & 18, this series: R.D. McDowall - *Ed.*].-

(a) Column-switching systems [1, 2], as routinely used for determining drug concentrations. Different link-up modes

---

**Abbreviations*.- ASPEC (system), Automatic Sample Preparation with Extraction Columns (Gilson); CBZ, carbamazepine; SPE *[Ed.'s term]*, solid-phase extraction [with cartridges, for which the term DEC's had been used by the authors, also the term liquid-solid extraction]; i.s., internal standard.

allow trace enrichment or multi-dimensional chromatography (heart-cutting technique).

(b) Robotic systems [3-5], which reproduce any desired number of manual-method steps such as liquid transfer, extraction by horizontal shaking, centrifugation or evaporation. They can perform the entire process from the initial sample aliquotting through the sample preparation steps to the final chromatographic separation and detection.

(c) Autosamplers with automatic sample preparation by SPE cartridge [6]: sample handling capability has been added to provide partial or full automation of SPE and HPLC.

Each of these approaches can be used to advantage in different situations, hence no particular one can be selectively recommended. We have not used robots but have experience of (a) and (c).

The 'automates' have recently been proposed. They are less complex and less expensive than robots. Since they perform SPE, they are akin to the trace-enrichment system but with the difference that the enrichment column is used once only. This can be considered as having two advantages: no memory effect such as may occur from one trace-enrichment to another, and no need to investigate the lifetime of the enrichment column with repeated use. Compared with the switching systems, the automates entail a more conventional approach: the sample preparation is separate from the chromatography. For these reasons, the automates are likely to gain favour as they fulfil certain of the aims already mentioned: good throughput, and ease of use and of method transfer from one laboratory to another. Here we relate some experiences with the ASPEC system.

MATERIALS AND EQUIPMENT

Ciba-Geigy (Basle) supplied CBZ, 10,11-epoxy-CBZ, 10,11-dihydro-10,11-*trans*-dihydroxycarbamazepine (dihydroxy-CBZ) and the i.s.: 5,6-dihydro-11-oxo-11H-dibenz(b,e)azepine-5-carboxamide. Solvents and reagents were of analytical grade from Carlo Erba France (Puteaux) or Merck (Darmstadt, F.R.G.). SPE cartridges (DEC's), 50 or 100 mg C-18 of 1 ml capacity, were from J.T. Baker (Deventer, The Netherlands),

The chromatographic system consisted of a pump (Model 303; Gilson, Villiers-le-Bel, France) and an ASPEC system (Fig. 1) which comprised an automatic sampling injector, a Model 401 diluter/pipetter and a set of racks and accessories for handling SPE cartridges and solvents. The SPE rack (Fig. 1) consisted of the cartridge holder (which is mobile), drain cuvette and collection rack.

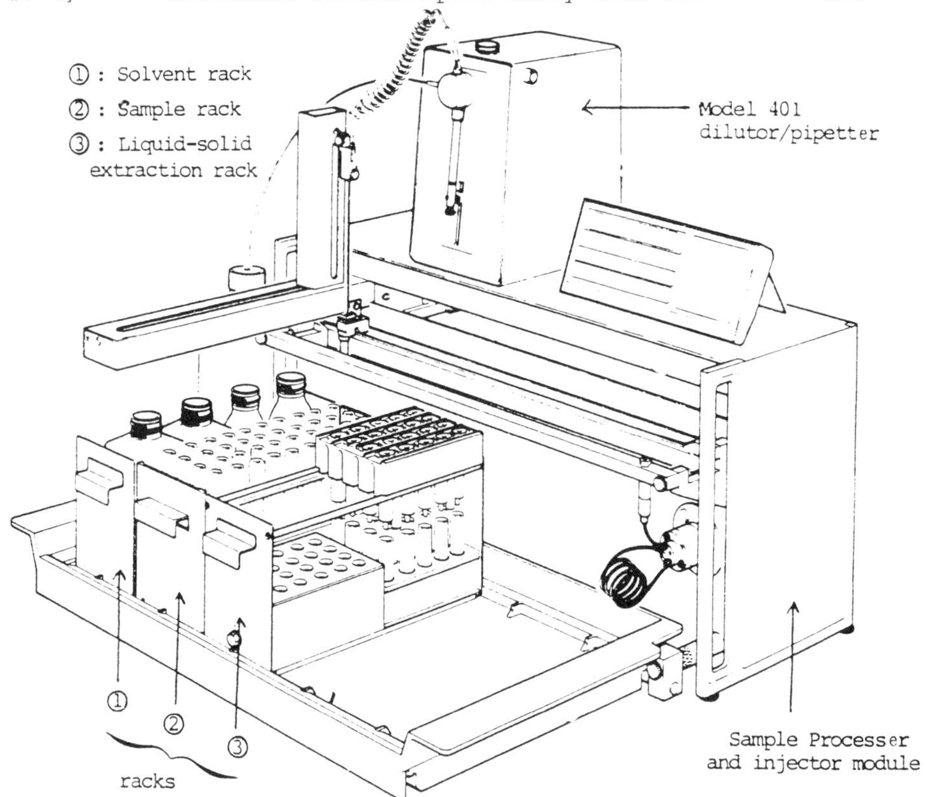

Fig. 1. The Gilson ASPEC system.

A variable wavelength Spectroflow detector (Model 783; Kratos, NJ) was used, at 210 nm (0.008 aufs, 1 sec response time), and an integrator-recorder (Model C-R3A, Shimadzu, Kyoto).

Columns and filter.- The 3 μm Supelcosil LC-18 column was from Supelco France (No. 5-8977; St-Germain-en-Laye), with a guard-column between the pump and injector. A filter (No. FL 01; S.F.C.C., Neuilly-Plaisance) with a replaceable 2 μm frit was inserted between the ASPEC and the column.

ANALYTICAL PROCEDURE

Sample preparation commenced with vortex-mixing, for a few seconds, of 600 μl plasma, 600 μl water and 60 μl i.s. in 1:1 (by vol.) methanol/water, introduced manually into a 5 ml polypropylene tube that was then placed on the rack of the ASPEC system which then performed the operations listed below. The liquids, once dispensed at the chosen flow-rate, were forced into the packing by pressurization with air at the same flow-rate; a special cap fitted on each cartridge assured air-tightness when the needle was dispensing liquid or air. Amplification is now given.

(1) Cartridge conditioning, by passage of 2 ml methanol and 2 ml water.

(2) Introduction of 1 ml diluted plasma was followed by washing with 1 ml 20 mM K_2HPO_4 and 1 ml 19:1 water/methanol.
The cartridge holder that had been located above the drain cuvette (Fig. 1) was now automatically positioned over the collection tubes for the following step, after which it was pulled back.

(3) Elution was performed with 250 µl methanol.

(4) After polarity adjustment by adding 950 µl water and bubbling with 2 ml of air to mix, 100 µl was injected by loop.

After each liquid transfer, the needle was rinsed with 1 ml water at 400 µl/sec. A 50 µl air segment was created before pipetting the liquid to be transferred, to obviate cross-contamination.

Chromatography, illustrated in Fig. 2, was with acetonitrile/methanol/50 mM KH_2PO_4 (15:5:80) at 2 ml/min. The throughput was one sample per 10.5 min, the time required for sample preparation and that for chromatography being similar.

RESULTS AND DISCUSSION

Sample throughput (~10 min for one determination of CBZ and metabolites) reflects the time-saving due to reduced manual interventions [7, 8]. Analysis times are comparable in two other newly established methods (for codeine and triclabendazole; unpublished data).

Accuracy and precision (reproducibility) were ascertained, with very good results, using 6 plasma samples spiked with the same amounts of the 3 analytes and assayed on the same day. For the 3 analytes individual recoveries were between 94% and 105% (n = 54) for 1-10 µmol/ml concentrations with C.V.'s between 0.4% and 1.5% (n = 6;9 series). Good behaviour of the system was also found for day-to-day reproducibility.

Sensitivity was the same as for manual methods [7, 8]: the quantitation limit was 10 ng/ml of CBZ and metabolites, adequate for pharmacokinetic purposes. Since almost all the sample eluted from the cartridge can be injected [6], losses due to extraction and liquid transfer are minimized. **Selectivity** was improvable by performing several rinsing steps to eliminate interfering compounds. Further improvement was obtainable by incorporating a switching valve operated by the automate to give column-switching capability.

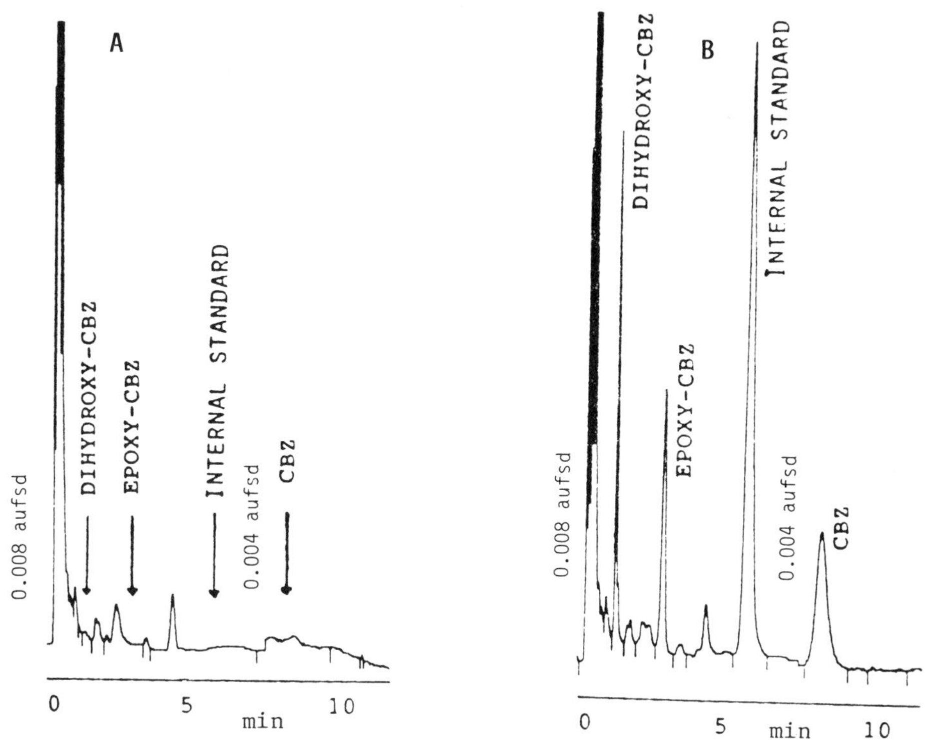

Fig. 2. Representative chromatograms for (**A**) a plasma blank, (**B**) plasma spiked with 0.5 µmol/L of CBZ, epoxy-CBZ and dihydroxy-CBZ. See text for conditions. Column: 3 µm Supelcosil C-18, 33 × 4.6 mm.

Use of the methods.- Although automated multiple operations call for familiarization, several technicians used the methods with confidence and ease, for compounds of different classes. The high reproducibility of the operations and the use of an i.s. allowed overnight running. Assays should be readily transferable from one laboratory to another, since all the automatic operations are precisely described in the method and the procedure is defined as a computer program.

CONCLUDING COMMENTS

Our experience with ASPEC automates during >1 year was positive. Such systems appeared to give an improved sample throughput along with ease of use. Reproducibility and accuracy were good. Several manufacturers are now proposing automates with new characteristics (e.g. in respect of mixing and of possibly weighing the liquids transferred), which would give greater scope for specific analytical tasks. Another positive factor is the cost of automates, which is close to that of an automatic injector.

References

1. Ramsteiner, K.A. (1988) *J. Chromatog. 456*, 3-20.
2. Lecaillon, J.B., Souppart, C., Dubois, J.P. & Delacroix, A. (1988) in *Bioanalysis of Drugs and Metabolites, Especially Anti-inflammatory and Cardiovascular* [Vol. 18, this series] (Reid, E., Robinson, J.D. & Wilson, I.D., eds.), Plenun, New York, pp. 225-233.
3. Luders, R.C. & Brunner, L.A. (1987) *J. Chromatog. Sci. 25*, 192-197.
4. Johnson, E.L., Pachla, L.A. & Reynolds, D.L. (1986) *J. Pharm. Sci. 75*, 1003-1005.
5. Grandjean, D., Beolor, J.C., Quincon, M.T. & Savel, E. (1989) *J. Pharm. Sci. 78*, 247-249.
6. Rouan, M.C., Campestrini, J., Lecaillon, J.B., Dubois, J.P., Lamontagne, M. & Pichon, B. (1988) *J. Chromatog. 456*, 45-51.
7. Menge, G.P., Dubois, J.P. & Bauer, G. (1987) *J. Chromatog. 414*, 477-483.
8. Chelberg, R.D., Gunawan, S. & Treiman, D.M. (1988) *Ther. Drug Monit. 10*, 188-193.

#C-5

IMPROVEMENTS IN THE HPLC MEASUREMENT OF DRUG AND METABOLITE LEVELS IN BIOLOGICAL FLUIDS

[1]A. Nicolas†, [1]P. Leroy, [2]D. Decolin and [2]G. Siest

[1]Laboratoire de Chimie Analytique

[2]Centre du Médicament - URA CNRS no. 597

Faculté des Sciences Pharmaceutiques et Biologiques, B.P. 403, 54000 Nancy, France

Requirement	*Sensitive assay methodologies for drugs and metabolites (examples below) with optimization of sample treatment and detection; applicability to plasma.*		
	Amoxicillin	***2-MPG* and its metabolite 2-MPA***	***Glucuronides produced* in vitro**
End-step	*RP-HPLC with MeOH-phosphate buffer; fluorimetric detection after post-column derivatization by fluorescamine.*	*Ion-pairing RP-HPLC with cetrimonium in acetonitrile-phosphate buffer; fluorimetric detection,* ***as on left*** *but using pyrenemaleimide.*	*RP-HPLC with MeOH-water or acetonitrile-TFA; direct UV detection, OR* ***as on left*** *but done pre-column with a coumarin, BrMmC.*
Sample handling	*Protein precipitation by TCA (to 10% w/v).*	*Reduction by TBP; protein precipitation by ethanol.*	*On-line clean-up or SPE.*
Comments	*Quantitation limit 0.1 μg/ml.*	*2-MPG and 2-MPA separated; sensitive and specific analysis.*	*Aglycone and glucuronide separated; incubation medium tolerated.*

The measurement of drugs and metabolites in biological fluids by HPLC can be improved in respects such as selectivity, sensitivity and automation by optimization of two critical steps: sample pre-treatment and detection. Protein precipitation is the simplest way to prepare plasma for HPLC analysis. Some problems occur with low-concentration measurement because of sample dilution and poor selectivity *vs.* endogenous compounds, especially with UV detection at low wavelengths. Liquid-liquid

†Addressee for any correspondence.

**Abbreviations*.- 2-MPG, 2-mercaptopropionylglycine; 2-MPA, 2-mercaptopropionic acid; BrMmC, 4-bromomethyl-7-methoxy-coumarin; MeOH, methanol; TBP, tributylphosphine; SPE, solid-phase extraction; TCA, trichloroacetic acid; TFA, trifluoroacetic acid.

and solid-phase (SP) extraction offer both better selectivity and the possibility of trace-enrichment. However, the former can be tedious and is unsuited to large sample batches, and solvent evaporation may cause drug degradation.

SPE requires the optimization of retention, washing and elution steps; moreover, different cartridge batches can vary in extraction efficiency, and recovery often decreases at very low concentrations. [Such points are considered in Vols. 16 & 18.-*Ed.*] The main advantage of SPE resides in the possibility of on-line processing with an HPLC switching system, offering both automation and trace enrichment.

Pre- and post-column derivatization has permitted many improvements in detection, and nearly all organic functional groups are amenable. Selectivity and sensitivity result from use of the near-UV and visible regions in spectrophotometry and fluorimetry or low potentials in amperometry. The post-column reaction mode looks more interesting because of its automatibility and high reproducibility.

The following examples demonstrate HPLC methodological improvements in relation to pharmacokinetic and biochemical studies.

ASSAY OF AMOXICILLIN IN PLASMA

The HPLC analysis of amoxicillin in biological fluids requires an efficient combination of a simple handling procedure with selective detection. This cephalosporin-type antibiotic has amphoteric properties (pKa's = 2.4 & 7.2), precluding liquid-liquid extraction. Reported HPLC techniques include use of SPE and UV detection [1], and of protein precipitation and a 2-column switching system with fluorescamine post-column derivatization [2], claimed to give a quantitation limit (0.05 μg/ml) at least 10-fold lower than with the former method using UV.

Amperometry has been tried as an alternative, but the high potential applied to the glassy carbon electrode generates many interferences from biological matrices [3]. *o*-Phthaldialdehyde has also been tried as a post-column fluorigenic reagent for amoxicillin, but failed to give fast condensation at room temperature in alkaline medium [4].

Initial trials.- Since protein precipitation and final UV detection did not permit the measurement of amoxicillin at concentrations <0.5 μg/ml, SPE with a C-18 cartridge was tried (Scheme 1) essentially as in the described procedure [1]. Recovery decreased from 92.5% to 85% in the concentration range 1.0-0.2 μg/ml, implying poor accuracy for the method and the need for an internal standard for its improvement.

PLASMA (1 ml) with
calibration standard (0.1 ml) added

Protein precipitation	SPE
Add 0.5 ml TCA (10% w/v) and centrifuge at 4000 rpm, 5 min	*Add 1 ml pH 4.5 20 mM phosphate buffer. Transfer onto SepPak C-18 cartridge pre-wetted with 4 ml methanol then 1 ml buffer; wash with 1 ml buffer and 1 ml water. Apply 2 ml methanol-water, 15:85 by vol.*
SUPERNATANT *To 0.5 ml add 0.5 ml pH 6.5 phosphate-citrate buffer*	ELUATE
Use 50 µl for HPLC	Use 50 µl for HPLC

Scheme 1. Plasma treatment for HPLC assay of amoxicillin: protein precipitation (the final buffer addition gives pH 4.8) **or** SPE.

Approach adopted.- The procedure finally arrived at hinged on protein precipitation and post-HPLC derivatization with fluorescamine. The results were very satisfactory (see below), without recourse to a column switching system as previously described [2]. TCA was the eventual choice for precipitating proteins (Scheme 1), rather than acetonitrile or ethanol, because of the lower dilution factor and better amoxicillin recovery (near 100% with TCA and <60% with organic solvents). However, amoxicillin is unstable in acidic media, and the pH has to be adjusted to 4.8 to prevent analyte loss in supernatants before HPLC analysis and to enable full automation with an autosampler injector.

HPLC and detection conditions, as given in the legend to Fig. 1.- The mobile phase composition was chosen for the following reasons:
- buffering at pH 7.2 provided fast condensation of amoxicillin with fluorescamine without further post-column base addition;
- no ion-pairing agent was necessary, and the methanol content was such as to give a short retention time for amoxicillin (<10 min; Fig. 1) and high frequency in sample analysis.

Derivatization and detection conditions were close to those previously elaborated [2].[⊗] The peak showed no significant broadening compared with direct UV detection when a slow flow rate for fluorescamine addition was used, with an open tubular reactor in a knitted configuration.

Characteristics of the assay for amoxicillin (capacity factor: 3.65).- Linearity prevailed (r = 0.9998) in the range tested, 0.1-10 µg/ml. There was good precision: C.V. (n = 5) 7.4% for 0.1 µg/ml and 2.4% for 1.0 µg/ml. There was excellent recovery: 99% for 0.2 µg/ml. For detection and quantification the limits (µg/ml) were respectively 0.025 (signal/noise = 5) and 0.1.

[⊗] The reactor was from Supelco; Hitachi-Merck #6 SS-A-13 3-way solvent delivery module.

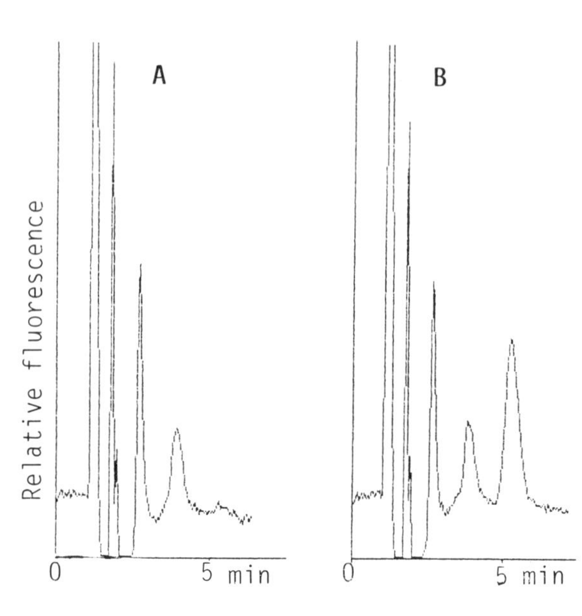

Fig. 1. Typical patterns for plasma after TCA deproteinization: **A**, blank; **B**, spiked with 1 µg/ml of amoxicillin. **Column:** LiChrospher RP-18e 5 µm (125 × 4 mm). **Elution** (35°): methanol: 10 mM phosphate pH 7.2 (7.5:92.5 by vol.), 1.5 ml/min. **Detection:** λ_{ex} 385 nm, λ_{em} ≥415 nm (long-pass filter) after passage through open tubular reactor (10 ft. × 0.5 mm) with 0.2 ml/min inflow of fluorescamine in acetonitrile, 0.2 mg/ml.

ASSAY OF 2-MPG AND ITS METABOLITE 2-MPA IN PLASMA

Methodological problems in measuring 2-MPG and 2-MPA in plasma arise in both sample preparation and detection. 2-MPG is a drug used in cystinuria and rheumatoid arthritis treatment. Several forms co-exist in plasma since its free sulphydryl group easily reacts to give disulphide bonds with itself, endogenous thiols and proteins. Moreover, its amide bond slowly hydrolyzes to release 2-MPA and, in turn, various oxidized forms. Detection modes applicable to these solutes mostly rely upon properties of free sulphydryl groups; hence -S-S- bonds require reduction prior to analysis.

2-MPG is too polar to permit use of an efficient partition technique (liquid-liquid or SP) for extraction. So protein precipitation remains the main approach for plasma sample preparation. The lack of a chromophore precludes sensitive UV spectrophotometric detection, and other modes have to be considered. A previously reported HPLC technique for the simultaneous measurement of 2-MPG and 2-MPA in plasma [5] included multi-step sample preparation, pre-column derivatization and fluorimetric detection. The main disadvantage of this method is the lengthiness of the analysis.

Method development.- A detection mode suitable for thiol compounds was selected for use in conjunction with a single protein-precipitation step following reductive treatment of the plasma. Among the three different techniques tested

Table 1. Different detection modes tested for HPLC analysis of 2-MPG and 2-MPA. S.C.E. denotes saturated calomel electrode.

MODE:-	Amperometry (glassy carbon electrode)		Fluorimetry
	Mode 1	Mode 2	
Derivatization reaction	no	post-column with *o*-phthaldialdehyde & taurine in borate buffer pH 9.5	post-column with pyrene-maleimide
Detection conditions	+1.2 V *vs.* S.C.E.	+0.7 V *vs.* S.C.E.	λ_{ex} = 342 nm λ_{em} ≥ 370 nm
Selectivity (plasma)	-	+	++

Scheme 2. Plasma treatment for HPLC of 2-MPG and 2-MPA.

PLASMA (1 ml) with calibration standard (0.1 ml) added

REDUCTION STEP: *Add 0.2 ml pH 8.0 0.2 M phosphate buffer & 0.2 ml TBP (10% w/v in $CHCl_3$). Heat 30 min at 50°.*

PROTEIN PRECIPITATION: *Add 2 ml ethanol; centrifuge at 4000 rpm, 10 min.*

↓

SUPERNATANT
Use 50 µl for HPLC

(Table 1), the third was chosen. The reaction with pyrene-maleimide, recently described for the determination of *N*-acetylcysteine in plasma [6], has now proved successful for detecting 2-MPG and 2-MPA. As alkaline reagent a triethylamine-Brij-35 solution was preferred to a borate buffer to prevent any blocking of the post-column module and detector flow-cell.

For plasma reduction (Scheme 2) TBP was chosen rather than dithiothreitol since no reagent peak appeared on the chromatogram and it gave complete reaction under milder pH conditions. Ethanol appeared better than TCA for protein precipitation in spite of greater sample dilution. Indeed, TCA lowers the pH of the supernatant, such that 2-MPG becomes unstable, and variations in retention time were observed for both analytes.

Chromatographic optimization.- The mobile phase being buffered at pH 7.0, the carboxyl group of each analyte was dissociated and hence capable of ion-pairing. Rather than tetrabutylammonium, a structurally 'bulky' ion, the chosen counter-ion was cetrimonium (hexadecyltrimethylammonium since, having a long alkyl chain, it gave better resolution (2.1) between 2-MPG and 2-MPA (capacity factors 5.5 and 6.3), both structurally linear. Fig. 2 shows typical HPLC patterns, the post-column reagents (each 0.17 ml/min) being (1) 30:70 acetonitrile-water containing (v/v) 2% triethylamine and 1% Brij-35, and (2) 5 $\times 10^{-4}$ M pyrenemaleimide in acetonitrile.

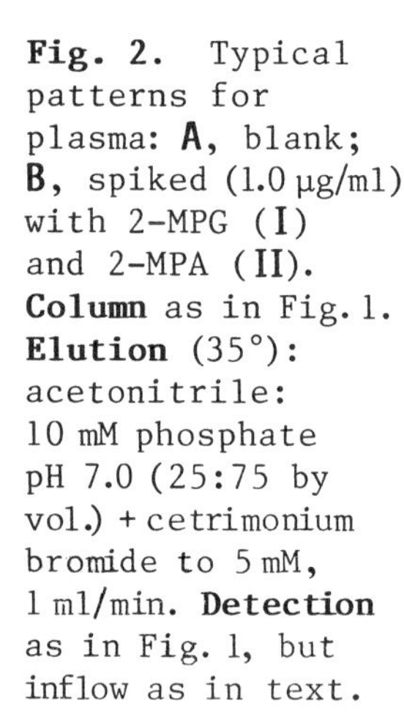

Fig. 2. Typical patterns for plasma: **A**, blank; **B**, spiked (1.0 µg/ml) with 2-MPG (**I**) and 2-MPA (**II**). **Column** as in Fig. 1. **Elution** (35°): acetonitrile: 10 mM phosphate pH 7.0 (25:75 by vol.) + cetrimonium bromide to 5 mM, 1 ml/min. **Detection** as in Fig. 1, but inflow as in text.

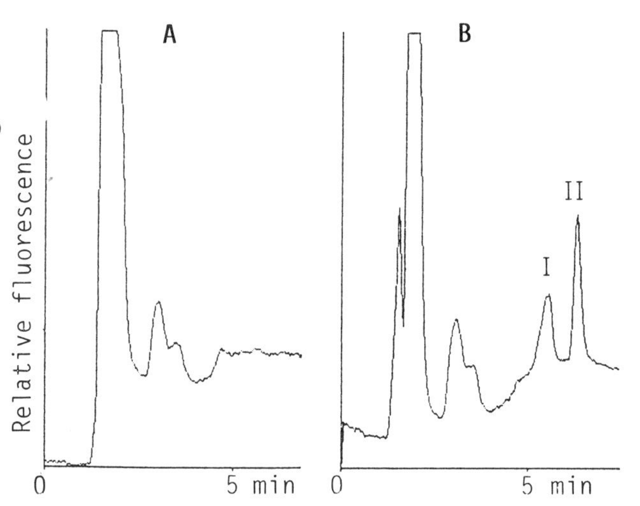

Characteristics of the assay.- For 2-MPG and 2-MPA respectively, linearity testing over the ranges 0.2-10 and 0.2-5 µg/ml gave r = 0.997 and 0.999. The respective precisions (n = 5) were 6.2% and 9.9% at 0.2 µg/ml, and 1.5% and 5.1% at 5 µg/ml. Mean recovery (1 µg/ml) was 95%, and the detection limit was 0.05 µg/ml.

MEASUREMENT OF GLUCURONIDES PRODUCED *IN VITRO*

UDP-glucuronyl transferase is a polymorphic enzyme; hence the study of its mechanism needs numerous *in vitro* assays involving various compounds as substrates. A direct HPLC method for measuring glucuronidation has to meet the following requirements:
- specificity *vs.* components of the incubation mixture (proteins, detergent, excess of aglycone and cofactor, i.e. UDP-glucuronic acid);
- sensitivity adequate to cope with low kinetic velocities or with aglycones that have poor detectability, e.g. menthol;
- low cost and simplicity, in view of the numerous assays.

Usual analytical approaches entail the isolation of glucuronides from microsomal suspensions by liquid-liquid extraction or SPE, and HPLC analysis with UV spectrophotometry. The main disadvantages are lengthy sample preparation and sometimes poor selectivity.

Method development.- Two methods have been devised.

(1) Acidified microsomal suspensions are injected directly into an HPLC switching system, for the analysis of glucuronides whose aglycones have suitable spectral properties (Fig. 3). Without the switching device, the lifetime of the analytical

Fig. 3. HPLC switching system (**a**, loading and cleaning; **b**, elution) used for on-line treatment of microsomal suspensions and analysis of *S*(+)-2-phenylpropionic acid glucuronide. The suspension, + 50 µl 0.2 M HCl, is injected directly onto the pre-column (LiChroprep RP-18, 25–40 µm, 25 × 4 mm, dry-packed) and flushed with 0.04% TFA in water (1 ml/min) for 3 min. The valve† was switched over and the sample was eluted onto the analytical column (LiChrosorb RP-18, 7 µm, 250 × 4 mm) with acetonitrile-water-TFA (19:81:0.04 by vol.), 1 ml/min. Detection was at 254 nm.

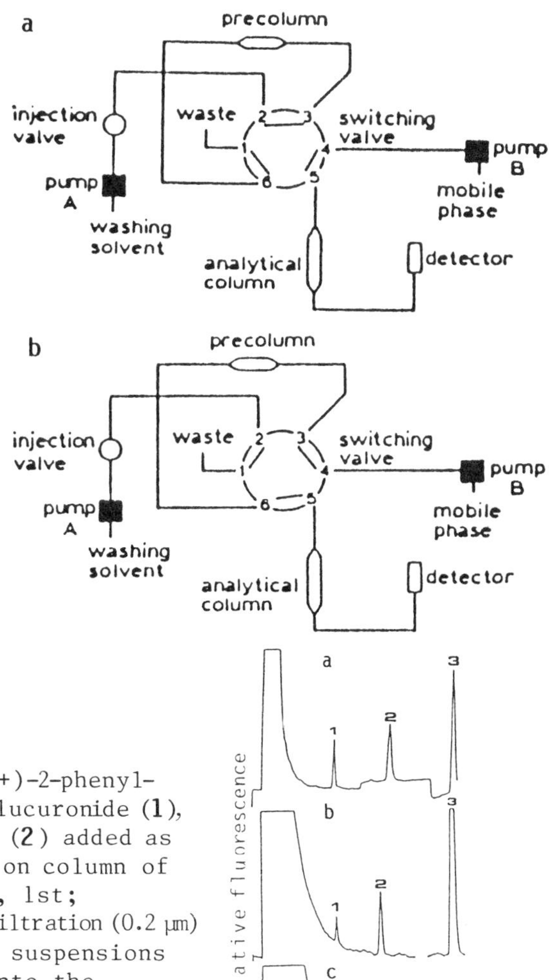

Fig. 4. Patterns for *S*(+)-2-phenylpropionic acid (**3**), its glucuronide (**1**), and 1-naphthol glucuronide (**2**) added as internal standard: effect on column of successive injections.- **a**, 1st; **b**, 10th; **c**, 15th. After filtration (0.2 µm) the acidified microsomal suspensions were injected directly onto the analytical column without the switching system.

†Two 6-way Rheodyne valves, #7010 and #7125, were used in combination.

column did not exceed 15 injections (Fig. 4), possibly because of binding of UDP-glucuronic acid onto free silanol groups. With the switching system and a pre-column change every 50 injections, the analytical column survived several hundred injections. Phenylpropionic acid glucuronides have been selectively measured by this method [7].

(2) SPE (C-18) is followed by pre-column derivatization with BrMmC (Scheme 3). This approach was applied to the measurement of terpene and steroid conjugates [8], and (Fig. 5) to microsomal incubates containing menthol glucuronide.

MICROSOMAL SUSPENSION (1 ml)
spiked with borneol glucuronide as i.s.

Scheme 3. Microsomal suspension treatment for HPLC measurement of menthol glucuronide.

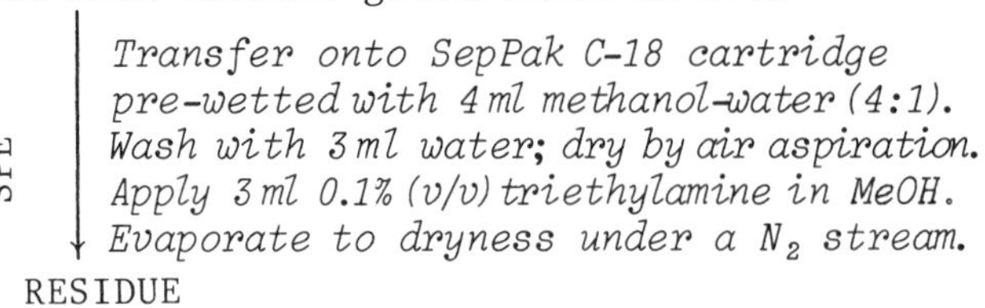

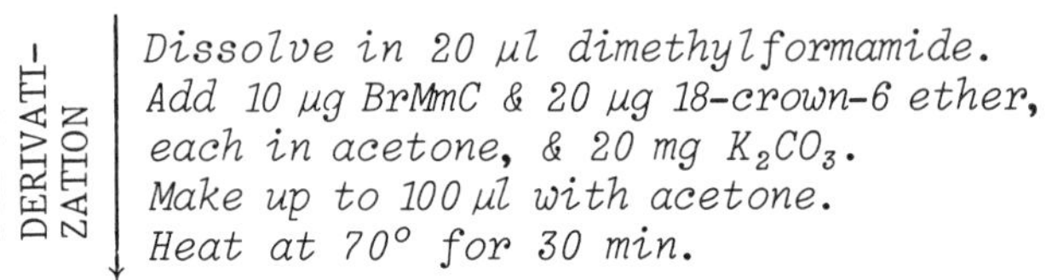

DERIVATIZED SOLUTES
Use 20 µl for HPLC

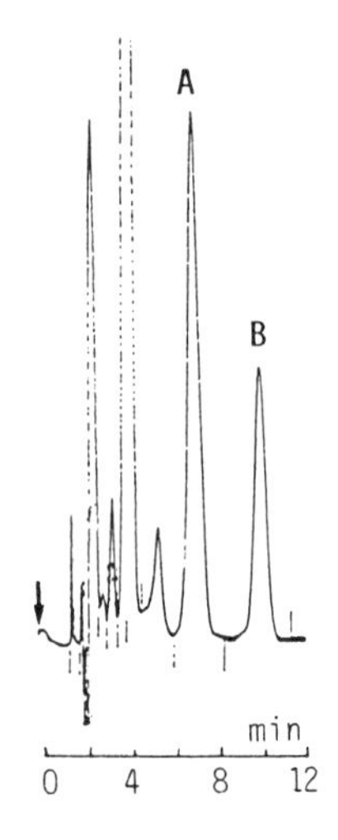

Fig. 5 *(left).* RP-HPLC pattern for glucuronides of (**A**) borneol (i.s.) and (**B**) menthol. **Column:** LiChrospher CH-18, 5 µm, 250 × 4 mm. **Elution:** methanol:water (75:25 by vol.), 1 ml/min. **Detection** at 328 nm.

Characteristics of the assay by Method (2).- Menthol glucuronide could be quantitated from 0.05 to 1 µg/ml with excellent linearity (r = 0.995) and precision (C.V. 4.44%; n = 5).

CONCLUDING COMMENTS

A simple pre-treatment such as deproteinization combined with selective post-column derivatization or 'on-line' clean-up with a column-switching system affords easy improvement in selectivity, specificity and automation in HPLC methods for drugs and their phase I and II metabolites in biological samples.

References

1. Lee, T.L. & Brooks, M.A. (1984) *J. Chromatog. 306*, 429-435.
2. Carlqvist, J. & Westerlund, D. (1985) *J. Chromatog. 344*, 285-296.
3. Brooks, M.A. & Hackman, M.R. (1981) *J. Chromatog. 210*, 531-535.
4. Rogers, M.E., Adlard, M.W., Saunders, G. & Holt, G. (1983) *J. Chromatog. 257*, 91-100.
5. Kagedal, B., Andersson, T., Carlsson, M., Denneberg, T. & Hoppe, A. (1987) *J. Chromatog. 417*, 261-267.
6. Johansson, M. & Westerlund, D.(1987) *J. Chromatog. 385*,343-356.
7. Fournel-Gigleux, S., Hamar-Hansen, C., Motassim, N., Antoine, B., Mothe, O., Decolin, D., Caldwell, J. & Siest, G. (1988) *Drug. Metab. Dispos. 16*, 627-634.
8. Chakir, S., Leroy, P., Nicolas, A., Ziegler, J.M. & Labory, P. (1987) *J. Chromatog. 395*, 553-561.

#C-6

HPLC-EC ANALYSIS OF HYCOSCINE (SCOPOLAMINE) IN URINE

Robin Whelpton and Peter R. Hurst

Department of Pharmacology, London Hospital Medical College, Turner Street, London E1 2AD, U.K.

Requirement *Assay of hyoscine (poor UV absorbance!), ~1 ng/ml urine.*
End-step *NP-HPLC with an i.s.*; EC detection. (GC: instability!)*
Sample handling *Toluene extraction at pH 10 (direct or after enzymic hydrolysis), then Bond Elut CN 'column'. Selectivity achieved.*
Comments *See* DISCUSSION.

Hycoscine (tropylscopine) is a potent anti-cholinergic alkaloid, chemically related to atropine (tropyltropine). It differs from atropine in having an epoxide group. Unlike atropine, which is a racemate, it is administered as the (-)-isomer. Its primary uses are as a pre-medicant, e.g. in Omnopon-Scopolamine, and in motion sickness, e.g. Kwells®. It has been estimated that steady-state plasma concentrations while transdermal patches are being worn are <100 pg/ml, posing analytical problems; another is that the compound is unstable.- It is heat-labile and tends to lose water to give apohyoscine on GC columns. At high pH values it tends to hydrolyze to scopoline (not scopine) and tropic acid (Fig. 1).

Bayne & co-workers [1] developed a GC-MS method in which hyoscine was first hydrolyzed to scopoline which was then derivatized with heptafluorobutyric acid anhydride (HFBA) and chromatographed. The only other approach that can assay plasma concentrations is radioreceptor assay (RRA) [e.g. 2]. Not having access to GC-MS, we decided to develop an HPLC assay. As the analyte absorbs UV light poorly (ε = 190 at 257 nm) and is not fluorescent, EC* detection was chosen

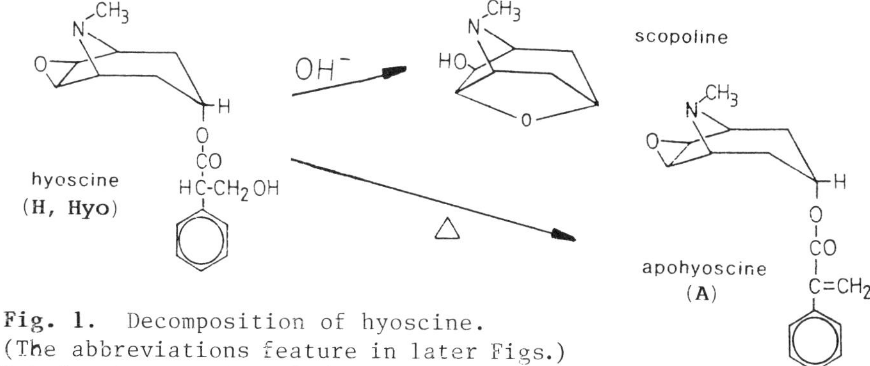

Fig. 1. Decomposition of hyoscine. (The abbreviations feature in later Figs.) Dehydration on GC packed columns produces **A**, whereas hydrolysis gives scopoline (by rearrangement of scopine) and tropic acid.

*EC, electrochemical; i.s. ['I.S.' in Figs.], internal standard.

even though the compound is not particularly electroactive. Considering these problems, we felt that HPLC-EC would be incapable of assaying plasma hyoscine. However, with a renal clearance of ~100 ml/min [3] concentrations in urine should be much higher. (When the urine flow rate is 1 ml/min a plasma concentration of 0.1 ng/ml should lead to a urine concentration of 10 ng/ml.)

MATERIALS AND APPARATUS

Hyoscine.HBr and scopine were from Sigma Chemical Co. Apohyoscine, norhyoscine, scopoline (metabolites/decomposition products), *N*-ethyl-norhyoscine and *N*-propylnorhyoscine (as potential i.s.'s) were prepared by Dr Ian Lawston (Chemical Defence Establishment, Porton Down, Salisbury). Norapohyoscine was obtained by pyrolyzing (120°) norhyoscine on a silica-gel TLC plate for 60 min and chromatographing ($CHCl_3$-methanol-ammonia, 80:10:1 by vol.) to separate the required product. Radioactive hyoscine and apohyoscine were prepared by treating the appropriate desmethyl compounds with methyl iodide; the products were >97% pure after TLC separation. HPLC-grade acetonitrile was from Fisons Scientific Apparatus, Loughborough. [^{3}H]Methyl iodide (85 Ci/mmol) was from Amersham International. β-Glucuronidase/aryl sulphatase (*Helix pomatia*) was from Boehringer. (The ^{3}H label was a recovery-assessment aid.)

Drug-free urine was from the authors, and samples containing hyoscine were from a male volunteer who took hyoscine on two occasions: (1) a 300 µg tablet was swallowed and the urine sample collected for 75 min thereafter; (2) the subject was water-loaded so that urine samples could be collected every 30 min, and the oral dose was 600 µg (2 Kwells® tablets).

HPLC conditions.- The pump was an ACS Model 300 or, with an additional dampener (Negretti, Southampton), an Altex 110A. The 3 µm silica column (Spherisorb, 150 × 4.6 mm) was from Phase Separations, Queensferry. Eluents were pumped at 0.8 ml/min. Peaks were detected with an ESA Coulochem dual-electrode detector, with the electrodes set at +0.9 V. UV detection was used when optimizing the chromatography as norhyoscine gave virtually no response with EC. Scopine and scopoline were detectable only by EC.

OPTIMIZATION OF PROCEDURES

The eluent was optimized (Fig. 2) after initial trials with that used previously for physostigmine [4]. Omitting the methanol and using acetonitrile-0.1 M ammonium acetate buffer, 90:10 by vol., furnished better column efficiencies than when any proportion of methanol was present. Increasing the pH of the buffer reduced the capacity factors. The best separation of the reference compounds was obtained with

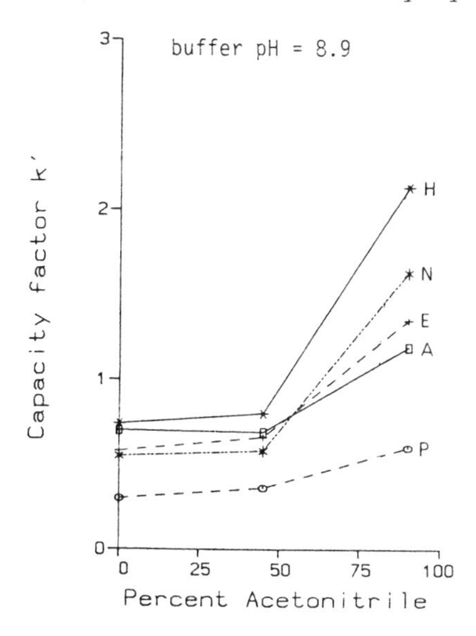

Fig. 2. NP-HPLC retentions as affected by % acetonitrile *(above, left)* and buffer pH. H, A: see Fig. 1; N = norH, E = *N*-ethylnorH, P = *N*-propylnorH.

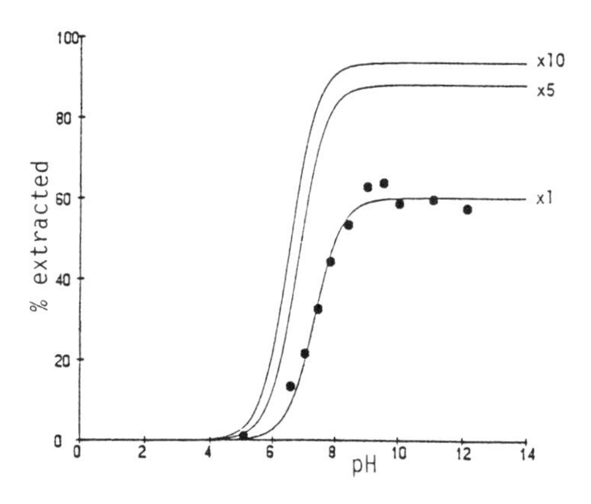

Fig. 3 *(right)*. Hyoscine extraction into toluene *vs.* buffer pH. Experimental points are for equal vols. of toluene and buffer; the associated curve (x1) represents the fitted data. The curves x5 and x10 are calculated for 5:1 and 10:1 toluene:buffer ratios.

pH 8.9 buffer (Fig. 2). *N*-Propylnorhyoscine chromatographed too near the solvent front for it to be suitable as an i.s. Scopine and scopoline, injected in admixture (not shown in the Fig.), had much greater k' values and chromatographed as a partially resolved broad peak.

Partition *vs.* pH.- Extraction characteristics were determined by shaking solutions of hyoscine or *N*-ethylnorhyoscine (chosen as i.s.) in toluene with equal volumes of Britton-Robinson buffers. The apparent partition coefficients were determined as a function of pH and the true partition coefficients calculated [5]. As shown in Fig. 3, using equal volumes the maximum extraction was ~60% for hyoscine, and calculation of extraction curves showed that a solvent:aqueous ratio of at least 5:1 was required to extract >90%. The i.s. being more lipophilic, it was more readily extracted.

The pH-extraction curve indicated that hyoscine decomposed at high pH. This was confirmed, as shown in Table 1. Urine presumably gives a lower pH than water because it acts as a buffer.

Table 1. Extraction recovery of hyoscine (**H**) and the i.s. *N*-ethylnorhyoscine (**E**) *vs.* vol. of 10 M NaOH added to 1 ml of aqueous phase.

ml NaOH	% recovery: **H**, urine	**H**, water	**E**, urine	**E**, water
0.01	83.4	26.4	100.4	39.2
0.02	83.8	10.8	98.4	23.1
0.04	51.4	3.4	65.0	3.9
0.10	10.0	0	17.4	0

Solid-phase extraction (SPE).- As evaporation of solvent extracts may cause problems, the toluene extracts were concentrated by SPE. Several phases were evaluated previously [6]. Acid-prepared nitrile columns seemed preferable. BondElut CN columns (100 mg, 1 ml) were prepared by sequentially drawing through, by vacuum, 1 ml portions of methanol (× 2), water (× 2), HCl, water, methanol and toluene. The toluene extract (4 ml) was drawn through, then 1 ml each of toluene and acetonitrile. The columns were removed from the vacuum manifold. Centrifugal elution was with 500 µl of HPLC eluent.

Urine samples >1 ml were pre-concentrated on BondElut LRC C-18 columns (100 mg; 10 ml reservoir), pre-washed with 1 ml portions of methanol (× 2), water (× 2) and 0.1 M K_2HPO_4. Urine loading was followed by 1 ml portions of water (× 3) and, for elution, methanol.

Urine preparation.- To 1 or 5 ml, i.s. was added (50 ng in 0.1 ml water), then 1 M K_3PO_4 to give pH 10. Toluene extraction as below was, for the 5 ml samples, preceded by centrifugation (to remove any particles) and, taking 4.5 ml, C-18 extraction as above: the methanol eluate was dried down (N_2, 60°) and the residue reconstituted in 1 ml pH 10 phosphate buffer. Toluene (5 ml) was added, and the tubes shaken for 15 min and centrifuged; to 4 ml of organic phase the above BondElut CN procedure was applied, and eluent aliquots (100 µl) chromatographed.

RESULTS

Mean recoveries found, ± C.V. (n = 5), for hyoscine [and apohyoscine] were as follows:- at 100 ng/ml, 84.1 ±2.9 [93.9 ±1.7]; 10 ng/ml, 88.2 ±2.1 [96.8 ±1.1]; 1 ng/ml, 86.2 ±2.0 [96.2 ±4.0]. As the % recovery was not concentration-dependent, adsorptive losses must have been negligible. The calibration curve was linear over the range 1-100 ng/ml (e.g. r = 0.9998). With

Fig. 4. Patterns for 1 ml urine, blank or with 100 ng hyoscine added; 50 ng i.s. added to each.

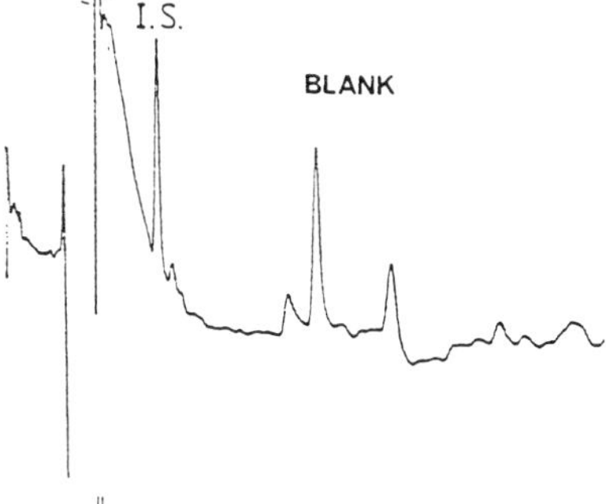

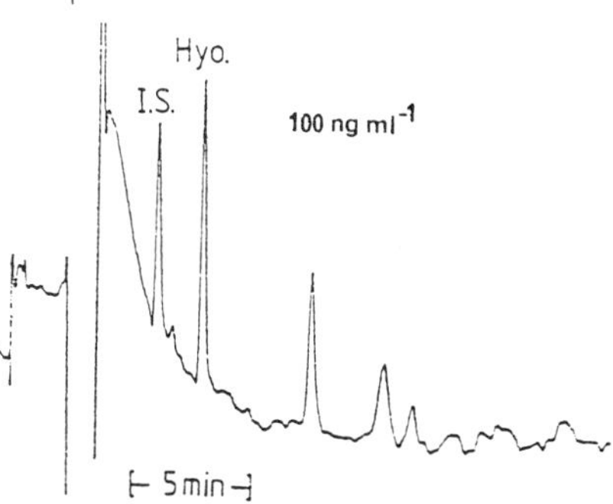

Fig. 5, *below.* Patterns for 5 ml urine; pre-dose, **A** & **B**, with hyoscine spike in **A** as in Fig. 4 (position of hyoscine marked in **B**); **C**, 0-75 min after 300 µg hyoscine.HBr orally. All spiked with 50 ng i.s.

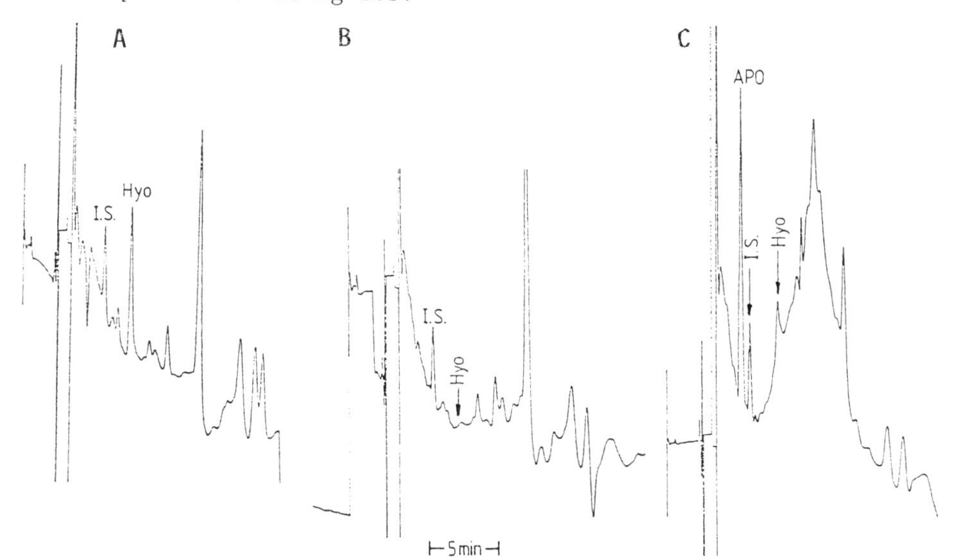

1 ml urine, 2 ng/ml could be detected with a signal:noise >3, improving only to 1 ng/ml with a 5 ml sample as 'noise' came mostly from urinary contaminants.

Fig. 4 shows patterns for blank and spiked 1 ml samples. Hyoscine was detectable in urine collected 75 min after 300 µg taken orally. With 5 ml (Fig. 5), the pattern post-dose (300 µg; **C**) showed a pre-i.s. peak in the position of apohyoscine.

After 600 μg, **A** was detectable in all samples at >30 min, and by 5 h 13 μg had been excreted. Free hyoscine was detected (7 μg) in the 30-60 and 60-90 min samples but not thereafter. Enzymic hydrolysis gave much higher values: 3 μg hyoscine after only 30 min, and altogether 130 μg after 5 h.

DISCUSSION

Hyoscine can be chromatographed by capillary GC, but we noted some pyrolysis even with on-column injection. Slight decomposition may be acceptable if MS detection is used - an isotopically labelled i.s. being employable - but not, we felt, for other GC techniques. Hence HPLC-EC was chosen as the only viable alternative (chemical) assay to GC-MS. By combining liquid-liquid extraction with SPE, we could avoid the problem areas commonly associated with these techniques. The preliminary liquid extraction obviated erratic recovery (due to protein binding) and blocked columns (due to particulate matter). A further bonus is that pH-controlled liquid extraction often imparts additional selectivity. The problems that often occur when extraction solutions are concentrated by evaporation were avoided by adsorbing the compounds onto SPE columns and eluting into a small volume of HPLC eluent. Hence we could use toluene, rather than benzene which can be evaporated more readily but is more toxic.

The bright yellow pigments, which elute with hyoscine, remained in the aqueous layer when extracted with toluene. When it became apparent that urine samples > 1 ml would be required, SPE was used to pre-concentrate the sample, rather than use large volumes of toluene or a more polar solvent. To maximize recovery, high-capacity columns (C-18) were used.

HPLC-EC can measure hyoscine in urine before and after hydrolysis and offers a means of monitoring the drug. For pharmacokinetic studies it may be better to assay total hyoscine, conjugated + non-conjugated, as some if not all the latter may arise from hydrolysis by urinary β-glucuronidase. The high proportion of apohyoscine we found (like hyoscine, it hydrolyzes to scopoline) casts doubt on assays by GC-MS [1], which give excretion values about twice those we found.

References

1. Bayne, W.F., Tao, F.T. & Crisologo, N. (1975) *J. Pharm. Sci. 64*, 288-291.
2. Metcalfe, R.F. (1981) *Biochem. Pharmacol. 30*, 209-212.
3. Shaw, J. & Urquhart, J. (1980) *Trends Pharm. Sci. 1*, 208-211.
4. Whelpton, R. & Hurst, P. (1986) *Meth. Surv. Biochem. Anal. 16*, 181-187 [Reid, E., Scales, B. & Wilson, I.D., eds.*].
5. Whelpton, R. (1989) *Trends Pharm. Sci. 10*, 182-183.
6. Whelpton, R. & Hurst, P.R. (1988) *Meth. Surv. Biochem. Anal. 18*, 289-294 [Reid, E., Robinson, J.D. & Wilson, I.D., eds.*].

*Plenum, New York.

#C-7

AN INVESTIGATION OF DIFFERENT ANALYTICAL TECHNIQUES FOR DETERMINING LOW LEVELS OF LACIDIPINE IN PLASMA

†G.L. Evans, †J. Ayrton, *P. Grossi, *M. Pellegatti, †J. Maltas and †A.J. Harker

*Glaxo Research Laboratories, Via Fleming 2, 36100 Verona, Italy

†Glaxo Group Research Ltd., Greenford, Middx. UB6 0HE, U.K.

*'Précis' by Ed.- This art. comprises an adapted Forum Abstract and (added near press-date) extra material from a late Ms. For routine assays of the drug (***'Ld'***; + a metabolite,* **LdPy***) at sub-ng/ml levels in plasma, the only tolerably satisfactory approach was HPLC-UV.*

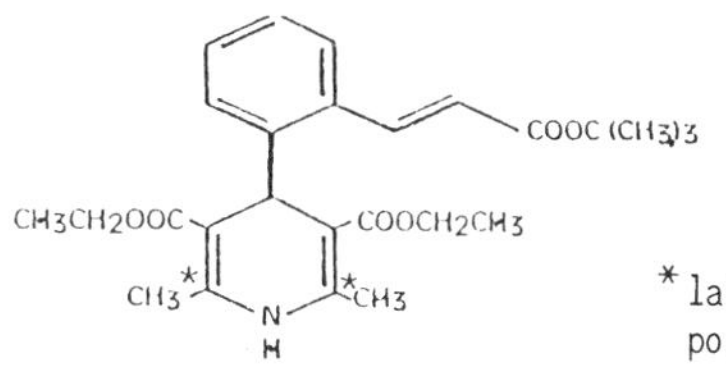

COOC(CH3)3
CH3CH2OOC
COOCH2CH3
CH3*
*CH3
N
H

*labelling position

LACIDIPINE

diethyl-4-{2-[(**tert**-butoxycarbonyl)-vinyl]phenyl}-1,4-dihydro-2,6-dimethyl-pyridine-3,5-dicarboxylate

Lacidipine is a dihydropyridine (DHP) calcium antagonist being developed as an antihypertensive drug, notably potent and long-acting (2 mg/day being the envisaged starting-dose in patients). An analytical method was required to allow the determination of pg levels in human plasma. As outlined in the first part of this article, various techniques (#**I**, #**II**,......) were investigated, with varying degrees of success. Amplification is then given for each technique, with comments.

Sample preparation, where applicable.- Lacidipine was extracted from plasma by precipitation with acetonitrile (1 vol.) followed by solid-phase extraction (SPE) using 500 mg C-18 cartridges. The method enabled the sample to be concentrated: 3 ml plasma was reduced to 100 µl after re-constituting the residue from drying down the eluate in the HPLC mobile phase.

(#**I**) **HPLC-UV**.- A RP-HPLC method was developed to assay plasma for lacidipine; it has a high UV absorbance, but an absorbance maximum (240 or 283 nm) could not be used because of interference by endogenous components. The wavelength used (300 nm) produces a cleaner trace but reduces sensitivity by ~20%. Originally the method had a limit of quantitation of 0.5 ng/ml, but improvements in the signal:noise ratio of detectors and reduction in reconstitution volume have improved sensitivity to 0.25 ng/ml (P = 0.05 for distinction from *nil*).

(#**II**) **HPLC with electrochemical detection**.- This technique was investigated as it had been used to assay nifedipine, another DHP calcium-channel blocker. A single-cell glassy carbon electrode was used with an EDT detector. A high potential (1.1 V) was required to achieve an oxidative response for lacidipine; at this potential the background response caused by endogenous plasma components prevented detection of lacidipine. When a lower potential was used, lacidipine was chromatographically resolved but the limit of detection was higher than that with UV detection.

(#III) **Capillary GC.**- Problems were experienced in attempts to assay lacidipine by capillary GC owing to decomposition of the parent drug to its pyridine analogue on the column. The stratagem adopted was to oxidize lacidipine to its pyridine analogue and chromatograph the latter. [See P.S.B. Minty's article in Vol. 18, this series.- *Ed.*] A maximum sensitivity of 0.35 ng/ml was in prospect; but this method was abandoned when the pyridine compound was found to be a metabolite in plasma. GC-MS of the pyridine analogue was also investigated, with a prospect of detecting 0.5 ng/ml.

(#IV) **Column-switching HPLC**, enabling large volumes of untreated plasma to be injected into a narrow-bore column, has been tried so as to increase assay sensitivity. Diluted plasma was injected into a short pre-column (10 × 4 mm) packed with pellicular material (Perisorb RP8, Merck). After washing, the pre-column was back-flushed into the analytical column (Spherisorb ODS-II, 100 × 2 mm) where the separation of the drug from residual matrix components took place. Detection was by UV absorption at 300 nm, or by amperometric oxidation at a potential of +1 V *vs.* the Ag/AgCl reference electrode.

Relatively good detection limits were achieved by increasing the volumes of the plasma samples: with 2.0 ml concentrations of lacidipine as low as 0.1 ng/ml were detectable. However, the efficiency of the analytical column was rapidly reduced by the large-volume injections. Moreover, dirty samples sometimes caused unpredictable pre-column blockages even with the use of stainless steel meshes (5 µm) as frits.

(#V) **Thermospray LC-MS** (HPLC linked to TSP-MS) has enabled a method to be developed for assaying lacidipine in plasma samples from human volunteer trials. After subjecting 3.0 ml plasma to SPE (see above), lacidipine and i.s. ($[^{13}C]_4$-lacidipine) were separated from endogenous components by HPLC. Intense pseudomolecular ions (ammonium adducts of lacidipine and i.s., m/z = 473 and 477 respectively) allowed the detection of lacidipine down to 0.1 ng/ml. Nevertheless, the upper limit for quantitation was $\not>$3 ng/ml because of non-linearity of response, thought to be due to the occurrence, at the higher concentrations, of absorption in the source. (Equipment: Finnegan MAT 4500; scan time 2 sec with ~1 sec for each ion.) [For a technique description, by T.J.A. Blake, see Vol. 18.- *Ed.*]

(#VI) **Radioimmunoassay.**- The setting-up of an RIA for lacidipine in plasma was first attempted using an immunogen obtained by coupling to bovine thyroglobulin the des-t-butyl analogue of the drug. The antiserum raised in rabbits was tested with three radiolabelled DHP's (^{3}H-nitrendipine, ^{3}H-PN 200/110 and ^{14}C-lacidipine). A comparison of the amounts of specific binding confirmed that the antibodies possessed an unexpected specificity for certain DHP's, such as to preclude their use in the assay of lacidipine. Because of the failure of this attempt, the development of a new RIA has recently been started using a different immunogen, and a method is now under development for assaying lacidipine in plasma.

(#VII) **Radioreceptor binding assay** (RRA).- Receptor preparations exhibiting high levels of specific binding for ^{3}H-nitrendipine were obtained routinely from rat cerebral cortex homogenates. Assay methods based on the displacement of this binding have been reported for DHP calcium-channel blockers.

============

Note by Ed. *- A follow-on para. on RRA that appeared in the Forum Abstract has been completely replaced by the up-dated version late in this article. For other techniques the foregoing wording (representing the mid-1989 position) has* ***not*** *been systematically altered to accord with what appears below.*

AMPLIFICATION AND UP-DATING (early-1990) (Lacidipine *denoted* **Ld**)

(#I) **HPLC-UV.**- Fig. 1 shows plasma chromatograms, with an explanation in the Legend of a stratagem to circumvent the problem that, although **Ld** elutes at 6.8 min, the run time has to be ~36 min because of late-eluting interferants. With ^{14}C-**Ld**, good extraction efficiency was demonstrated; ~10% of the label was not retained in the SPE step (with either C-8 or C-18). With 1-16 ng/ml the calibration line showed good linearity (R.S.D. typically 1.8%).

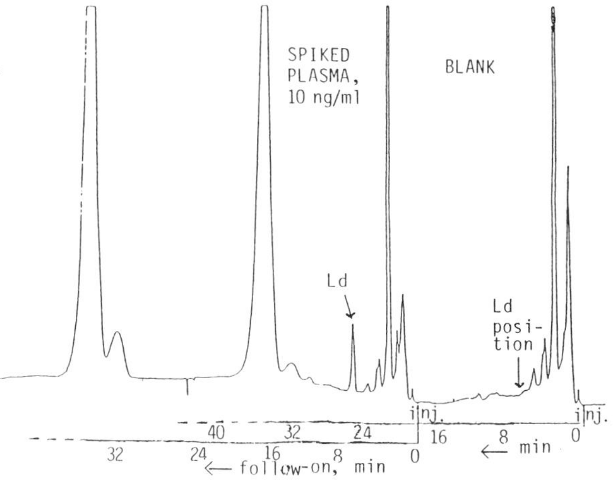

Fig. 1. HPLC-UV pattern for **Ld** added to human plasma. Taking advantage of a flat baseline between 8 and 28 min, the second run was started at 18 min when late-running endogenous material had not yet eluted.
Column: 5 µm Hypersil ODS, 100 × 5 mm, kept at 40°; 100 µl injected. Mobile phase: methanol/water/acetonitrile, 66:28:6 by vol.; 1 ml/min.

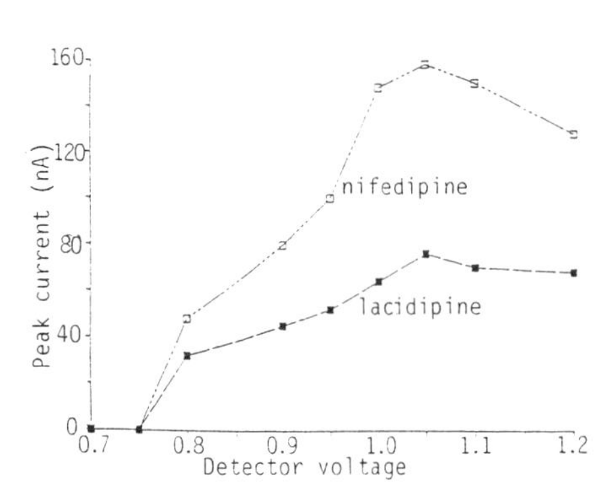

Fig. 2. Voltammograms for **Ld** and nifedipine (see foregoing text).

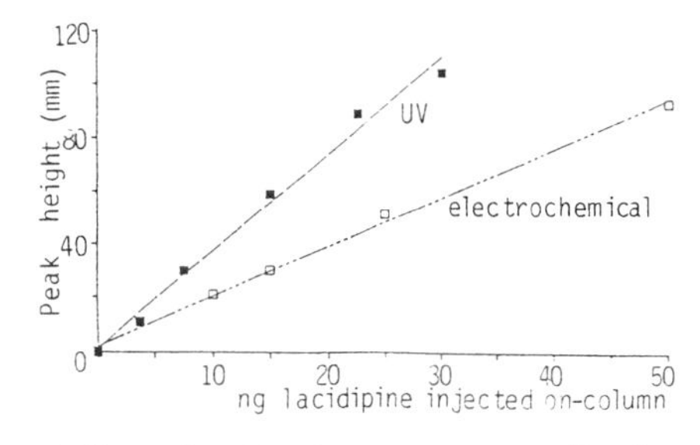

Fig. 3. HPLC-UV assay of **Ld** compared with HPLC-EC (at 0.95 V), which gave poorer sensitivity. Interferants precluded 1.1 V (optimal).

(#**II**) **HPLC with EC detection** (voltammogram: Fig. 2) is compared with HPLC-UV in Fig. 3. A dual-cell instrument has been tried, so far unsuccessfully, with the higher sensitivity gain in the hope of reducing endogenous interferences.

(#**III**) **Capillary GC, with trial of MS.**- After extraction from plasma, **Ld** was oxidized by $NaNO_2$ to the pyridine analogue (**LdPy**); this was extracted, and the residue from drying-down reconstituted in toluene for GC. Detector comparisons showed that sensitivity with ECD was ~100-fold that with FID and ~12-fold that with NPD, allowing an on-column load of ~0.1 ng to be detected. However, marked endogenous interference (Fig. 4) showed a need for better sample clean-up.

In MS trials on **LdPy** with methane as reagent gas, positive-ion CI gave a poor response. SIM chromatograms at m/z 234 or 453 showed similar signal-to-noise for EI and negative-ion CI (NICI), but the general background was less with NICI, and it was tried with spiked plasma as shown in Fig. 5 for 0.5 ng/ml - the probable detection limit.

(#**IV**) **HPLC with column switching** was performed (see above, and Fig. 6) on plasma centrifugal supernatants with 0.1 vol. of acetonitrile added to disrupt **Ld** protein binding. Although 0.1 ng/ml was detectable (Fig. 7), detrimental effects of injecting large volumes of plasma (see above) precluded routine use.

(#**V**) **Thermospray LC-MS** entailed protein precipitation on 3 ml plasma (1 vol. acetonitrile), SPE on the supernatant with drying-down of the acetonitrile (2 ml) eluate, reconstitution using 120 µl of mobile phase, and injection of 100 µl (NovaPak C-18 column, 75 × 4 mm, 50°). **Ld** and i.s. were eluted, at 2.1 min, with 72:28 methanol/50 mM pH 5.5 ammonium acetate (1.5 ml/min). SIM: see above.

(#**VI**) **RIA.**- As mentioned, use of desbutyl-**Ld** as hapten (to allow thyroglobulin attachment) was abortive; **Ld** did not displace

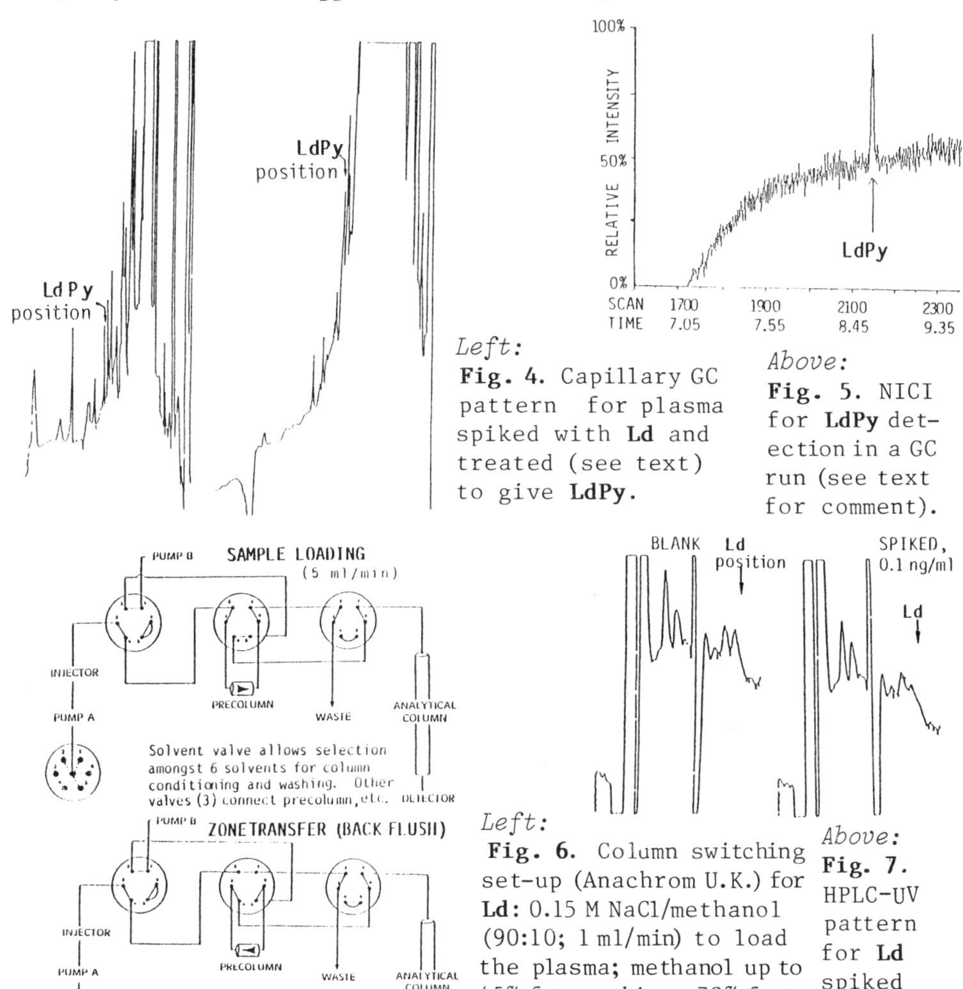

Left: **Fig. 4.** Capillary GC pattern for plasma spiked with **Ld** and treated (see text) to give **LdPy**.

Above: **Fig. 5.** NICI for **LdPy** detection in a GC run (see text for comment).

Left: **Fig. 6.** Column switching set-up (Anachrom U.K.) for **Ld**: 0.15 M NaCl/methanol (90:10; 1 ml/min) to load the plasma; methanol up to 45% for washing, 70% for back-flushing and elution on analytical column.

Above: **Fig. 7.** HPLC-UV pattern for **Ld** spiked into plasma.

the radiolabel. Immunoreactivity with certain DHP's appears to depend on the substituent on the phenyl moiety. Trial of $-CH=CHCO_2C(CH_3)_3$ at its 3-position is in hand; if Ab's showing affinity for **Ld** are achieved, metabolites will need checking.

(#**VII**) **RRA.**- Fig. 8 illustrates the binding characteristics. Freezing the preparation caused ~10% loss of specific binding, but thereafter no further deterioration occurred. However, there was batch-to-batch variability and general assay irreproducibility: for the mean of quadruplicates the S.D. was often 16-17%. Yet an assay was developed for **Ld** in plasma, the extract from SPE being reconstituted in the incubation mixture. Specific binding of **Ld**, as % of ^{3}H-nitrendipine binding in controls lacking **Ld**, was linear if plotted on a

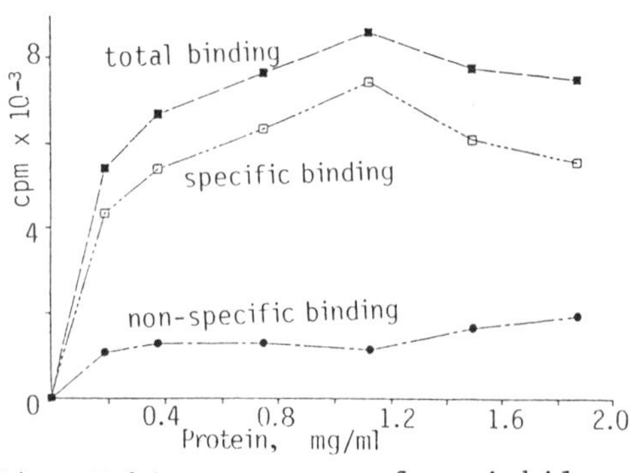

Fig. 8. Testing the viability of the receptor preparation with ^{3}H-nitrendipine.

Note on the calibration line (not shown): variability was so high that any **Ld** concentration that inhibited the binding by <30% or >80% would be outside the linear range.

Log:Logit basis (0.1-1.1 ng/ml). Taking account of variability, the detecion limit with 1 ml plasma is at present ~0.5 ng/ml. Four plasma pools from dosed volunteers, with set levels (ng/ml) based on assays by HPLC-UV, were assayed by RRA also (shown **bold**):- 0.5, **1.0**; 1.0, **8.0**; 1.5, **8.5**; 2.0, **>10**. The higher values by RRA probably reflect metabolite interferences; in an attempt to circumvent these by pre-RRA HPLC separation, isolation losses were unduly great.

DISCUSSION

None of the methods investigated has achieved the required pg sensitivity, and several have notable disadvantages. RRA is affected by metabolites which, being structurally similar, would not be readily separable by a simple extraction method. RIA failed because the Ab's had unexpected specificity for certain DHP's.

If **LdPy**, a metabolite in dog and rat plasma, is present in human plasma too, it will be co-measured with **Ld** if generated by oxidizing assay **Ld**, as in GC-ECD and GC-MS which are in any case rather insensitive. If, however, their sensitivity could be improved, levels of **LdPy** itself might be measurable in human plasma.

HPLC-UV, giving 0.5 ng/ml sensitivity with 3 ml plasma, has proved very robust and has enabled >5000 samples to be successfully assayed. Column switching does not provide a suitable alternative for routine assay because of the problems with blockages. Although a detection limit of 0.2 ng/ml seemed to be attainable, its reproducibility has not been assessed and it might be higher in practice. LC-MS provides a potential sensitivity of 0.1 ng/ml, but may lack the reproducibility and robustness needed for routine assay, especially with large numbers of samples.

#C-8

SOME APPLICATIONS OF MASS SPECTROMETRY IN DRUG DETECTION AND METABOLISM STUDIES IN THE HORSE

E. Houghton, P. Teale, M.C. Dumasia, A. Ginn, †D. Marshall and †D.B. Gower

Horseracing Forensic Laboratory Ltd.,
P.O. Box 15, Snailwell Road, Newmarket CB8 7DT, U.K.

and †U.M.D.S., Division of Biochemistry,
Guy's Hospital, London SE1 9RT

The prominence of our Laboratory in the area of drug detection in racing thoroughbreds has been maintained through ongoing research and technological updating. As is outlined in this article, drug screening, usually on urine with initial SPE, is presently based on GC with nitrogen-specific detection, HPLC and immunoassay. For the requisite confirmation of any positive response in a screening procedure, there is an important role for MS, notably GC-MS, for certain steroids and drugs in general. In steroid and other areas GC-MS has also served to support studies of* in vivo *and* in vitro *metabolism. MS-MS has recently been adopted for drug screening, confirmatory procedures and metabolism studies; it allows quite crude extracts to be analyzed rapidly.*

In 1988 the Horseracing Forensic Laboratory completed 25 years of service to horseracing. During this period, through the rapid introduction of state-of-the-art technology into routine screening and commitment to extensive research, the Laboratory has remained at the forefront in drug detection in thoroughbred horseracing.

A positive response to a screening procedure - initial SPE, then GC, HPLC or immunoassay - only indicates the possible presence of a drug or metabolite in the biological fluid under investigation: it is mandatory in sport to chemically identify the specific drug present, and confirmatory methods must be developed. In the latter context MS has played an important role.

**Abbreviations.*- HCG, human chorionic gonadotrophin; SPE, solid-phase extraction; MS, mass spectrometry (CAD, collisionally activated dissociation; CI, chemical ionization; EI, electron impact; SIM, selected ion monitoring); TBDMS, t-butyldimethylsilyl; i.s., internal standard.

GC-MS has been used extensively in the development of confirmatory analysis procedures for drugs in general ([1], and arts. #ncC-1 & #ncC-2, this vol.), besides the corticosteroids [2, 3] and anabolic steroids [4-8]. The technique has also been used to support drug metabolism studies [9-17] in the horse, both *in vivo* and *in vitro*, and for steroid biogenesis [18] and profile studies [19]. In metabolism studies stable isotopes have been used to facilitate metabolite identification [18]. Recently the principle of threshold values has been introduced into the Rules of Racing to overcome the problems of detecting small quantities of substances of dietary origin which, due to their pharmacological action, are regarded as prohibited substances, and to allow for the detection of administration of endogenous steroid hormones. In establishing these threshold values, quantitative MS has played an important role [20-23].

The recent introduction of a triple quadrupole instrument into the Laboratory has resulted in the application of MS-MS in drug screening, confirmatory procedures and metabolism studies. Some of these applications are reviewed here.

GC-MS AND MS-MS analysis

The instruments available for our use and their major applications are as follows:

- FINNIGAN 1020: support of the routine screening programme;
- FINNIGAN MAT INCOS 50: as for FINNIGAN 1020;
- HEWLETT PACKARD 5790: GS-MS screening for anabolic steroids by SIM, and
  quantitative GC-MS for steroid endocrinology studies;
- FINNIGAN MAT TSQ 70: development of confirmatory analysis methods for drugs,
  support of drug metabolism studies,
  CI/negative-ion CI studies,
  MS-MS applications in drug screening/confirmatory analysis, and
  thermospray studies - and for the MS-MS studies also;
- FINNIGAN MAT ITD: steroid profile studies in biological fluids and tissues, and
  in vivo/in vitro metabolism studies in the horse.

All GC-MS studies were carried out using fused silica capillary columns with helium as carrier gas. For both steroid analysis and general drug analysis OV1 columns were used.

DEVELOPMENT OF A QUANTITATIVE GC-MS METHOD FOR URINARY TESTOSTERONE

In the field of thoroughbred horseracing, as in human athletics, the detection of the abuse of anabolic steroids

presents a challenging analytical problem. The anabolic formulations used in veterinary practice are based upon four parent steroids: nandrolone, testosterone, 1-dehydrotestosterone and trenbolone. In the development of screening and confirmatory analysis procedures for detection of abuse in horseracing, problems stem from a number of sources:-
(1) the complex metabolism of these steroids in the horse,
(2) the endogenous presence of some of these steroids and their metabolites in normal horse urine, and
(3) interference from other endogenous compounds in normal urine.
The situation is complicated further by the fact that there are 3 categories of racing thoroughbred: the gelding (castrated male), the colt (intact male) and the filly, each showing a characteristic urinary steroid profile.

Urinary steroid profile studies in the colt have demonstrated significant levels of testosterone present primarily as a sulphate conjugate. Metabolism studies on testosterone and its proprietary anabolic preparations in the gelding have also demonstrated that the major testosterone metabolite is the sulphate conjugate. Thus, in order to detect the administration of this endogenous hormone to the colt two approaches were considered. — Firstly, the measurement of the ratio of the urinary testosterone level to a second steroid which is not a metabolite of testosterone [24] (the method adopted for athletics [25]); secondly, the development of a quantitative method for urinary testosterone and utilization of this method to determine the normal range of values in a large population of colts. From statistical analysis of these data it might be possible to establish a value above which a particular sample is declared positive.

For simplicity and speed in developing a urinary testosterone assay, steroid isolation was based on SPE and it was aimed to analyze the testosterone present as sulphate. Following addition of the i.s. [(16,16,17-$^2H_3$)-testosterone] the steroid conjugates were extracted on a C-18 SepPak cartridge and eluted with solvolysis solvent: ethyl acetate/methanol/H_2SO_4, 50 : 10 : ~0.2 (by vol.). The sulphate conjugates were then cleaved by incubation of the eluate either overnight at 37° or for 2 h at 50°. The neutral steroids were isolated by washing the ethyl acetate with 2 M NaOH and, after drying, by removing the solvent under N_2. The residue was derivatized for GC-MS analysis.

In the development of quantitative GC-MS methods for steroid analysis, careful consideration must be given to the choice of derivative. Ideally it should show significant ions in the high-mass region of the spectra, thus enhancing

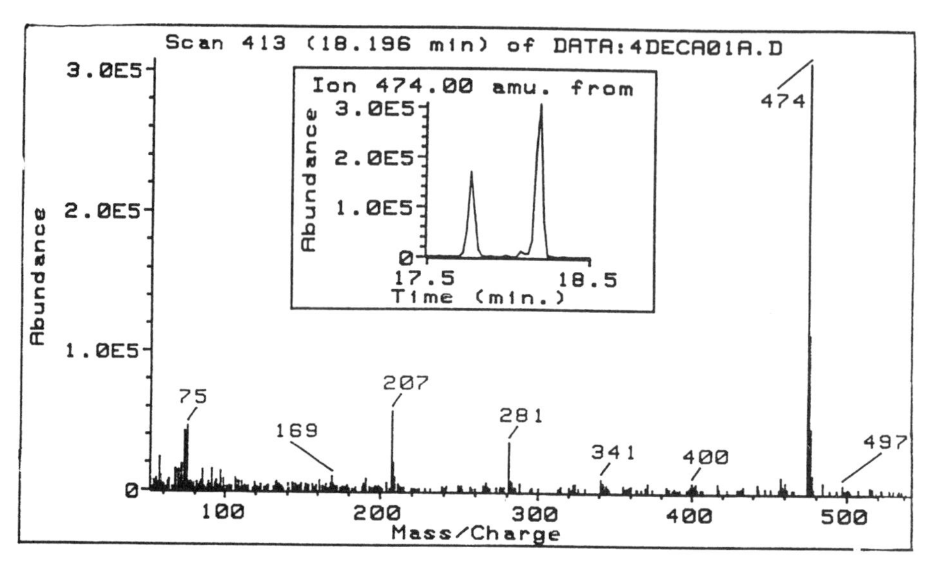

Fig. 1. EI mass spectrum of the oxime-*bis*TBDMS derivative of testosterone (*syn* and *anti* isomers: *INSET* shows their separation).

both sensitivity and specificity. For testosterone this requirement is met by the oxime-*bis*TBDMS derivative for which the major portion of the total ion current is concentrated in the ion at m/z 474 ($M^{+\cdot}$-57; Fig. 1). Also, due to the bulky nature of the TBDMS group on the oxime function at C-3, the *syn* and *anti* isomers of the derivative are clearly separated by GC analysis (Fig. 1) and quantitation can be based upon either peak [26].

Using this method, calibration lines for testosterone were established, over the range 0-200 ng/ml, in urine (from a gelding) which was initially passed through an XAD-2 column to remove any endogenous steroids. Linear calibration lines were obtained (Fig. 2) based on m/z 474 (testosterone) and m/z 477 (the i.s.). The method was subsequently used to determine the effect of administered HCG on urinary testosterone levels in the colt. HCG produced a marked effect, the testosterone peak value (3.7 μg/ml) 4 days after dosing being 100-fold greater than the normal value in the horse studied. The quantitative method is now being used to determine the range of normal values for urinary testosterone in the colt.

USE OF STABLE ISOTOPES AND GC-MS IN STEROID BIOSYNTHESIS STUDIES IN THE HORSE

In steroid biosynthesis studies, the obvious need to distinguish between endogenous and substrate-derived material is well met by the use of stable isotopes in conjunction with GC-MS [27]. Provided that these labels (the introduction of which is not difficult) are in metabolically stable positions, the characteristic mass shift or isotopic clusters produced in GC-MS allow unequivocal identification of metabolites. This technique has been used to investigate the *in vitro* biosynthesis of C_{18} neutral steroids in horse testes [18] and more recently in studies related to steroid biosynthesis in the pregnant mare.

In an attempt to gain a more detailed insight into enzymic activities within the foeto-placental unit in the mare, *in vitro* incubations of various foetal tissues have been done with various steroid substrates. In a dual-labelling study involving $4\text{-}^{14}C$ and ^{2}H -testosterone, aromatase activity has been shown to be concentrated in the allantochorion, the foetal side of the placenta. Following incubations with fresh tissue homogenates the metabolites were isolated by either solvent extraction or SPE and the neutral and phenolic steroids separated by chromatography on triethylamino-hydroxypropyl Sephadex LH-20 [28]. The steroid derivatives (methyloxime-TMS ethers) were analyzed by GC-MS. The use of ^{14}C-testosterone allowed careful monitoring of the efficiency of the analytical procedures and the % conversion to phenolic steroids, whilst in the GC-MS analysis of the derivatized metabolites the ^{2}H label provided unequivocal evidence that the phenolic steroid oestradiol was substrate-derived. This is clearly demonstrated in Fig. 3 which shows the twin ion peaks at m/z 416 and 420 for the molecular ions of oestradiol and its ^{2}H-analogue. The usefulness of stable isotope techniques with GC-MS in metabolic studies is also exemplified by our identification studies on urinary propranolol metabolites (#ncC-1, this vol.) [and by various arts. in earlier vols.- *Ed.*].

SOME APLICATIONS OF MS-MS IN CONFIRMATORY ANALYSIS AND DRUG METABOLISM STUDIES

In sport the mandatory unequivocal identification of the particular drug or metabolite poses an analytical challenge where concentrations are low because of prolonged excretion, extensive metabolism or, with potent drugs, low dosage. In horseracing, matrix complexity (urinary extracts) further complicates the situation.

Over the past few years, MS-MS has been used increasingly in the analysis of trace components in complex biological matrices [29, 30]. The technique provides the analyst with an extremely powerful tool applicable to direct analysis of complex mixtures, to the provision of structural information and to the elucidation of MS fragmentation processes. High

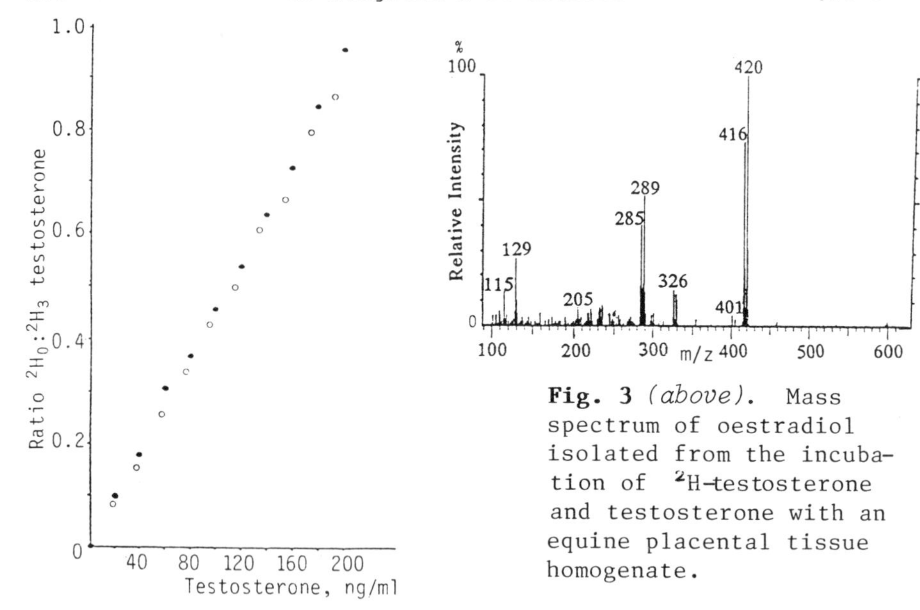

Fig. 3 *(above)*. Mass spectrum of oestradiol isolated from the incubation of ^{2}H-testosterone and testosterone with an equine placental tissue homogenate.

Fig. 2. Calibration lines for testosterone based on the *syn*- and *anti*-isomers of the oxime-*bis*-TBDMS derivatives *vs.* $^{2}H_{3}$-testosterone as i.s.: r = 0.9987 and 0.9991 for the minor (●) and major (o) peaks respectively. (To be settled: *syn* & *anti* assignments.)

specificity of the MS-MS system, low detection limits and a minimal need for sample pre-treatment allowing a high sample throughput are advantageous in the analysis of targeted compounds in complex biological matrices. Inclusion of chromatographic separation and/or the use in combination with soft ionization techniques enhance the probability of identification.

We have used the daughter-ion mode to obtain spectra for confirmatory analyses where normal MS techniques have failed, and to elucidate fragmentation processes. The parent-ion mode has proved useful in metabolite detection and for targeted screening procedures for specific drug groups.

MS-MS data were generated on a triple-quadrupole instrument, the Finnigan MAT TSQ 70. In the daughter-ion mode a single ion is transmitted through the parent set of quadrupole rods Q1 (Fig. 4) and subjected to CAD in the collision cell Q2, which is pressurized with argon (1-3 mtorr). Q2 acts as an ion transmission device and transmits all the fragment ions to Q3 which is scanned over a specific mass range yielding the daughter-ion spectrum. In the parent ion-mode (Fig. 4) Q1 is scanned over a specific mass range and the range of ions is subjected to CAD; a single ion is monitored

Fig. 4. Schematic diagram for production of a daughter ion and a parent ion spectrum using a triple stage quadrupole MS.

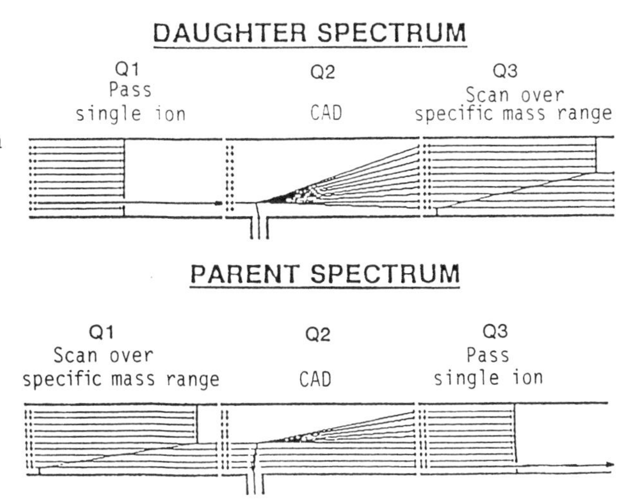

by Q3. The output in the latter mode thus consists of all the ions transmitted through Q1 which fragment in the collision cell to yield the common daughter ion transmitted through Q3. This technique is thus ideal for detecting members of a specific group of drugs which in MS show a common fragment ion.

We screen for diuretics by HPLC of a Tox Elut extract following removal of the basic drugs. The method provides an effective screen for the 'neutral drugs', e.g. the methyl-xanthines and a number of diuretics, hydrochlorthiazide, tri-chlormethiazide and methazolamide.

Following detection of hydrochlorthiazide in a urine sample by HPLC, confirmatory analysis proved difficult due to the complexity of the urinary extract. Tox Elut extraction and methylation followed by GC-MS analysis failed to yield satisfactory confirmatory data, and the sample was subjected to MS-MS. Fig. 5A shows the direct-insertion probe EI spectrum of methylated hydrochlorthiazide and of the urinary extract. The latter, due to the complexity of the extract, showed little evidence for the presence of the methylated drug; the molecular ion, M^+ of m/z 353, is of low intensity. The specificity of MS-MS was demonstrated when the same extract was analyzed in the daughter-ion mode for m/z 353: the crude urinary extract gave comparable data to those of the standard (Fig. 5B). As the analysis was carried out in the EI mode, complementary data were rapidly obtainable by switching the instrument to monitor daughter ions of the fragment ion m/z 310. Daughter-ion spectra for the sample and the standard were virtually superimposable.

Opposite: **Fig. 5.** MS studies on hydrochlorthiazide, following its detection in a Tox Elut extract of urine.

Hydrochlorthiazide

A & **B**. Methylated hydrochlorthiazide: direct-insertion probe EI mass spectra (**A**), and daughter-ion spectra for m/z 353 (**B**). **C** shows daughter-ion spectra m/z 269 for underivatized hydrochlorthiazide.

As shown in Fig. 5C, following SPE re-extraction on Tox Elut it was also possible to generate the daughter-ion spectrum for underivatized hydrochlorthiazide, monitoring daughters of the molecular ion m/z 269.

Due to the complexity of the extract it was felt that it would have been extremely difficult to obtain confirmatory data by any other means than MS-MS without resorting to extensive purification.

THERMOSPRAY MS-MS SCREENING FOR PROPRANOLOL

The cardioselective β-adrenergic antagonists (β-blockers) have recently been added to the list of doping agents banned by the International Olympic Committee. Screening methods for β-blockers are difficult to develop because of the number of drugs available, the dosage and routes of administration, extensive species-dependent metabolism and the instability of some of the metabolites.

The *in vivo* metabolism of propranolol in the horse was studied in order to define a specific analyte upon which to develop a method of confirmatory analysis.* After administration of a mixture of $^2H_0$-, $^2H_2$- and ^3H-propranolol hydrochloride, 62% of the dose was recovered in the urine in the first 24 h. After mixed-enzyme hydrolysis of the 0-24 h pooled urine (10 ml), the metabolites were extracted at pH 9.6 with ethyl acetate. Preliminary TLC analysis indicated that the major metabolite had an R_f value correspon-ding to authentic 4-hydroxypropranolol. However, due to instability of this metabolite, difficulty was experienced in confirming its identity in the TLC eluate by GC-MS after derivatization. Therefore an aliquot of the crude extract (= 100 µl urine) was injected *via* a Rheodyne loop and analyzed by thermospray MS-MS (filament on) in the parent-ion mode, monitoring the parents of m/z 72. [In Vol. 18, this series, T.J.A. Blake surveyed thermospray MS, in the HPLC context.-*Ed.*]

* Propranolol features in a companion art., #ncC-1.

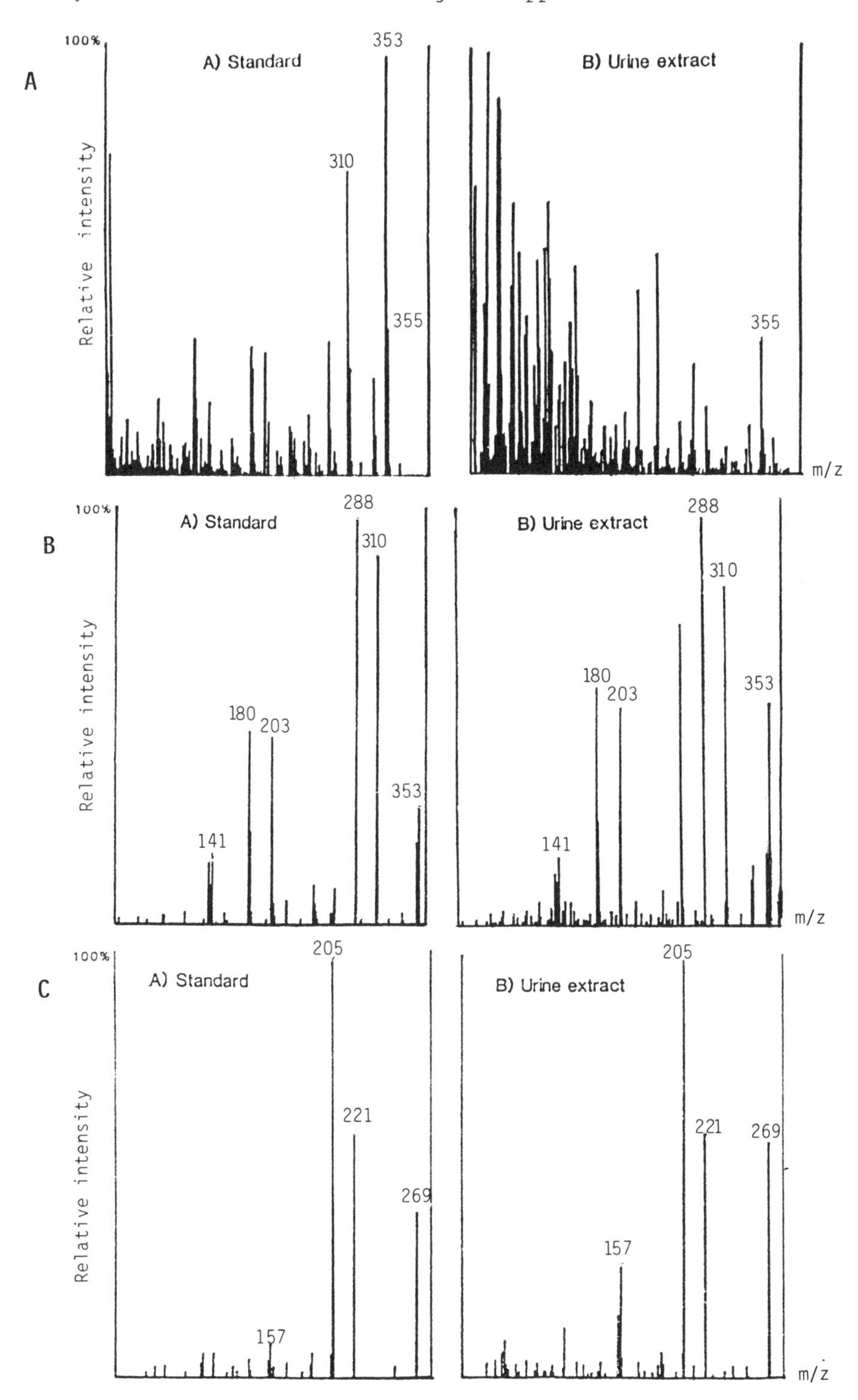
A
100%
A) Standard
B) Urine extract
Relative intensity
353
310
355
355
m/z
B
100%
A) Standard
B) Urine extract
Relative intensity
288
310
180
203
141
353
288
310
180
203
353
141
m/z
C
100%
A) Standard
B) Urine extract
Relative intensity
205
221
269
157
205
221
269
157
m/z

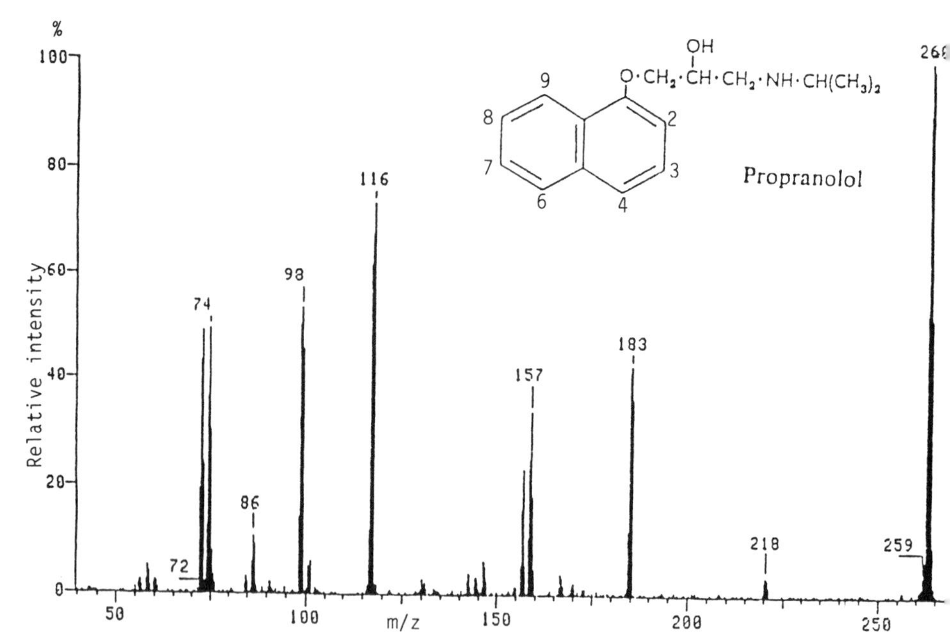

Fig. 6. Daughter-ion spectra of m/z 260 $(M+H)^+$ for propranolol using the thermospray inlet (filament-on mode).

Fig. 6 shows the thermospray MS-MS daughter-ion spectrum of authentic propranolol [$(M+H)^+$ m/z 260]. The ion at m/z 72 is characteristic of the isopropylaminopropan-2-ol side chain. The fragment ion m/z 72 should be common to all metabolites retaining the side chain and also to other drugs possessing a similar side chain. Thus monitoring for parent ions of m/z 72 on crude extracts should provide a profile for these metabolites or drugs.

Analysis of the crude urine extract in the parent-ion mode gave the ions at m/z 276 and 278 corresponding to $(M+H)^+$ of $^2H_0$- and $^2H_2$-4-hydroxypropranolol shown in Fig. 7, and confirmed the identity of the metabolite. Retention of the two ^2H atoms demonstrated that the hydroxylation had occurred *via* the 'NIH shift mechanism'*.

CONCLUSIONS

Due to the advances in drug development leading to increased potency, the administration of endogenous steroids and their corresponding trophic hormcnes and the use of products of biotechnology, drug detection in sport is an ever-increasing analytical challenge. It can be met only

*named after the labs. where it was discovered: formation of arene oxides, the proton being retainable by a shift to an adjacent C atom.

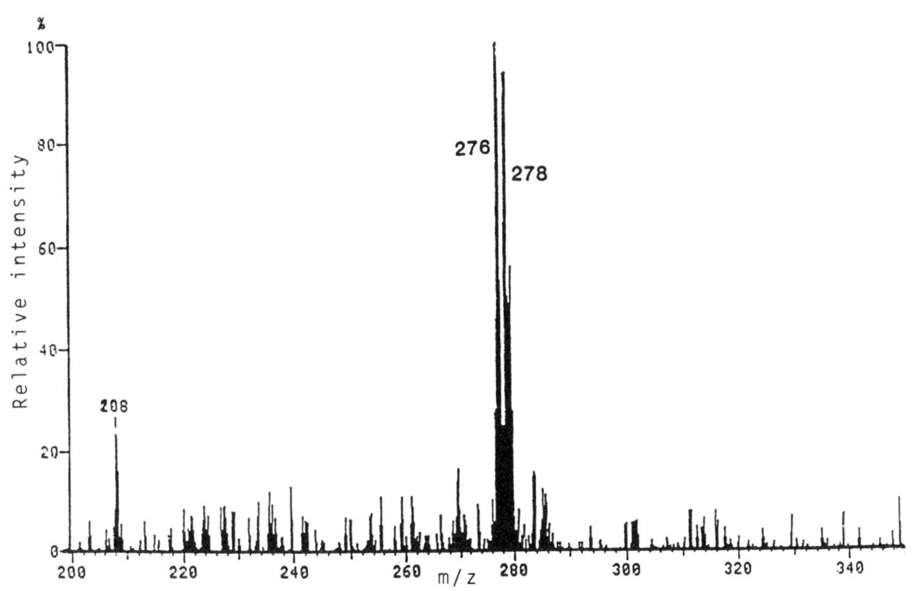

Fig. 7. Parent-ion spectrum of m/z 72 showing the presence of the major metabolite (4-hydroxypropranolol) in a horse urine extract following administration of propranolol and $^{2}H_2$-propranolol.

by the application of state-of-the-art technology and related research programmes.

MS is an area where major advances have been made in the past decade with the introduction of new ionization techniques, extension into the high mol. wt. area, new and improved integrated systems, and the introduction of low-cost and multiple-sector instruments. These advances have broadened the areas of application of the technique to cover many disciplines. This laboratory has been fortunate in being able to take advantage of some of these advances with the introduction of automated GC-MS systems and triple-quadrupole technology in the areas of drug screening, confirmatory analysis, drug metabolism and drug-related programmes.

References

1. Houghton, E. (1982) *Biomed. Mass Spectrom.* *9*, 103-107.
2. Houghton, E., Teale, P., Dumasia, M.C. & Wellby, J.K. (1982) *Biomed. Mass Spectrom.* *9*, 459-465.
3. Houghton, E., Teale, P. & Dumasia, M.C. (1984) *Analyst* *109*, 273-275.
4. Houghton, E., Oxley, G.A., Moss, M.S. & Evans, S. (1978) *Biomed. Mass Spectrom.* *5*, 170-173.
5. Dumasia, M.C., Houghton, E. & Sinkins, S. (1986) *J. Chromatog.* *377*, 23-33.

6. Houghton, E., Dumasia, M.C., Teale, P., Moss, M.S. & Sinkins, S. (1986) *J. Chromatog. 383*, 1-8.
7. Dumasia, M.C., Houghton, E. & Teale, P. (1987) *Proc. 6th Int. Conf. Racing Analysts & Vets.*, Hong Kong, 1985 (Crone, D.L., ed.), Macmillan Publishers (HK), pp. 225-229.
8. Houghton, E., Dumasia, M.C. & Teale, P. (1988) *Analyst 113*, 1179-1187.
9. Houghton, E. (1977) *Xenobiotica 7*, 683-693.
10. Houghton, E. & Dumasia, M.C. (1980) *Xenobiotica 10*, 381-390.
11. Dumasia, M.C. & Houghton, E. (1984) *Xenobiotica 14*, 647-655.
12. Houghton, E. & Dumasia, M.C. (1979) *Xenobiotica 9*, 269-279.
13. Dumasia, M.C. & Houghton, E. (1981) *Xenobiotica 11*, 323-331.
14. Dumasia, M.C., Houghton, E., Bradley, C.V. & Williams, D.H. (1983) *Biomed. Mass Spectrom. 10*, 434-440.
15. Dumasia, M.C. & Houghton, E. (1988) *Biomed. Environ. Mass Spectrom. 17*, 383-392.
16. Marsh, M.V., Caldwell, J., Smith, R.L. Horner, M.W., Houghton, E. & Moss, M.S. (1981) *Xenobiotica 11*, 655-663.
17. Marsh, M.V., Caldwell, J., Hutt, A.J., Smith, R.L., Horner, M.W., Houghton, E. & Moss, M.S. (1982) *Biochem. Pharmacol. 31*, 3225-3230.
18. Smith, S.J., Cox, J.E., Houghton, E., Dumasia, M.C. & Moss, M.S. (1987) *J. Reprod. Fert. Suppl. 35*, 71-78.
19. Dumasia, M.C., Houghton, E. & Jackiw, M. (1989) *J. Endocrinol. 120*, 223-229.
20. Moss, M.S., Blay, P., Houghton, E., Horner, M.W., & Teale, P. (1987) *as for* 7., pp. 97-99.
21. Houghton, E., Ginn, A., Teale, P., Dumasia, M.C. & Moss, M.S. (1987) *as for* 7., 229-232.
22. Houghton, E., Ginn, A., Teale, P., Dumasia, M.C. & Moss, M.S. (1986) *Equine Vet. J. 18*, 493-497.
23. Haywood, P.E., Teale, P. & Moss, M.S. (1990) *Vet. Record 21*, in press.
24. Houghton, E., Ginn, A., Teale, P. & Dumasia, M.C. (1990) in *Proc. 7th Int. Conf. Racing Analysts & Vets.*, Louisville, Kentucky, 1988, in press.
25. Donike, M., Barwald, K.R., Klostermann, K., Schanzer, W. & Zimmermann, J. (1983) in *Sport Leistung und Gesundheit* [Sportarztekongress, Köln, 1982] (Heck, J., Hollmann, W., Lisen, H. & Jost. R., eds.), Deutscher Artz-Verlag, Cologne, pp. 293-298.
26. Gould, V.J., Turkes, A.G. & Gaskell, S.J. (1986) *J. Steroid Biochem. 24*. 563-567.
27. Braselton, W.E., Orr, J.C. & Engel, L.L. (1973) *Anal. Biochem. 53*, 64-84.
28. Marshall, D.E., Gower, D.B., Houghton, E. & Dumasia, M.C. (1989) *Biochem. Soc. Trans. 17*, 1018-1019.
29. Covey, T.R., Lee, E.D. & Henion, J.D. (1986) *Anal. Chem. 58*, 2453-2460.
30. Lee, M.S. & Yost, R.A. (1988) *Biomed. Environ. Mass Spectrom. 15*, 193-204.

#ncC

NOTES and COMMENTS relating to

APPROACHES FOR VARIOUS DRUGS AND METABOLITES

#ncC-1

A Note on

SOLID-PHASE EXTRACTION AND GC-MS IDENTIFICATION OF PROPRANOLOL AND ITS METABOLITES FROM HORSE URINE

M.C. Dumasia, E. Houghton and P. Teale

Horseracing Forensic Laboratory Ltd.,
Soham House, Snailwell Road, Newmarket CB8 7DT, U.K.

PR* is an important β-adrenergic antagonist used in the treatment of cardiovascular disorders. It is extensively metabolized in man and other animal species: reported pathways include *N*- and *O*-dealkylation, *N*- and *O*-methylation, aromatic hydroxylation and conjugation, giving rise to acidic, basic and neutral metabolites [1-5]. Evidence also exists for enantiomeric differences in PR metabolism in animal species [6, 7].

The use of β-blocker drugs in sports to reduce sympathetic activity and thus aid sporting performance has been reported, and several methods are available for screening and detecting such drugs in biological matrices [8-10]. In order to develop screening and confirmatory methods to detect the use of β-adrenergic drugs in horseracing, the *'in vivo'* biotransformation of (±)-PR after oral or i.v. administration was investigated. Here we describe a rapid procedure for identifying PR and its basic metabolites in horse urine and blood using SPE, the formation of a cyclic derivative of the β-hydroxyamine side-chain and analysis by GC-MS. Thereby a novel dihydrodiol metabolite of PR was identified in urine [11].

Biological samples.- Thoroughbred horses were dosed i.v. (0.2 mg/kg) or orally (1.0 mg/kg) with a mixture of unlabelled and $^2H_2$-labelled (±)-PR hydrochloride plus ^{3}H-labelled† drug. Blood samples were collected during 24 h and urine samples for 72 h after dosing. The plasma was immediately separated and both plasma and urine samples were stored at -20° until analyzed. **Conjugates** in urine samples, adjusted to pH 4.8, were hydrolyzed at 37° overnight with mixed glucuronidase/arylsulphatase (*Helix pomatia* juices).

Sample preparation.- Bond Elut Certify® cartridges used with a vacuum manifold (Analytichem International) were pre-conditioned by washing with 2 ml methanol and 2 ml 0.1 M

**Editor's abbreviation.*- PR, propranolol: 1-(isopropylamino)-3-(1-naphthoxy)-2-propanol. *Others.*- DMS, dimethylsilylmethylene; MS, mass spectrometry; SPE, solid-phase extraction.

†The ^{3}H served to monitor excreted dose and recoveries in analysis.

pH 6.0 phosphate buffer. Plasma (3.0 ml) was diluted with 4 ml of buffer. Hydrolyzed urine samples (5.0 ml) were adjusted to pH 6.0 and 2.0 ml buffer added. The samples (total vol. 7.0 ml) were applied to the activated cartridges which were then rinsed with 1.0 ml 1 M acetic acid, dried under full vacuum for 5 min, washed with 6.0 ml methanol, and re-dried for a further 2 min. The basic metabolites were recovered with 5.0 ml ethyl acetate containing conc. ammonia (2.0% v/v). The extract was taken to dryness under N_2 at 40°.

Cyclic DMS derivatives.- To 2.0 ml hexane in a screw-capped vial were added 150 µl *N*-diethylamine and 150 µl chloromethyldimethylchlorosilane. The mixture was vortexed and centrifuged at 3000 rpm for 5 min. The supernatant (100-150 µl) was added to the urine/plasma extracts dissolved in 25 µl toluene. After 20 min at 50°, the reagents were removed under N_2 and the derivatized residue redissolved in 30-50 µl toluene for GC-MS analysis.

Capillary GC-MS.- The mass spectra were recorded on a Finnigan-MAT TSQ-70 instrument in the EI mode (mass range 80-650 amu). An OV-1 column (18 m × 0.3 mm i.d., 25 µm film thickness) was used with helium as carrier gas (linear velocity 40 cm/sec).

RESULTS AND DISCUSSION

In these pilot studies using PR as a model drug, an SPE procedure using Bond-Elut Certify cartridges was developed which rapidly and efficiently isolated the basic compounds from both urine and plasma. The formation of cyclic DMS derivatives of β-hydroxyamines was first reported by Hammar [12] using a slightly different procedure. In the present study the cyclic derivatives of PR and its basic metabolites were rapidly and quantitatively formed by an intramolecular cyclization with the concomitant loss of HCl and were stable in solution for up to a week when stored at 4°.

The mass spectra of these derivatives showed universal and specific ions derived from the electron-induced fragmentations of the cyclized side-chain. For the cyclized basic metabolites of PR, the base peak at m/z 186 arose from the cleavage of the isopropylaminopropanol side-chain, and the ions at m/z 100 and 128 were formed from cleavage of the heterocyclic ring. These ions can also be used specifically for selected ion monitoring studies. The metabolites of PR identified, using the above procedures, in horse urine after both oral and i.v. administrations are listed in Table 1. The dihydrodiol metabolite of PR has not been identified before in any other species.

Table 1. Metabolites of propranolol (PR) in horse urine after intravenous or oral administration.

	Metabolite	Regioisomers	Intravenous	Oral
#1	PR (parent drug)	-	-	√
#2	4-hydroxy	-	major	major
#3	Other monohydroxy	2	√	√
#4	Hydroxymethoxy	-	√	√
#5	Dihydrodiol	-	√	√
#6	Dihydroxy	3	√	√

Propranolol O·CH_2·CH(OH)·CH_2·NH·$CH(CH_3)_2$

(ring positions: 9, 8, 7, 6, 4★, 3, 2★)

The major metabolite #2 was identified by comparison with authentic standard, and the others (very minor) by GC-MS. Not yet settled: the ring positions of the OH's of isomers (#3 & #6) and the structures of #5 and the catecholic hydroxymethoxy metabolite #4.

Applicability of the procedures.- SPE is gaining acceptance in doping analysis as a rapid technique for sample pre-treatment giving optimum recovery and selectivity. It is felt that the method described here can be exploited both for selectively extracting and for derivatizing this drug class. The method is being investigated for general applicability in screening horse urine samples for the presence of other β-hydroxyamines, both β-agonists and β-antagonists, following their administration.

References

1. Bourne, G.R. (1981) *Prog. Drug. Metab. 6*, 77-110.
2. Walle, U.K., Wilson, M.J. & Walle, T. (1981) *Biomed. Mass Spectrom. 8*, 78-84.
3. Talaat, R.E. & Nelson, W.L. (1988) *Drug Metab. Dispos. 16*, 212-216.
4. Bargar, E.M., Walle, U.K., Bai, S.A. & Walle, T. (1983) *Drug Metab. Dispos. 11*, 266-272.
5. Walle, T., Walle, U.K. & Olanoff, L.S. (1985) *Drug. Metab. Dispos. 13*, 204-209.
6. Walle, T., Wilson, M.J., Walle, U.K. & Bai, S.A. (1983) *Drug Metab. Dispos. 11*, 544-549.
7. Lindner, W., Rath, M., Stoschitzky, K. & Semmelrock, H.J. (1989) *Chirality 1*, 10-13.
8. Maurer, H. & Pfleger, K. (1986) *J. Chromatog. 382*, 147-165.
9. Leloux, M.S., De Jong, G.E. & Maes, R.A.A. (1989) *J. Chromatog. 488*, 357-367.

10. Musch, G., Buelens, Y. & Massart, D.L. (1989) *J. Pharm. Biomed. Anal.* *7*, 483-497.
11. Dumasia, M.C. & Houghton, E. (1989) *Biomed. Environ. Mass Spectrom.* *18*, 1030-1033.
12. Hammar, C.G. (1978) *Biomed. Mass Spectrom.* *5*, 25-28.

#ncC-2

A Note on

THE USE OF GC-MS-MS TO DETECT METABOLITES OF DETOMIDINE IN THE HORSE

M.A. Seymour, P. Teale and M.W. Horner

Horseracing Forensic Laboratory Ltd.,
P.O. Box 15, Snailwell Road,
Newmarket, Suffolk CB8 7DT, U.K.

Detomidine (DET*), a potent α-adrenoreceptor agonist used in veterinary practice as an analgesic sedative, is of interest to us because of concern that it might be used as a 'stopping drug' in horseracing. The following study relates to the need for urinary screening procedures.

Drug administration.- After i.v. administration of ^{3}H-DET to a thoroughbred colt (19 μg/kg), urine samples were collected for 3 days, and were stored frozen until analyzed. Only 60% of the ^{3}H dose was recovered in the urine, >50% being excreted during the first 16 h.

Extraction of urine samples.- Ethyl acetate extracted ~2% of the radioactivity from post-administration urine. After hydrolysis by *Helix pomatia* juice, this increased to 10-15%, depending on pH; other, less polar, solvents extracted <10%. In an alternative approach, a range of non-polar SPE sorbents were conditioned by washing with methanol, then PB. Hydrolyzed urine was applied to the columns, which were washed again with methanol, then eluted with PB. Several sorbents (C-18, C-8, C-2, cyclohexyl, Ph) extracted >50%, the best being cyclohexyl (65%); however, the extracts were very dirty. Trials were then performed with two cation-exchange sorbents, SCX (propylbenzenesulphonic acid) and PRS (propylsulphonic acid).

The cation-exchangers were conditioned with methanol, then PB at pH 2, 3 or 4. Hydrolyzed urine at the same pH was applied to the columns, which were then washed with PB followed by methanol, and eluted with methanolic HCl. Extraction on SCX rose from 25% at pH 4 to 51% at pH 2, whilst PRS extracted only ~1% at each pH. SCX, unlike

*Abbreviations.- DET, 4-(2,3-dimethylbenzyl)-imidazole; DET-COOH, detomidine carboxylic acid; OH-DET, hydroxydetomidine; MS, mass spectrometry (CI, chemical ionization; EI, electron impact); MSTFA, *N*-methyl-(trimethylsilyl)-trifluoroacetamide; PB, 0.1 M sodium phosphate buffer (pH 7 unless otherwise stated); SPE, solid-phase extraction.

PRS, is quite lipophilic, and retention probably involves both ionic and hydrophobic interactions. However, the non-polar interactions were not strong enough to prevent elution of all the retained radioactivity with a pH 10 PB wash. The SCX extracts were cleaner than any of the non-polar extracts.

Identification of metabolites.- Analysis of SCX extracts by TLC and, after clean-up and derivatization (Scheme 1), by GC-MS(EI) identified DET-COOH (27.5% of the excreted dose) and OH-DET (6.1%, present as a glucuronide) as metabolites of DET in the horse. Little or no unconjugated OH-DET or free DET is excreted. The metabolism of DET in the rat is qualitatively similar [1, 2].

DEVELOPMENT OF A ROUTINE EXTRACTION PROCEDURE

Our first procedure was as shown (Scheme 1, left). DET-COOH was the analyte of choice; the main problems were its poor chromatography and the presence of interfering compounds, probably endogenous amines. In attempting to overcome these, three strategies were tried. The first was to improve the selectivity of the extraction by optimizing the washing and eluting solvents, but this was largely unsuccessful.

The second strategy was to derivatize the analyte, in order to shift its retention time away from the interferences and to improve chromatography. Esterification of the carboxyl group with a variety of alcohols was tried, but only the methyl ester was formed in high yields; chromatography was also improved by acetylation of the imidazole ring. In the hope that negative-ion CI-MS could have been used with benefit to both selectivity and sensitivity, attempts were made to form various perhaloacyl derivatives using the appropriate anhydrides, *N*-methyl-bis(trifluoroacetamide), heptafluorobutyryl imidazole or pentafluorobenzoyl chloride. However, these were unsuccessful because the halogenated compounds appeared to be unstable, decomposing on-column. The third strategy was to use MS-MS (EI) in the daughter-ion mode, to increase the selectivity of the detector (as in art. #C-8; Finnigan TSQ 70 triple quadrupole instrument).

At around this time, Bond Elut Certify® cartridges, which contain a mixture of non-polar and cation-exchange sorbents, became available. Using these and optimizing the washing and eluting solvents (Scheme 2), extracts clean enough for GC-MS-MS analysis without further processing were obtained (Fig. 1). Methylation was necessary in the SCX procedure in order to facilitate extraction of the analyte from pH 10 PB. However, since there is no solvent extraction step in the Certify® procedure, there is no need to derivatize at an

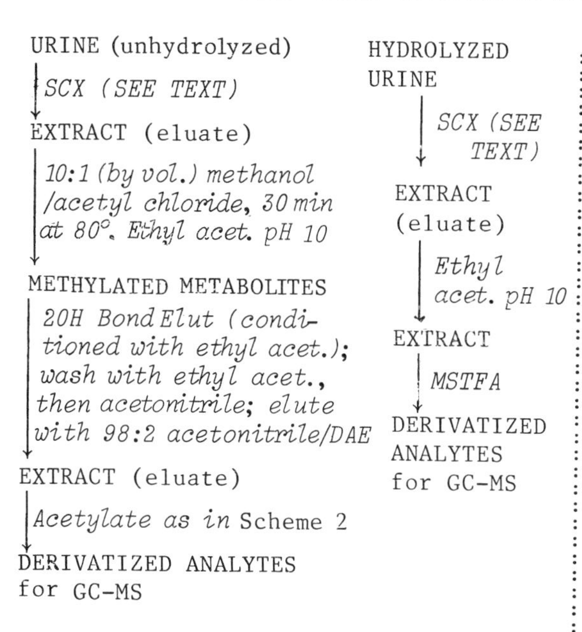

Scheme 1. Preparation of samples for GC-MS analysis (full scan EI). *DAE*, 1,2-diaminoethane; '*2OH*', propoxypropan-1,2-diol; *acet.* is abbreviation for *acetate*.

URINE (10 ml, pH 3)

↓ *Certify® Bond Elut*; wash with 1 ml 1 M acetic acid, 6 ml methanol, 5 ml 98:2 (by vol.) ethyl acet./ammonia & elute with 3 ml 9:1 butan-2-ol/ammonia*

EXTRACT (eluate)

↓ *4:1:1 ethyl acet./acetic anhydride/acetic acid, 2 h at 80°*

ACETYLATED ANALYTES

↓ *Redissolve in 20 µl ethyl acet.; add 5 ml MSTFA*

DERIVATIZED ANALYTES for GC-MS-MS: *scan daughters of m/z 198*

Scheme 2. Routine extraction procedure. 'Ammonia' is s.g. 0.88 NH_4OH.

**conditioned with 2 ml methanol, then 2 ml pH 3 PB*

early stage. Instead the carboxyl group remains available for simple on-column derivatization with MSTFA to form the trimethylsilyl ester, which has excellent chromatographic properties.

Acknowledgement

We are grateful to Farmos Group Ltd. for providing DET, ^{3}H-DET and authentic metabolite standards.

References

1. Salonen, J.S. & Suolinna, E-M. (1988) *Eur. J. Drug Metab. Pharmacokin.* *13*, 53-58.
2. Salonen, J.S., Vuorilehto, L., Eloranta, M. & Karjalainen, A. (1988) *Eur. J. Drug Metab. Pharmacokin.* *13*, 59-65.

[Fig 1: OVERLEAF

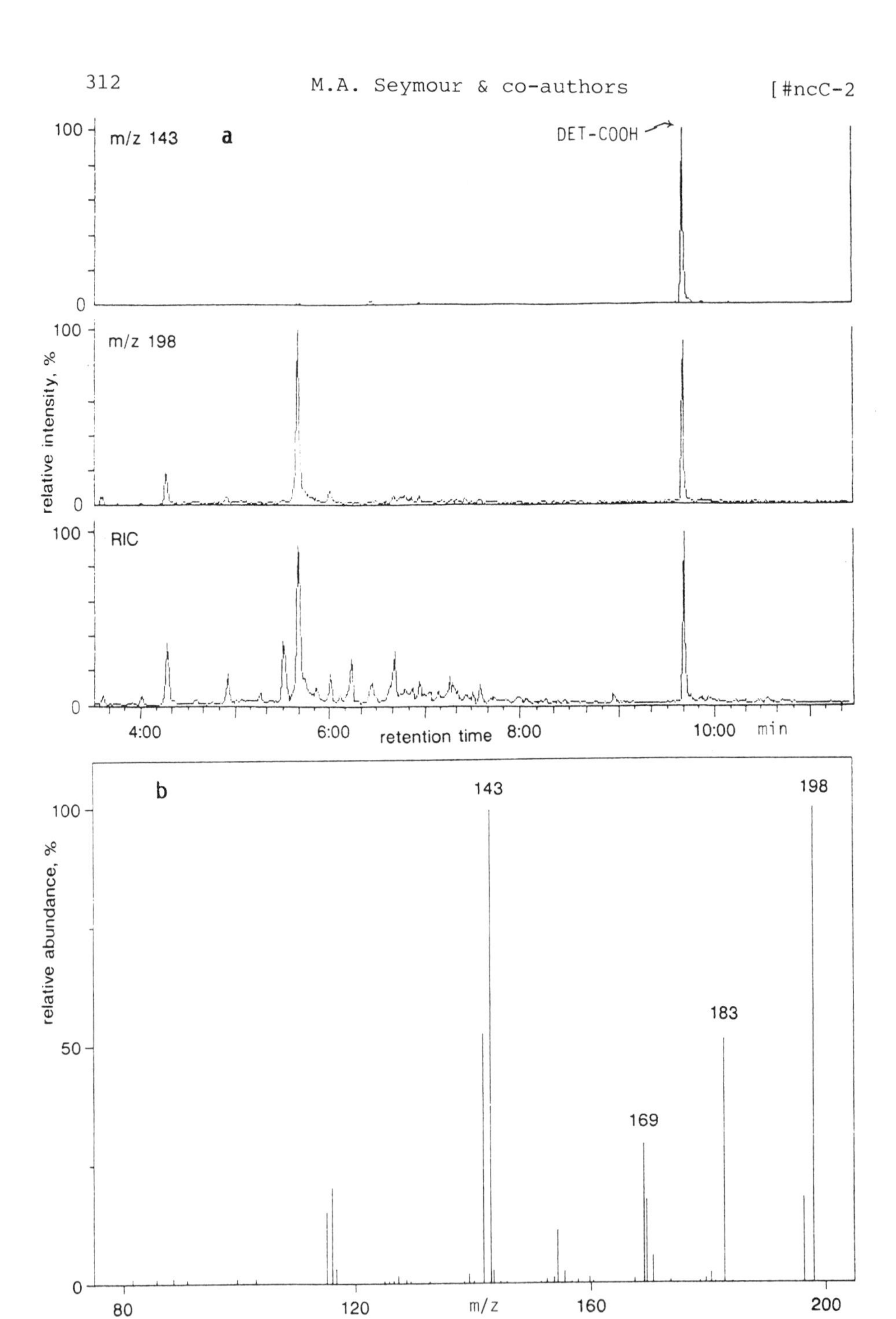

Fig. 1. GC-MS-MS of Certify® extract of colt urine after DET administration: **a**, ion chromatogram; **b**, daughter ion spectrum. RIC = reconstructed ion chromatogram (the combined trace of all ions over the scanned range, m/z 80-200). The spectrum shows all the fragments of the m/z 198 parent ion.

#ncC-3

A Note on

SOME PAST FORUM PRESENTATIONS, ESPECIALLY ON METABOLITES

Eric Reid

Guildford Academic Associates,
72 The Chase, Guildford GU2 5UL, U.K.

On the 21st anniversary of the updating of a book which, at least for endogenous constituents, is a classic [1], it is worth recalling that in the era before the first Forum was held (1975), prediction of drug-metabolite structure rested on a body of knowledge derived partly from functional-group colour tests on spots found after developing sheet chromatograms (paper sheets in pre-TLC days!). Such simple approaches, mentioned by Martin & Reid [2], should not be overlooked even though powerful instrumental techniques are available nowadays. An example of the latter, touched on by F.A.A. Dallas at the 1985 Forum,* is TLC plate-scanning by a linear analyzer, in the context of the radiolabel approach which, whilst now classical, rightly remains popular.

Analyte extraction.- An approach which is now classical, and still useful in cases of metabolite diversity, is selective solvent extraction with judicious choice of polarity (metabolites are usually more polar than the parent compound) and of pH, as described for fluphenazine at the 1975 Forum by R. Whelpton & S.H. Curry (cf. #**C-6**, this vol.). Too high a pH may suppress extraction of phenolic metabolites; this can be a nuisance but can be turned to advantage. Surveys by E. Reid ([3], and arts. in Vols. 10 & 12) include consideration of solvent extraction in respects such as efficiency, polarity, adventitious contaminants, trace-supplementation with an alcohol (which may minimize adsorptive losses; cf. #**B-8**, this vol.), and drying-down without loss or detriment. 'Old hands' in the bioanalytical field are steeped in the requisite lore. For solvent extraction and many other aspects, younger members of the analytical fraternity could advantageously become acquainted with relevant lore, nowadays easier to acquire by reading (e.g. in [4], besides the present book series including E. Reid's art. in Vol. 12).

For solid-phase extraction (SPE), near-supplanting solvent extraction in routine assays although not without pitfalls, there have been indications in recent Forum presentations

*See end of art. for the book corresponding to each Forum.

(summarized in [5]) of some selectivity in respect of metabolite elution, e.g. with 4-trifluoromethylaniline as studied in J.K. Nicholson's laboratory (Vol. 18; cf. # **ncC-5**, this vol.). However, only for a glucuronide (R.D. McDowall, 1987 Forum) has SPE given a sharp separation from the parent drug and Phase I metabolites; 'chromatography' in the SPE stage ('SPEC') has not yet really materialized.

Role of GC.- At the 1981 Forum where the focus was on metabolites (cf. the literature tabulation by E. Reid in Vol. 12), contributions such as that of J.A.F. de Silva and co-authors (on morphinan-type compounds) testified to the continued usefulness of GC. GC in conjunction with MS can, with interpretive knowledge which computerization has not yet supplanted, facilitate metabolite identification even if authentic compounds are unavailable. By the time of the 1983 Forum, MS had become much cheaper, and capillary columns had near-superseded packed columns. Examples of capillary-GC methodology presented in 1987 (as recapitulated in [5]) included assay of diclofenac and 5 metabolites, following an extractive alkylation step (P.H. Degen; temperature-programmed GC-ECD), and assay of a trace-level antispasmodic compound and a desmethyl metabolite (H. de Bree): with use of a thermostable bonded phase that allowed GC operation at >300° (on-column injection) there was no need to derivatize. Insofar as derivatization will still have a useful role, in HPLC as well as GC, there are two excellent sources of guidance [6, 7], supplementing personal lore.

Role of HPLC.- As a pre-GC step in the above-mentioned difficult assay, de Bree and co-workers performed sample clean-up by NP-HPLC (cf. G.G. Skellern's survey at the 1981 Forum). Normally, of course, the role of HPLC is for the final separation. Forum contributions over the past decade reflect the dominance of HPLC and the trough in GC popularity for investigating as well as assaying metabolites. Thus, A. Jørgensen and co-workers (1985 Forum) separated a thioxanthene and its desalkyl metabolite (distinguishing *cis* and *trans* isomers) by normal-phase HPLC, although late in their study they obtained promising results with capillary GC using a phenylmethylsilicone film. Elaborate sample preparation [3] fell out of favour with the advent of HPLC, but recent Forum contributions indicate that the pendulum has swung: there seems to be a waning of the practice of performing HPLC on raw plasma or urine, maybe with a guard-column as a palliative. Another trend is exploitation of switching modes.

Chirality of drugs and of drug metabolism has come to the fore. — Consult 'Chiral' Index entries in Vols. 14 (only 1 entry), 16 & 18. Chiral columns, especially for HPLC, are superseding the stratagem of diastereoisomer formation.

Conjugates.- For conjugate isolation, not readily achievable by solvent extraction, the advent of RP-HPLC, which allowed early rather than miserably late elution, proved a boon as evidenced by the contributions of G.G. Skellern and of J. Caldwell to the 1981 Forum. At the 1985 Forum a remedy for the problem of distinguishing eluted glucuronides was described by A. Hulshoff: they can be rendered fluorescent by pre-column derivatization. For glucuronide types ([2], & tabulation by E. Reid in Vol. 12), past confusion has shrunk. For ion-pairing, G. Schill's survey (1975 Forum) remains apt.

Reference compounds.- A notable feature of Caldwell's contribution is the guidance on synthesizing reference compounds, rarely given in bioanalytical literature. Bioanalysts are, however, daunted by synthetic chemistry, and it would be a boon if non-company bioanalysts having neither willing chemists at hand nor funds to commission syntheses could procure authentic specimens of known or putative metabolites besides those which, for psychoactive drugs, happily are procurable from the N.I.H. in Bethesda. This comment has some applicability to internal standards (i.s.'s) also. Nevertheless their use may arguably be of the nature of a luxury - especially in the context of important metabolites, each of which should, strictly speaking, have its own i.s. A good point concerning i.s.'s has been made (not at a Forum) by Marten [8], bearing on i.s. choice and HPLC conditions: an early-eluting *O*-desmethyl active metabolite of a radiosensitizer, mesonidazole, might have been missed if the i.s. had eluted before rather than after the parent drug.

Increasing role of MS and NMR.- Already the conjoining of HPLC and MS [regrettably termed 'LC-MS'], which had sample-introduction problems as discussed by L.E. Martin at the 1983 Forum, has been eased by the thermospray approach that was discussed in 1985 and 1987. Another innovation, evidenced by a 1987 Forum contribution (I.D. Wilson), is facile conjoining of MS with TLC. In recent volumes, NMR has featured as a tool which, if high sensitivity is not imperative, allows screening for metabolites, notably conjugates, to be performed with minimal sample preparation on urine or other biological specimens. NMR has, of course, long served to help identifications on isolated metabolites, as in the above-mentioned presentation by J. Caldwell. A striking example was presented by C.W. Vose in 1987 (see [5] for a 'visual aid'): ^{1}H-NMR on an HPLC component showed the absence of a phenolic group that had been deduced from direct-probe MS because of artefactual dehydration of a ring-diol (vicinal hydroxyls, forming a phenol) in the probe. With pre-derivatization and GC-MS, the dehydration was obviated and there was accord with the NMR results. This tale is salutary.

Concluding comments

A general 'moral' concerning published assays *(Journal referees:- please note!)* is that authors should cater for the reader who wants to know the rationale and blind-alleys underlying the choice of sample-preparation and chromatographic conditions. Thus, through a lapse in editorial vigilance, HPLC conditions are produced 'out-of-a-hat' in an above-mentioned presentation (Jørgensen; Vol. 16). It is based on a *J. Chromatog.* paper which has the same shortcoming, too prevalent in the justifiably escalating assay literature related to drug development. Prevalance in the Forum-based books is generally lower than in the literature at large.

Past Forum presentations that featured metabolites have mostly concerned the assay of known metabolites. Metabolite discovery and identification has suffered from some neglect — possibly remediable at a 1991 Forum; novices might welcome guidance, not hinging on access to sophisticated instruments.

The Forum series, and corresponding books, *comprising the 'Analytical' subseries of METHODOLOGICAL SURVEYS IN BIOCHEMISTRY AND ANALYSIS:* ed. E. Reid OR E. Reid *et al.* *- full titles of recent vols. appear at start of* ***this*** *vol.*

#1975 Forum: Vol. 5, 'Assay of Drugs......' *(now stocked only at address in heading to present article)* (publ. 1976)
#1977 Forum: Vol. 7, 'Blood Drugs ......' (1978)*
#1979 Forum: Vol. 10, 'Trace-Organic......' (1981)*- *Out of print*
#1981 Forum: Vol. 12, 'Drug Metabolite Isolation......'(1982)†
#1983 Forum: Vol. 14, 'Drug Determination...... ' (1984)†
#1985 Forum: Vol. 16, 'Bioactive Analytes......'(1986)†
#1987 Forum: Vol. 18, 'Bioanalysis of Drugs......' (1988)†

*Ellis Horwood, Chichester (distrib. J. Wiley/Halsted Press)
†Plenum, New York

Other references

1. Smith. I.D.. ed. (1969) *Chromatographic and Electrophoretic Techniques*, Vol. 1: *Chromatography*, 3rd edn., Heinemann Medical, London, 1080 pp.
2. Martin, L.E. & Reid, E. (1981) *Prog. Drug Metab. 6*, 197-248.
3. Reid, E. (1976) *Analyst 101*, 1-18.
4. Chamberlain, J. (1985) *Analysis of Drugs in Biological Fluids*, CRC Press, Boca Raton, FL, 320 pp.
5. Reid, E. (1989) in *Biologie Prospective* (Galteau, M., Henny, J. & Siest, G., eds.), John Libbey, Paris, pp. 687-694.
6. Blau, K. & King, G., eds. (1977) *Handbook of Derivatives for Chromatography*, Heyden, London, 576 pp.
7. Knapp, D.R. (1979) *Handbook of Analytical Derivatization Reactions*, Wiley, New York, 741 pp.
8. Marten, T.R. (1985) *Chem. Br. 21*, 745-748. *Cites* Marten, T.R. & Ruane, R.J. (1980) *Chromatographia 13*, 137-140.

#ncC-4

A Note on

THE METABOLISM OF A NEW PERIPHERAL BLOOD FLOW ENHANCER

I. Szinai, K. Ganzler, J. Hegedus-Vajda, E. Gacs-Baitz and S. Holly

Central Research Institute for Chemistry of the Hungarian Academy of Sciences, P.O. Box 17, H-1525 Budapest, Hungary

Whereas eburnamine derivatives such as vincamine are cerebral vasodilators, the new compound (-)-1β-ethyl-1α-(hydroxymethyl)-1,2,3,4,6,7,12bα-octahydroindolo(2,3α)quinolizine (RGH 2981H ®, Fig. 1) is a good peripheral vasodilator [1, 2]. Its disposition and metabolism have now been studied in rats.

EXPERIMENTAL

6-^{14}C-RGH 2981H (0.48 GBq/mmol) was synthesized in our Institute, and given orally (10 mg/kg) in 1% DL-tartaric acid to male Wistar rats (180-200 g). The urine and faeces were collected daily for 5 days. Bile taken during 6-8 h from cannulated rats was pooled. Radioactivity in samples was measured directly as $^{14}CO_2$ by LSC* after combustion [3].

Fig. 1. Structure of RGH 2981H. Metabolites referred to later: **I**, HO- replacing H at †; **II**, =O replacing 2 H at ‡.

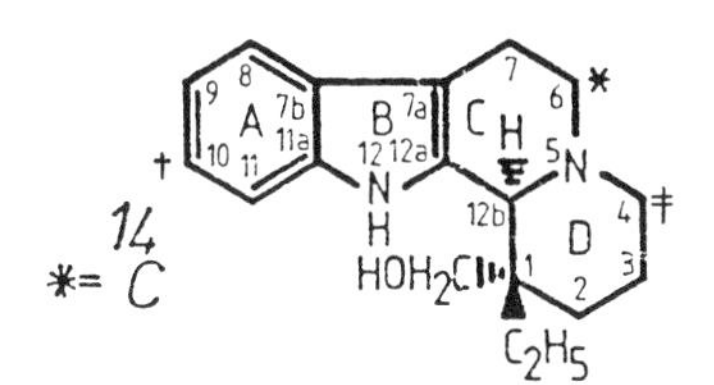

Sample preparation

Microwave extraction of faeces: samples (3-5 g) were suspended in 30 ml methanol-water-acetic acid (5:4.75:0.25 by vol.) and homogenized with an Ultra-Turrax homogenizer. The suspension was irradiated for 5 × 30 sec with short cooling intervals in a microwave oven (2500 MHz, 1140 W; Toshiba, ER 638ET D model) [4] and then centrifuged (11,000 **g**, 10 min). The supernatant was then further purified.

Purification and identification (Scheme 1) of RGH 2981H and its metabolites in urine, bile and faeces extracts.- For the first step use was made of an XAD-2 column (Serva, GFR; 80 mesh, 500 × 20 mm). The eluted analytes were treated

®Gedeon Richter Co. (Hung. patents 194221, 198207).

Abbreviations.- DEA, diethylamine; LSC, liquid scintillation counting; MS, mass spectrometry; TEA, triethylamine.

URINE (24 h) [15.1 ±7.7%] | BILE (6 h) [22.5±5% & 2 ±0.3%] | FAECES (48 h) [34.8±94%]
Microwave extr^n.; centrifuge
SUPERNATANT (80 ±9%)

Analytes on Amberlite XAD-2 column
Water wash (discard), then methanol
Eluate (94-96%)
Enzymatic hydrolysis, then diethyl ether extr^n., pH 8.5 with DEA (90-95%)

Organic phase (URINE, 54.8%; BILE, 43.1%)
Aqueous phase (URINE, 40.1%; BILE, 54.3%)
Analytes on Florisil column
Stepwise elution: 8 fractions (93.6%) *(see legend)*
#1; #2-4; #5; #6; #7 & 8
(nil) (51.1%) (18.45%) (14.1%) (9.9%)
(no products identified)

Analytes on TLC plate | Analytes on TLC plate
see legend for development

RGH 2981H (URINE, 16.5%*; BILE, 23.3%*)
Metabolites **I** & **II** (Fig. 1) (URINE: **I**, 48.6%*; **II**, 0.3%*; BILE: **I**, 43.9%*; **II**, 2.7%*)

Scheme 1. Isolation of RGH 2981H and metabolites. Values shown [ ] are % of administered dose, the second value for bile being enterohepatic circulation; those shown ( ) are the recovery in the purification step, * signifying % of radioactivity applied to the TLC plate. Elution from Florisil [5] was with 100-200 ml cyclohexane (#1), with diethyl ether alone (#2) or plus ethyl acetate (1:1; #3) then latter alone (#4); with methanol (#5) and then methanol:water (9:1, #6; 7:3, #7; 1:1, #8). The TLC development was with benzene-methanol-DEA (9.5:0.5:0.5); the R_f's were: RGH 2981H. 0.6; **I**, 0.36; **II**, 0.4. S.D.'s (n = 15) shown ±.

with β-glucuronidase/arylsulphatase [5]. After ether extraction, the aqueous phase yielded successive fractions (Scheme 1) by stepwise elution from a Florisil column (Floridine Co.; 100-300 mesh). For identification approaches, see below.

TLC was performed with pre-coated (0.25 mm) silica gel plates having a concentration zone (Merck, FRG). **HPLC** was performed with an Isco #2350 pump, Valco injector (20 or 100 μl), and UV detector - Isco V^4 at 284 nm or Hewlett Packard 8452 diode array, 190-350 nm. The analytical column was 5 μm Hypersil C-18 (Shandon; 250 × 4.5 mm i.d.); similarly the pre-column (50 × 4.6 mm); the mobile phase (1.0 ml/min) was acetonitrile-water (7:3 by vol.) adjusted to pH 8.5 with DEA. Runs were also done with a 10 μm Spherisorb C-18 column (260 × 10 mm; 10 × 4.6 mm pre-column) using acetonitrile-water (6:3), pH 8.5 TEA (3 ml/min). Peak purity was checked by diode array detection and rechromatography. The spectroscopic equipment used for verifying the structures of isolated metabolites was: EI-MS, AEI 902 (70 eV); ^{1}H-NMR, Varian X1-400 (400 MHz); FT-IR, Nicolet 170SX (in KBr micro-pellets).

Table 1. Occurrence and relative intensities (I, %) of EI-MS peaks. For RGH 2981H, M is $C_{18}H_{24}N_2O$. Values are for e/z. The HPLC R_t was 10 min for RGH 2981H, 6.1 min for metabolite **I**; **II** was not separated by HPLC, being identified from the TLC plate.

Ion:	M	M-1	M-15	M-17	M-31	M-47	$C_{13}H_{12}N_2O$:	$C_{11}H_{10}N_2$:
RGH 2981H	284 (83)	283 *(100)*	269 (18)	267 (60)	253 (7.5)			
I	300 (80)	299 *(100)*	285 (14)	283 (60)	269 (7)			
II	298 (50)			281 (3.8)	267 (5)	251 (9)	212 (30)	170 *(100)*

RESULTS AND DISCUSSION

As the radioactive compounds could not be extracted without enzymatic treatment, we assumed that RGH 2981H and its metabolites were excreted as conjugates (glucuronides or sulphates). The HPLC separation proved to be selective enough ($\alpha_{1\,2}$ = 3) to monitor the drug and metabolite **I**. The parent compound and metabolites **I** (aromatic-hydroxylated RGH 2981H) and **II** (=CO instead of CH_2 in D ring at C-4 position; Fig. 1) [unpublished observations by J. Tamas] were identified by comparative spectroscopic methods. Table 1 summarizes the EI-MS results. The C-10 position of the OH group in **I** was established from the ^{1}H-NMR data [H8: 7.12 ppm J_{ortho} = 8.5 Hz; H9: 6.72 ppm J_{ortho} = 8.5 Hz, J_{meta} = 2.2 Hz; H11: 6.85 ppm J_{meta} = 2.2 Hz] and from FT-IR data [OH+NH 3500 cm^{-1}, γC_{ar}-C_{ar} 1625,1590!,γC-O(H) 1120,γC_{ar}-O(H) 1200,γC_{ar} H (9- or 10-OH) 912,840,800 cm^{-1}].

References *overleaf*

References

1. Kalaus, Gy., Malkieh, N., Katona, I., Kajtar-Peredy, M., Koritsanszky, T., Kalman, A., Szabo, L. & Szantay, Cs. (1985) *J. Org. Chem. 50*, 3760-3767.
2. Nogradi, M. (1986) *Drugs of the Future 11*, 853-856.
3. Gacs, I., Vargay, Z., Dobis, E., Dombi, S., Payer, K. & Otvos, L. (1982) *J. Radioanal. Chem. 68*, 93-98.
4. Ganzler, K., Salgo, A. & Valko, K. (1986) *J. Chromatog. 371*, 299-306.
5. Ledniczky, M., Szinai, I., Ujszaszi, K., Holly, S., Kemeny, V., Mady, Gy. & Otvos, L. (1978) *Arzneim.-Forsch./ Drug. Res. 28*, 673-677.

#ncC-5

A Note on

^{19}F- AND ^{1}H-NMR STUDIES OF THE METABOLISM OF 4-TRIFLUOROMETHYLBENZOIC ACID IN THE RAT

[1] **Farida Y.K. Ghauri**, [2] **Ian D. Wilson and** [1] **Jeremy K. Nicholson**

[1] Chemistry Department, Birkbeck College, Gordon House, 29 Gordon Sq., London WC1H OPP, U.K. and [2] Safety of Medicines, ICI Pharmaceuticals, Alderley Park, Macclesfield, SK10 4TG, U.K.

There is considerable interest in the application of NMR techniques to obtain information on the fate and disposition of foreign compounds without the need for radiolabelled compounds. ^{19}F-NMR shows particular promise for studies on the metabolism of fluorinated compounds [1, 2]. There is also negligible interference from endogenous fluorinated compounds. In addition the large chemical shift range seen with ^{19}F-NMR also gives high sensitivity to changes in the molecule at sites up to at least 8 bonds distant from the ^{19}F nucleus. Thus, the presence of a single ^{19}F atom can provide a useful 'handle' for the metabolism of the whole molecule [3]. The ^{19}F nucleus also has a high detection sensitivity (83% of ^{1}H) and 100% natural abundance. Here we report studies on the major urinary metabolites of TFMBA* following i.p. administration to the rat, using single-pulse ^{19}F- and ^{1}H-NMR methods for identifying metabolites. Also illustrated is the application of chemical modification and/or enzymic hydrolysis to the biofluid for assignment of ^{19}F resonances, and the identification of conjugated metabolites. Partial balance studies have also been performed without recourse to radiolabelled compounds.

Dosing and sample collection.- TFMBA was from Fluorochem Ltd. Male Sprague-Dawley rats (200-250 g) were dosed with TFMBA in corn oil (100 mg/kg i.p.), and over the following 48 h were kept in metabolism cages whilst urine was collected. Urinary volumes and pH were recorded and samples stored at -40° until analyzed.

NMR measurements.- Samples (0.5 ml) with 0.1 ml of $^{2}H_2O$ added as an internal field frequency lock were placed in 5 mm glass NMR tubes for measurement. Alternatively, urine samples (typically 0.6-1.0 ml) were freeze-dried and redissolved in 600 µl $^{2}H_2O$ to allow concentration of the samples before acquiring the spectra. Measurements were made on Varian VXR400

**Abbreviations.*- FID, free induction decay; TFM, trifluoromethyl; TFMBA, *as in title.*

and Bruker AM400 spectrometers operating at 9.4 T field strength (376 MHz ^{19}F frequency) and on a Jeol GSX 500 spectrometer operating at 11.75 T field strength (500 MHz ^{1}H frequency).

^{1}H-NMR spectroscopy.- ^{1}H-NMR spectra were measured at 500 mHz using a 45° pulse (3 µsec) over a 6000 Hz sweep width. Typically, 500 FID's were collected into 32,384 k computer points with a 2.5 sec acquisition time. A further delay of 2.5 sec between pulses was used to ensure that the spectra were fully T_1-relaxed. The FID's were multiplied by an exponential function corresponding to 0.2 Hz line broadening prior to Fourier transformation. A continuous secondary irradiation field at the water resonance frequency was applied in order to suppress the intense signal resulting from water protons. Chemical shifts were referenced externally to sodium 3-(trimethylsilyl)-1-propanesulphonate (δ = 0 ppm).

^{19}F-NMR spectroscopy.- Proton-coupled ^{19}F spectra were measured at 376 MHz using a 45° pulse (15 µsec) over a 8000 Hz sweep width. Typically, 64 FID's were collected into 8192 k computer points zero-filled to 64 k; acquisition time was 0.5 sec. A further delay of 4 sec between pulses was used to ensure that the spectra were fully T_1-relaxed. The FID's were multiplied by an exponential function corresponding to 0.5 HZ line-broadening prior to Fourier transformation. Chemical shifts were referenced externally to $CFCl_3$ (δ = 0 ppm) or 4-fluorobenzoic acid (δ =-107.7 ppm).

RESULTS AND DISCUSSION

Following i.p. dosing with TFMBA urine was collected from rats and analyzed both by ^{19}F-NMR and, as exemplified in Fig. 1B (1C = control), by ^{1}H-NMR: in the aromatic region there are discernable resonances which clearly are due to or relate to TFMBA. However, it is not obvious how many metabolites of TFMBA are present, due to overlap and the presence of signals from endogenous compounds. On the other hand the ^{19}F-NMR spectrum (cf. Fig. 2A) showed a surprisingly complex pattern of ^{19}F signals for the TFM group, one being the parent compound (confirmed by standard addition), including two sets of ^{19}F resonances which are similar to doublets but cannot be the result of spin-coupling. The complexity of the ^{19}F spectrum, unequivocally indicating 7 metabolites, is perhaps surprising since benzoic acid itself mainly yields glucuronides and glycine conjugates.

Accordingly, we further investigated the metabolism of TFMBA, still using ^{19}F-NMR. Reasoning that at least some of the signals were attributable to glucuronide conjugates, we tried both chemical and enzymic hydrolysis. Incubation with β-glucuronidase caused rapid disappearance of the peak at δ = -64.8 ppm (Fig. 2C) and a consequent increase in that

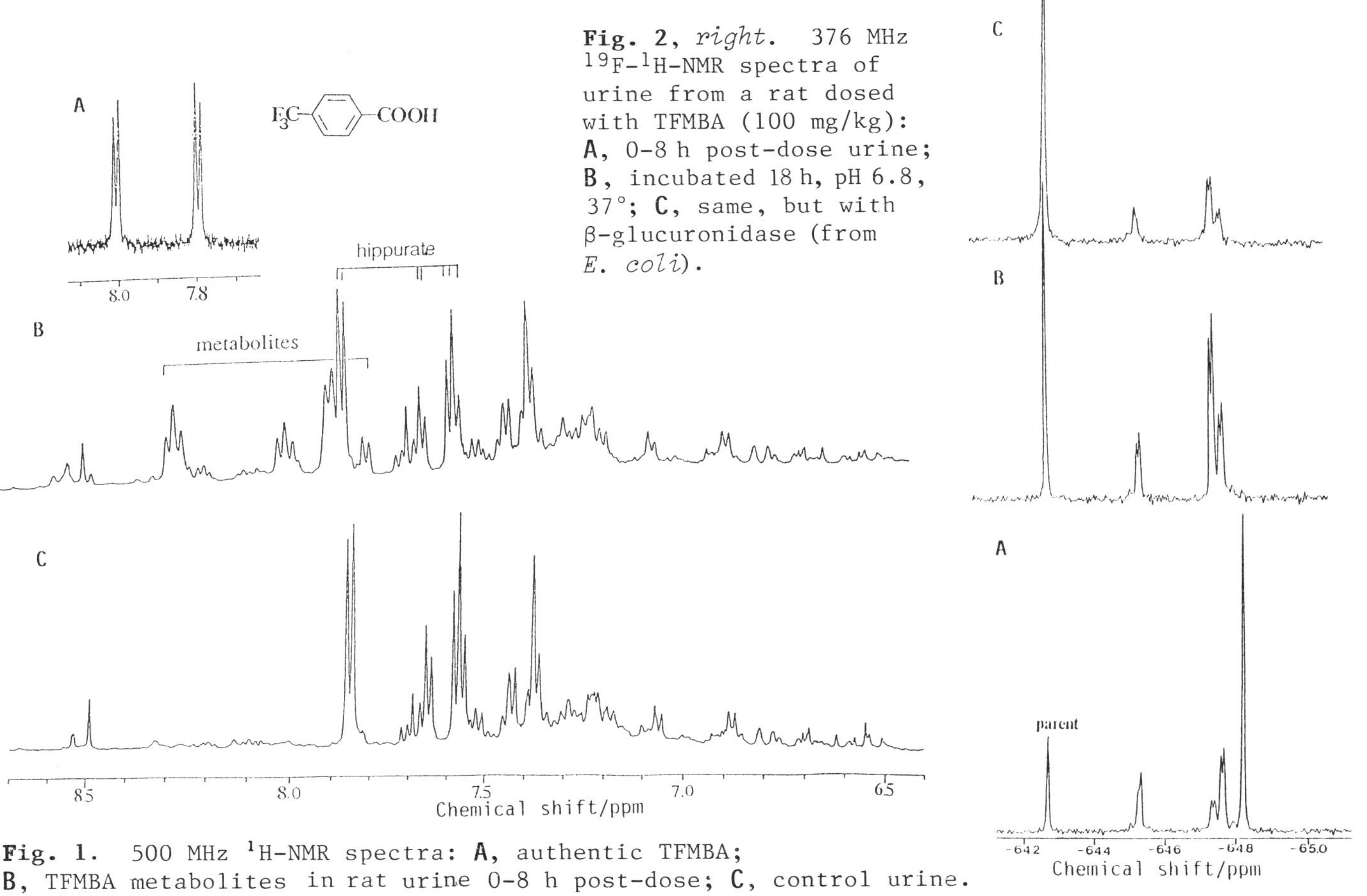

Fig. 2, *right*. 376 MHz ^{19}F-^{1}H-NMR spectra of urine from a rat dosed with TFMBA (100 mg/kg): **A**, 0-8 h post-dose urine; **B**, incubated 18 h, pH 6.8, 37°; **C**, same, but with β-glucuronidase (from *E. coli*).

Fig. 1. 500 MHz ^{1}H-NMR spectra: **A**, authentic TFMBA; **B**, TFMBA metabolites in rat urine 0-8 h post-dose; **C**, control urine.

of the parent, indicating that the compound was a glucuronide of the parent (δ = -64.2 ppm). Interestingly, however, the same signal loss and increase in parent signal occurred at pH 6.8 even without the enzyme (Fig. 2B). Such behaviour is typical of that of ester glucuronides, which undergo increasingly rapid hydrolysis with increasing pH. Support came from ^{19}F-NMR results after hydrolysis at pH 10 (with sodium deuteroxide at 20°). With incubation at pH 6.8 not only did the signal for the presumed 1-β-D-glucuronide disappear, but the ratios of the other peaks at δ = -64.52, -64.76 and -64.72 changed significantly with time. Such behaviour is explicable in terms of the now well known propensity of ester glucuronides to undergo pH-dependent transacylation from the 1-*O*-acyl to the β-glucuronidase-resistant 2,3- and 4-*O*-acyl forms [4] (& arts. - J. Caldwell, I.D. Wilson - in Vol. 12, this series - *Ed.*). The splitting of the resonances for these glucuronide forms can be rationalized by postulating the formation of both α- and β-anomers by mutarotation during transacylation.

Concluding comments.- Having already shown [e.g. 5] the effectiveness of ^{19}F-NMR in studying and assaying other compounds, we have extended this to TFMBA, whose metabolic fate in the rat was hitherto unknown, and identified, with simple methodology, the ester glucuronide as the main urinary metabolite. Unexpectedly, ^{19}F-NMR not only distinguished different transacylated forms but revealed the presence of α- and β-anomers. This subtle discrimination is striking, especially as the ^{19}F 'handle' is 8 bonds away from the site of metabolic modification, serving as an exquisitely sensitive probe:

'handle' → F_3C–C_6H_4–C(=O)–O-gluc ← metabolic site

We are continuing to study the effect of the TFM substituent on how simple aromatic compounds are metabolized.

Acknowledgements

We are grateful to Dr H.G. Parkes for technical expertise, and to Analytichem International for part-supporting F.Y.K.G.

References

1. Malet-Martino, M.C., Bernadou, J., Martino, R. & Armand, J.P. (1988) *Drug Metab. Disp. 16*, 78-84.
2. Bernadou, J., Armand, J.P., Lopez, A., Malet-Martino, M.C. & Martino, R. (1985) *Clin. Chem. 31*, 846-848.
3. Everett, J.R., Jennings, K. & Woodnutt, G. (1985) *J. Pharm. Pharmacol. 37*, 869-873.
4. Paul, H., Illing, A. & Wilson, I.D. (1981) *Biochem. Pharmacol. 30*, 3381-3384.
5. Wade, K.E., Troke, J.A., Macdonald, C.M., Wilson, I.D. & Nicholson, J.K. (1988) *Bioanalysis of Drugs and Metabolites* [Vol. 18, this series] (Reid, E., *et al.*, eds.), Plenum, N.Y., pp. 383-388.

#ncC-6

A Note on

THE USE OF ^{15}N-NMR IN STUDYING THE METABOLISM OF ^{15}N-LABELLED XENOBIOTICS, EXEMPLIFIED BY ^{15}N-ANILINE

[1]K.E. Wade, [2]I.D. Wilson and [1]J.K. Nicholson

[1] Chemistry Department, Birkbeck College, 20 Gordon Street, London WC1H OPP, U.K.

and [2]Safety of Medicines, ICI Pharmaceuticals, Alderley Park, Macclesfield SK10 4TG, U.K.

In our investigations [1] on multinuclear NMR techniques especially to determine the metabolic fate and disposition of foreign compounds, we have favoured ^{1}H-NMR because few organic molecules lack protons in their structure. Moreover, it is non-invasive and rapid, and needs only 0.4 ml of sample, with little sample preparation, and no pre-selection of instrumental conditions for different classes of compounds. However, ^{1}H-NMR is handicapped in metabolism studies by endogenous compounds which may mask signals due to metabolites. We have shown the utility of ^{19}F-NMR for following the metabolic fate of fluorinated xenobiotics ([2, 3] and #ncC-5, this vol.). Here we have investigated the urinary excretion of ^{15}N-labelled aniline in the rat. The main metabolite of aniline is *N*-acetyl-*p*-aminophenyl sulphate [4-6]. ^{15}N as an NMR nucleus has a large chemical shift range and hence good specificity, but is only 3.85×10^{-4}% as sensitive as ^{1}H-NMR and has a natural abundance of only 0.37%. However, ^{15}N-enriched compounds can be used, with the added advantage that there is no background interference from endogenous metabolites.

^{15}N has a negative nOe* which in terms of observing NMR resonances means that signals may often be lost. Quantification of spectra is also difficult because of this, and for quantitative analysis standard compounds with the same nOe as the drug metabolites must be used. ^{15}N relaxation times are long, resulting in lengthy analysis for single-pulse experiments. However, ^{15}N inverse polarization transfer experiments can be performed which utilize the coupling between ^{15}N and ^{1}H spectra to show correlations only from protons coupled to ^{15}N nuclei. The proton relaxation time is used; hence accumulation of spectra is faster and, if a compound is enriched with ^{15}N, spectra can be obtained rapidly. NMR spectral editing techniques have been investigated using ^{15}N-^{1}H-

**Abbreviations.*- nOe, nuclear Overhauser effect; FID, free induction decay; SPE(C), solid-phase extraction (chromatography). For others, see text.

shift correlation approaches which correlate ^{15}N nuclei and protons because of the coupling between them (hence only correlations from xenobiotic metabolites are observed). This is, then, a useful spectral editing technique as it 'removes' resonances due to endogenous metabolites and produces a clear picture of the number and structure of metabolites of the compound. The position of the metabolite protons can be pinpointed in the proton spectrum; hence information on sites of metabolism can be obtained.

To investigate the utility of ^{15}N-NMR for metabolism studies we have here used ^{15}N-aniline as a model compound. Fig. 1 shows the metabolic fate of aniline. We have been able to identify several aniline metabolites in untreated urine by ^{1}H-NMR using the methods of standard addition and consideration of chemical shift and couplng constant [1]. In addition, identification was aided by conjoining NMR with SPEC [7, 8]. The same samples have now been studied by ^{15}N-NMR.

Dosing and sample collection.- Male Sprague-Dawley rats (200-250 g) were dosed with ^{15}N-aniline (99%; Cambridge Isotope Labs.) in corn oil (200 mg/kg i.p.). For the next 48 h the rats were kept, with food and water *ad lib* , in metabowls and urine was collected. Samples were stored at -40° until analysis.

NMR measurements.- Samples (0.4 ml) containing 0.1 ml $^{2}H_2O$ added as an internal field-frequency lock were placed in 5 mm glass NMR tubes. Alternatively, urine samples (2-5 ml) were freeze-dried and redissolved in $^{2}H_2O$ to concentrate them before spectroscopy. Measurements were made on Bruker AM400 and WH400 spectrometers operating at 9.4 T field strength (400 MHz ^{1}H frequency or 40.52 MHz ^{15}N frequency).

^{1}H-NMR spectroscopy.- The spectra were collected with a sweep width of 6000 Hz, acquisition time 3.2 sec and a 2.5 sec delay between pulses to allow T_1 relaxation, 52° pulse (6 µsec), 16,384 computer points and 64 acquisitions. The FID's were multiplied by an exponential function correspon-ding to 0.5 Hz line broadening prior to Fourier transformation (FT). Water suppression was obtained by continuous secondary irradiation at the water resonance frequency. Chemical shifts were referenced internally to sodium 3-(trimethylsilyl)-1-propanesulphonate (TSP; δ = 0 ppm).

^{15}-NMR spectroscopy.- Urine samples (typically 2 ml) were freeze-dried and resuspended in H_2O/D_2O, 90:10 by vol. ^{15}N-NMR spectra were obtained using inverse gated proton-decoupling (to eliminate nOe's during the acquisition). The

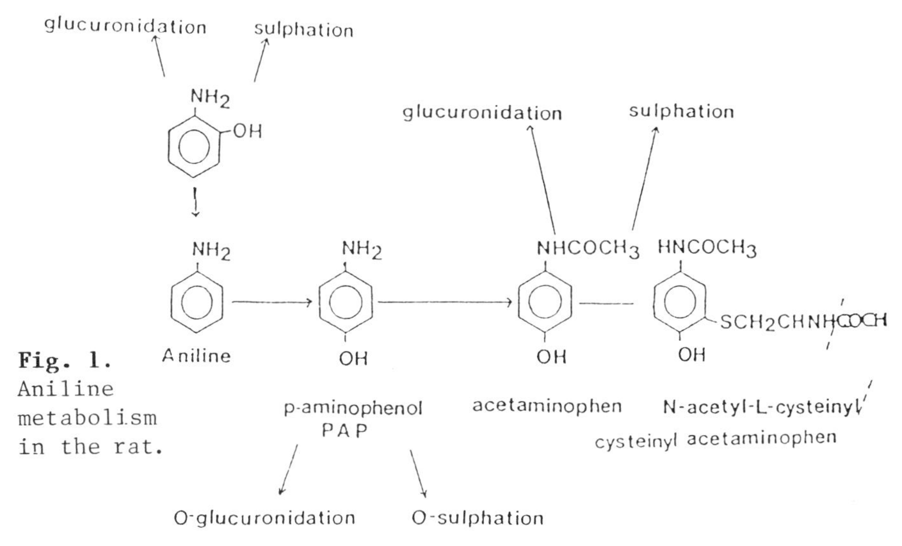

Fig. 1. Aniline metabolism in the rat.

parameters were: sweep width 15.015 Hz, acquisition time 1.091 sec, 2 sec delay between pulses, 30° pulse angle, and 32,768 computer points. The pre-FT multiplier (see above) was 4 Hz. Chemical shifts were referenced externally to $NH_4{}^{15}NO_3$.

^{1}H-^{15}N 2D shift correlation NMR spectroscopy.- Spectra were obtained using the polarization transfer approach of Bax & Morris [9] with broadband ^{1}H decoupling during ^{15}N acquisition. The pulse sequences were:

^{1}H channel:

D_0 [90°- D_0 - 90°- D_3 - 180°- D_3 - 90° - D_0 - D_3 -90°- D_4 - broadband decoupling]

^{15}N channel:

D_0 [D_1 - 180°- D_1 -90°- D_4 - collect FID]

where D_1 was 1-5* T1 (the relaxation time for protons), D0 was the preparation time for decoupler pulsing, D_3 was the polarization time of $\frac{1}{2}$ J (NH), and D_4 was the delay needed for refocussing the ^{15}N resonances equal to $\frac{1}{4}$ J (NH). The **inverse version** of this approach was as for the normal version except that the ^{1}H nucleus was observed.

SPE of ^{15}N-aniline metabolites.- In order to characterize the metabolites further, we tried the SPEC-NMR approach [7, 8] for isolating and purifying them. Urine was loaded onto a C-18 BondElut 'column' (3 ml size; Analytichem) at pH 2 and the metabolites were eluted first with water (pH 2) and then methanol. Each fraction was loaded onto a fresh C-18 column and eluted with a step-gradient of methanol (acidified with 1% formic acid to pH 2) from 0% to 100% in 20% increments. Solvent was removed by a N_2 stream and freeze-drying. Residues were redissolved in ^2H_2O and analyzed by ^{1}H-NMR.

RESULTS AND DISCUSSION

^{1}H-NMR of urine.- Fig. 2 shows typical ^{1}H-NMR spectra for post-dose and control urine. Numerous resonances can be detected from both endogenous compounds and aniline metabolites. Certain aniline metabolites were identified by standard addition and consideration of chemical shift (Table 1) as paracetamol glucuronide, paracetamol sulphate, *p*-aminophenol glucuronide and paracetamol. Aniline toxicity was manifested by changes in endogenous metabolite concentrations; thus lactate was elevated, indicating nephrotoxicity [10].

SPE-^{1}H-NMR of aniline metabolites.- Paracetamol glucuronide and *p*-aminophenol glucuronide were not retained on the C-18 'column' and passed straight through in the water wash of the initial loading. ^{1}H-NMR spectra of the fractions obtained by stepwise elution of the originally obtained '100% methanol' fraction, when applied to a fresh column and re-run, showed partial separation of the metabolites (Fig. 3). Paracetamol sulphate was eluted by 20:80 methanol:water and a further metabolite of aniline was found in the 60:40 and 80:20 fractions, and from its ^{1}H-NMR spectrum was identified as the 3-cysteinyl conjugate of paracetamol. This was confirmed by adding the authentic metabolite. These fractions also showed a small amount of the mercapturate of paracetamol as evidenced by the singlet from proton **a** at $\delta = 1.84$ ppm. In many species mercapturate/cysteinyl formation is a process of detoxification for the reactive hepatotoxic metabolite of paracetamol, *N*-acetyl-*p*-benzoquinoneimine (NAPQI). NAPQI reacts with glutathione and is excreted in the urine as 3-mercapturic paracetamol and 3-cysteinyl-paracetamol. It is unclear from the literature whether the latter is a metabolite of paracetamol in the rat. Our study suggests that both compounds are excreted in rats treated with aniline, which may be pertinent to understanding its metabolism.

^{15}N-NMR of whole urine.- Fig.4 shows a typical spectrum, manifesting 21 ^{15}N resonances. The identification of these metabolites is continuing, but from the relative proportions of the major metabolites detected in the ^{1}H-NMR spectrum some of the resonances have been tentatively assigned. In brief, paracetamol sulphate was identified after isolation by SPE and analysis by ^{15}N-NMR. There are 4 groups of resonances, indicating 4 different types of aniline metabolite, possibly conjugates of paracetamol, *p*-, *o*- and *m*-aminophenol. The resonances at ~40 ppm are probably *p*-aminophenol conjugates since this is the applicable ^{15}N chemical shift. Previously

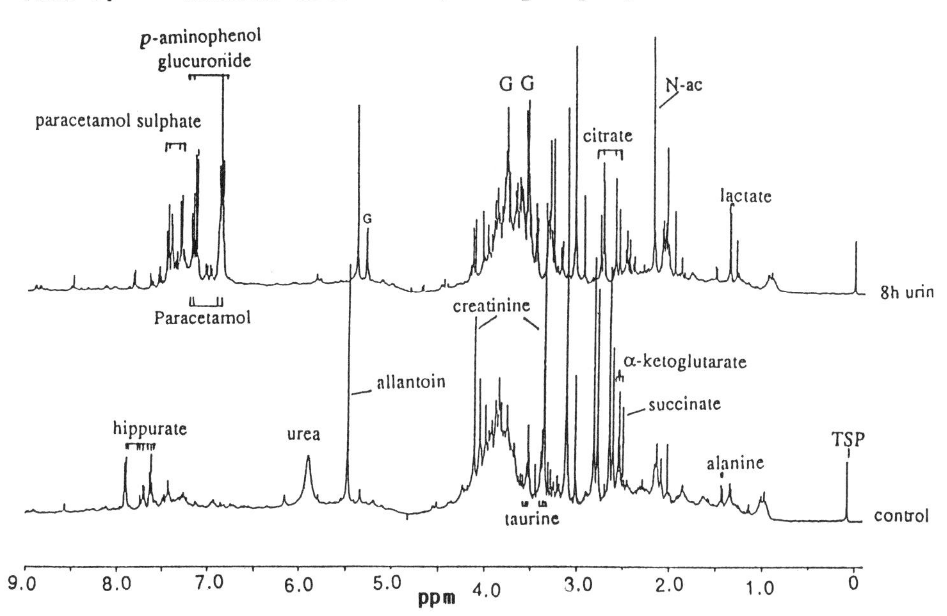

Fig. 2. ^{1}H-NMR spectra for rat urine - ^{15}N-aniline, 0-8 h post-dose, and control. For conditions see text. G, glucuronide resonances from *p*-aminophenol and paracetamol glucuronides; N-ac, *N*-acetyl resonances from paracetamol metabolites.

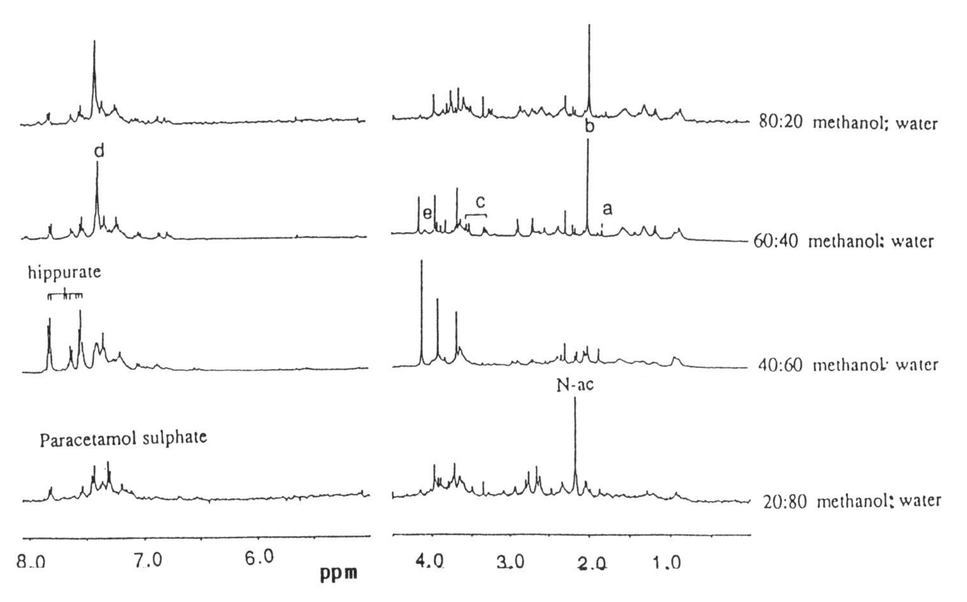

Fig. 3. ^{1}H-NMR spectra for urine, subjected to SPE, from aniline-dosed rats: successive SPEC eluates in re-run of material originally eluted with 100% methanol ('re-chromatography'). For conditions see text.

Fig. 4. ^{15}N-NMR spectra for ^{15}N-aniline metabolites in freeze-dried urine. For conditions see text, which alludes to the 21 resonances indicated.

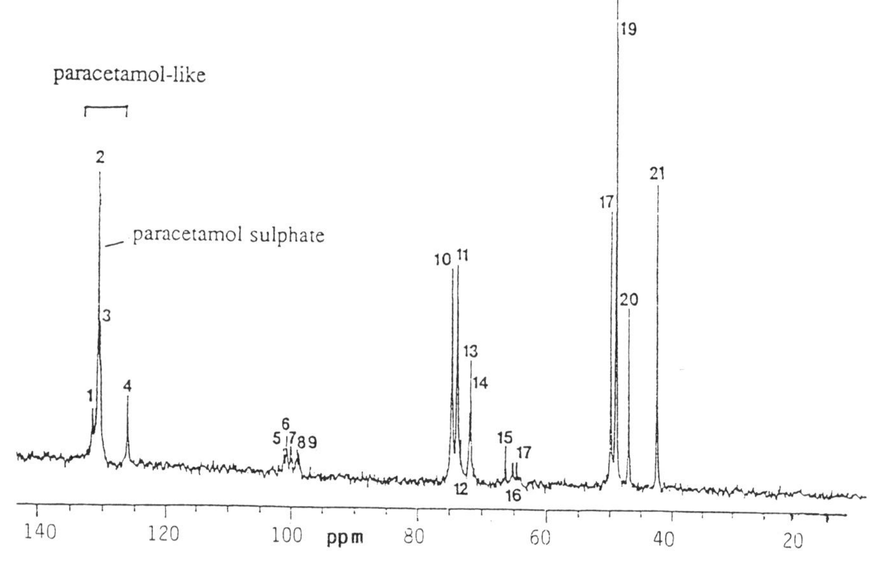

Table 1. ^{1}H-NMR chemical shifts and coupling constants (J/Hz) for aniline metabolites. *Data from ref. [1].* (s) = singlet, (d) = doublet.

R group	Chemical shift/ppm					J/Hz, H_a-H_b (& H_a-H_c)
	H_a	H_b	H_c	H_d	R	
H	7.25(d)	6.90(d)	2.15(s)		-	8.7
$SO_3^-K^+$	7.45(d)	7.31(d)	2.17(s)		-	8.9
I	7.34(d)	7.13(d)	2.16(s)		5.11(d), 3.60-3.94	8.9
II	7.26	6.98	-		-	8.8
III	7.02	6.83	-		-	8.8
IV	7.26(dd)	6.99(d)	7.51(s)	2.15(s)	3.35(ABX), 3.99 (X)	8.7
V	7.23(dd)	6.93(d)	7.42(s)	2.14(s)	1.8(s), 3.28(ABX), 4.3(X)	8.7 (2.1)

Particular **R** groups in the phenol

HNCOCH$_3$ (d); ring positions a, a, b, b; OR

I: glucuronide ring (O, CO$_2$H, H, OH, H, OH, H, OH, H)

II, R' = H; **III**, R' = glucuronide — HN, ring positions a, a, b, b; OR'

HNCOCH$_3$ (d); ring positions a, b, c; OH; SCH$_2$CHNR' — **IV**, R' = H; **V**, R' = COCH$_3$

The compounds, in order of listing, are: paracetamol and its sulphate and glucuronide; *p*-aminophenol and its glucuronide; 3-cysteinyl and 3-*N*-acetylcysteinyl conjugates of paracetamol.

[4-6] only 11 urinary metabolites have been found: paracetamol and its glucuronide and sulphate, acetanilide, *p*- and *o*-aminophenol glucuronides and sulphates, nitrosobenzene, aniline *N*-glucuronide and 2,4-diphenylamine. Some metabolites of paracetamol, itself an aniline metabolite, might also be present, such as 3-methylsulphinyl- and 3-thiomethyl-paracetamol, *m*-aminophenol glucuronide and sulphate, and 5-(*S*-acetamido-2-glucuronosidophenyl)cysteine.

Concluding comments.- The experiments outlined here clearly illustrate the potential of ^{15}N-NMR when a ^{15}N-labelled compound is available for study. The contrast between the ^{1}H-NMR spectrum and that obtained for the same sample using ^{15}N-NMR is that the latter shows a much clearer picture of the number and relative proportions of the metabolites. Using the two approaches together, we have been able to identify rapidly some metabolites and, in conjunction with the SPEC-NMR technique, have been able to isolate and identify a 3-cysteinyl paracetamol metabolite of aniline which has not previously been observed in the rat, whilst already having been encountered as a paracetamol metabolite in man [1].

References

1. Bales, J.R., Sadler, P.J., Nicholson, J.K. & Timbrell, J.A. (1984) *Clin. Chem. 30*, 1631-1636.
2. Wade, K.E., Troke, J., Macdonald, C.M., Wilson, I.D. & Nicholson, J.K. (1988) in *Bioanalysis of Drugs and Metabolites especially Anti-inflammatory and Cardiovascular* [Vol. 18, this series] (Reid, E., Robinson, J.D. & Wilson, I.D., eds.), Plenum, N.Y., 383-388.
3. Wade, K.E., Wilson, I.D., Troke, J.A. & Nicholson, J.K. (1989) *J. Pharm. Biomed. Anal.*, in press.
4. Kao, J., Faulkner, J. & Bridges, J.W. (1978) *Drug. Metab. Disp. 6*, 549-555.
5. Grossman, S.J. & Jollow, D.J. (1986) *Drug. Metab. Disp. 14*, 689-691.
6. Parke, D.V. (1960) *Biochem. J. 77*, 493-503.
7. Wilson, I.D. & Nicholson, J.K. (1987) *Anal. Chem. 59*, 2830-2832.
8. Wilson, I.D. & Nicholson, J.K. (1988) *J. Pharm. Biomed. Anal. 6*, 151-165.
9. Bax, A. & Morris, G. (1981) *J. Magn. Res. 42*, 501-505.
10. Gartland, K.P.R., Timbrell, J.A., Nicholson, J.K. & Bonner, F.W. (1986) *Human Toxicol. 5*, 122-123.

#ncC-7

A Note on

BIOSENSORS, AN APPROACH TO DRUG ANALYSIS? - PRELIMINARY STUDIES ON THE DEVELOPMENT OF A WARFARIN ASSAY

R. Hyland, J. McBride, G.W. Hanlon, A.J. Hutt and ⊗C.J. Olliff

Department of Pharmacy, Brighton Polytechnic, Moulsecoomb, Brighton BN2 4GJ, U.K.

Biosensors are analytical devices which combine the unique specificity of a biological sensing element with a physical transducer that can convert a biological event into an appropriate signal often with the aid of a mediator. The biological component of the system may be an enzyme or enzyme system, antibody, tissue section, membrane fraction or microorganism. The interaction of the biocatalyst with an appropriate substrate is monitored by the transducer, which measures changes in the system's optical, electrical, calorimetric or mechanical properties. Such measurements may be based on potentiometry, amperometry, conductance, absorption, fluorescence, total internal reflectance, surface plasmon resonance spectroscopy, ellipsometry or field-effect transduction [1]. The most frequently used biosensor combination involves the coupling of enzymes with either potentiometric or amperometric transducers; the glucose sensor is the most widely studied. Which combinations of components are used depends on factors such as selectivity, sensitivity, stability and cost.

Biosensors have potential advantages over some existing analytical technologies in that they may be both highly selective and sensitive, have relatively rapid response times and can be used for the analysis of material present in complex mixtures with minimal sample clean-up. Biosensors may also be prepared cheaply and in miniaturized form. An additional advantage is that biosensors should be capable of enantiodifferentiation and therefore offer a novel alternative approach for enantioselective analysis. Few of the enantiospecific analytical methods at present available utilize the inherent chirality of nature for enantiodifferentiation, except for RIA and radioreceptor techniques [2].

⊗ addressee for any correspondence

One disadvantage of enzyme- or antibody-coupled systems is their short working life. More stable systems may be prepared using whole microbial cells since the enzyme systems are present in their optimal environment. In addition they may be prepared relatively cheaply and simply, and in certain cases have the capacity for regeneration through immersing the sensor in a suitable medium and allowing the cells to reproduce. Bacterial biosensors, using a range of transducers and mediators, have been developed for the analysis of a wide range of substrates, e.g. amino acids, sugars, alcohols and cholesterol [3, 4].

Biosensor technology therefore sits at the interface between the rapidly developing fields of miniaturized electronics and cell biology. At the present time biosensors are used in the clinical biochemistry area, and in the future they will become important in the food and pharmaceutical industries as well as for environmental monitoring. However, little attention has hitherto been paid to the application of biosensors in drug analysis although we have previously reported the construction of a 'penicillin electrode' by coupling the enzyme penicillinase to a glass electrode [5].

The present work is based on the observation [6] that the bacterium *Nocardia corallina* (ATCC 19070) reduces the coumarin anticoagulant drug warfarin to the corresponding warfarin 'alcohol' with complete substrate stereoselectivity:

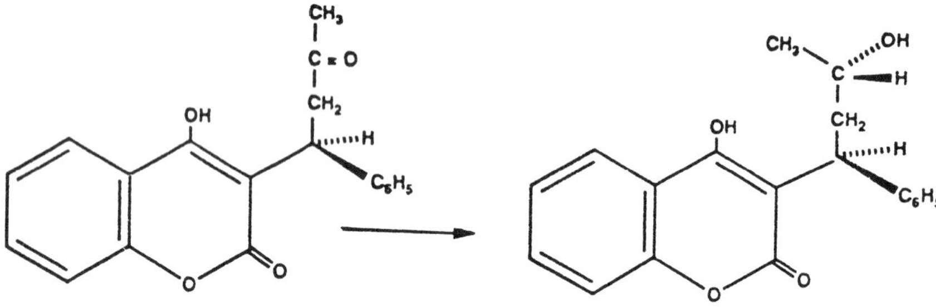

(*S*)-warfarin (*S*,*S*)-warfarin 'alcohol'

We have previously reported the construction of an amperometric electrode by coupling *N. corallina* to glassy carbon rods in the presence of the mediator ferrocene [7]. The electrode produced exhibited a linear response to racemic warfarin over the range 30-300 µM. We now report on methods to increase the sensitivity of the system and the stereoselectivity of the electrode produced.

EXPERIMENTAL

An increase in electrode response would be expected by increasing the surface area and also by platinizing the carbon surface, as we have observed (unpublished) with enzyme-coupled electrode systems. An area increase was achieved by replacing the glassy carbon with carbon felt. For platinization the felt, serving as one electrode with Pt as the other, was placed in a solution of chloroplatinic acid (3.5%, w/v) and lead trihydrate (0.005%); a 25 μA current was passed between the electrodes for 10 min and then reversed for 10 min, and then the entire process was repeated. The presence of Pt attached to the carbon felt was confirmed by both SEM and electron-probe microanalysis. To the platinized electrode after treatment with nitric acid, then ferrocene [7, 8], the bacterium was coupled by initial immersion in 1-cyclohexyl-3-(2-morpholinoethyl)carbodiimide-*p*-methyltoluene sulphonate (0.15 M) in acetate buffer (pH 5.2) for 90 min; a wash with water was followed by immersion in a thick suspension of *N. corallina* in phosphate-buffered saline (PBS; pH 7.3) at 4° for 2 h. Electrodes which, on examination by cyclic voltammetry (5 mV/sec voltage sweep, 0 to 600 mV), gave <1 μA background current at 400 mV were used for the warfarin analysis. To examine the response of the biosensor the 3-electrode mode was used (Ag-AgCl reference electrode and Pt auxiliary electrode): the system was placed in a 50 ml reservoir kept at 27° in a water-bath. Samples were examined by adding appropriate drug solutions in PBS to the reservoir. Electrodes were routinely stored in PBS at 4°.

RESULTS

The range of electrode response *vs.* concentration depended on the type of carbon surface and its treatment. The current

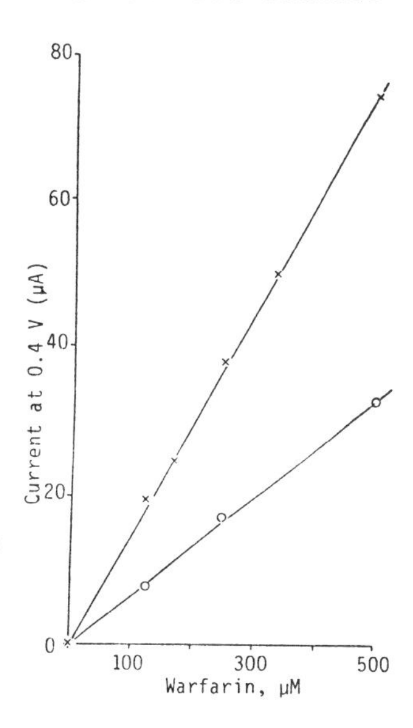

Fig. 1. Relationship between current produced and the concentration of the individual isomers of warfarin: *S*-(-)-, ×; *R*-(+)-, o. The slopes of the calibration curves are 0.162 and 0.065 $A.M^{-1}$ respectively; that for the racemate (not shown) was 0.112 $A.M^{-1}$, within 5% of that calculated from the individual isomer responses. Throughout the pH was 8.6.

response was increased 4-fold using carbon felt compared to a glassy carbon electrode of the same diameter (3 mm). Platinization of the felt gave an increase of 5-fold, or 8- to 12-fold with the mediator ferrocene present; then current *vs.* concentration was linear over the range 5-500 μM for racemic warfarin (data not shown).

The pH profile with 100 μM racemic warfarin showed a maximal response at pH 8.6, as used thereafter. Some enantio-selectivity was demonstrated (Fig. 1; amplification in legend). It is of interest that the *R*-isomer gave some response, which may indicate a non-stereospecific response within the system. In support of this view, an electrode response value of 0.064 $A.M^{-1}$ was given by the related compound 7-hydroxy-coumarin (umbelliferone), an analyte which obviously cannot undergo the bacterially mediated reduction.

CONCLUSIONS

This preliminary report indicates that biosensors may be successfully constructed for the analysis of drug molecules at sensitivities matching the concentrations found in biological fluids. In addition these systems are capable of significant enantiodifferentiation.

Acknowledgements

The enantiomers and racemic warfarin were generously gifted by Duncan Flockhard, Greenford. *N. corallina* was supplied by Prof. P.J. Davis, Austin, Texas, and the carbon felt was gifted by Prof. C. Bourdillon, Compiègne, France.

References

1. Owen, V.M. & Turner, A.P.F. (1987) *Endeavour 11*, 100-104.
2. Hutt, A.J. & Caldwell, J. (1989) in *Xenobiotic Metabolism and Disposition* (Kato, R., Estabrook, R.W. & Cayen. M.N., eds.), Taylor & Francis, London, pp. 161-169.
3. Corcoran, C.A. & Reichnitz, G.A. (1985) *Trends in Biotechnology 3*, 92-96.
4. Delaney, G.M., Bennetto, H.P., Mason, J.R., Roller, S.D., Stirling, J.L.& Thurston, C.F. (1984) *J. Chem. Tech. Biotechnol. 34B*, 13-27.
5. Olliff, C.J., Williams, R.T. & Wright, J.M. (1979) *J. Pharm. Pharmacol. 30*, 45P.
6. Davis, P.J. & Rizzo, J.D. (1982) *Appl. Environ. Microbiol. 43*, 884-890.
7. Fitzgerald, Y.J., Hanlon, G.W., Hutt, A.J. & Olliff, C.J. (1987) *J. Pharm. Pharmacol. 39*, 149P.
8. D'Costa, E.J., Higgins, I.J. & Turner, A.P.F. (1986) *Biosensors 2*, 71-87.

#ncC-8

A Note on

FACILE PREPARATION OF CHIRAL HPLC COLUMNS BY INJECTION OF CHIRAL ISOCYANATES DIRECTLY ONTO AMINOPROPYL COLUMNS

Robin Whelpton and ⊗Dennis G. Buckley

Department of Pharmacology,
London Hospital Medical College,
Whitechapel, London E1 2AD

and ⊗Department of Chemistry, Queen Mary College,
Mile End, London E1 4NS

Several years ago we were asked by Phase Separations Ltd. to evaluate a chiral column ('S5') that had been prepared by refluxing (*R*)-*N*-(α)-phenethyl-*N*'-triethoxysilylpropyl-urea with silica. We used it to resolve thioridazine (TDZ) and some of its metabolites [1, 2] but, as it was not clear whether the column would be available commercially, we considered making our own. We chose a simpler approach, reacting (*R*)-1-phenethylisocyanate with aminopropyl-modified silica.

Spherisorb S5 NH_2 columns (250 × 4.6 mm i.d.) were obtained from Phase Separations Ltd. (Queensferry, N. Wales), and the above isocyanate and (*R*)-1-naphthylethylisocyanate were from Aldrich Chemicals Ltd. (Gillingham, Dorset). An Altex Model 110A pump and Rheodyne 7125 injector with a 1 ml loop were used. The columns were equilibrated with HPLC-grade dichloromethane (DCM; from Fisons Scientific Apparatus, Loughborough; pentene present as stabilizer) by pumping ~60 ml at 1 ml/min. The flow was adjusted to 0.2 ml/min, and ~0.5 g isocyanate injected. A further ~0.5 g (residue in the ampoule) was injected 30 min later. After 30 min the flow was increased to 1 ml/min for 60 min to remove excess reagent.

The columns were tested using the following test compounds (racemic):
- (a) valine, phenylalanine and alanine (as their *N*-3,5-dinitrobenzoyl methyl esters);
- (b) 1-phenylethylamine (as its *N*-3,5-dinitrobenzoyl derivative);
- (c) 2',2',2'trifluoro-1-(9-anthryl)ethanol ('Pirkle test compound').

Hexane-DCM-methanol (50:10:1 by vol.) was used as eluent for (a) and (b), and 1.5% (v/v) propan-2-ol in hexane for (c).

RESULTS AND DISCUSSION

Generally the company-supplied ('S5') chiral column performed better than the home-made equivalent (phenyl), but the results with the latter were considered acceptable: the k' values and the separation factors using (a) were inferior with the phenyl column but greatly superior with the naphthyl column (Fig. 1).

The mechanism for chiral recognition on the types of phase we used is suggested to be a 3-point interaction of dipole-dipole stacking of -CO-NH- groups, Π-Π bonding and steric interaction [3]. The vastly increased resolution of 3,5-DNB-phenylethylamine on the naphthyl column (Fig. 2) can be explained by increased Π-bonding. However, separation of the Pirkle compound (c) does not fit the model: the best separation was achieved with the phenyl column (Fig. 3). The resolution of TDZ metabolites [2] does not fit the model either. Although the systems are not particularly stable in respect of TDZ, the enantiomers of the desmethyl metabolite could be resolved on the company-supplied chiral column and our phenyl column using a variety of eluents, including 100% methanol. The differences in resolution between the tertiary (TDZ) and secondary (norTDZ) amine suggest that hydrogen bonding may be important. Like the Pirkle compound, norTDZ chromatographed as a distorted, incompletely resolved, peak on the naphthyl column.

Direct injection of chiral isocyanates onto amino columns offers an inexpensive and convenient way of producing urea-bonded chiral columns. The isocyanates cost us, respectively, <£7/g and <£30/g, and no additional apparatus such as a packing pump was required. The 1 ml injection loop is a luxury: in our early experiments we used an empty HPLC column to introduce the reagents. Pirkle and colleagues [4] have re-circulated isocyanates through amino columns, but we disliked the idea of passing isocyanates through our analytical pumps. With the current growth of interest in optical isomers, it is likely that more chiral isocyanates (or the amines from which to prepare them) will become available. Isocyanates react smoothly without giving any other products; hence a chiral column can be prepared in ~3 h.

In summary, urea-bonded chiral columns are readily prepared using commercially available isocyanates, injected directly onto an aminopropyl column. The column properties compare favourably with those of a company-supplied column.

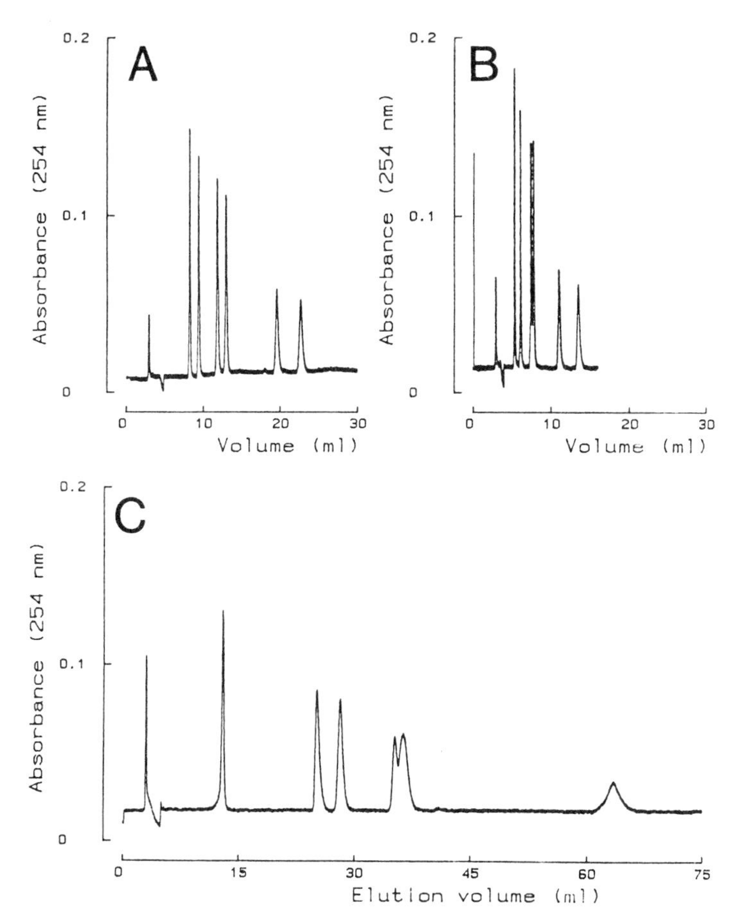

Fig. 1. Test mixture of amino acid derivatives as in (a). Elution order: valine, phenylalanine and alanine. Columns: **A**, company-supplied, 'S5 Chiral 1'; **B**, phenyl; **C**, naphthyl. Eluent: hexane-DCM-methanol, 50:10:1 by vol.

References

1. Watkins, G.M., Whelpton, R., Buckley, D.G. & Curry, S.H. (1986) *J. Pharm. Pharmacol. 38*, 506-509.
2. Whelpton, R., Jonas, G. & Buckley, D.G. (1988) *J. Chromatog. 426*, 222-228.
3. Wainer, I.W. & Doyle, T.D. (1984) *J. Chromatog. 284*, 117-124.
4. Pirkle, W.H. & Hyun, M.H. (1985) *J. Chromatog. 322*, 295-307.

[Figs. 2 & 3: OVERLEAF

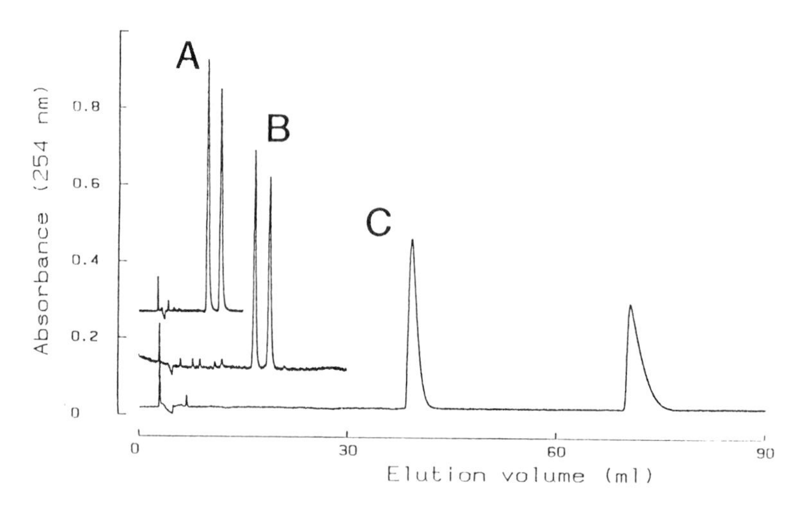

Fig. 2. 1-Phenylethylamine, (b) in text, tested for separation of stereoisomers (as 3,5-DNB derivatives). Columns: **A**, phenyl; **B**, company-supplied, 'S5 Chiral 1'; **C**, naphthyl. Conditions as for Fig. 1.

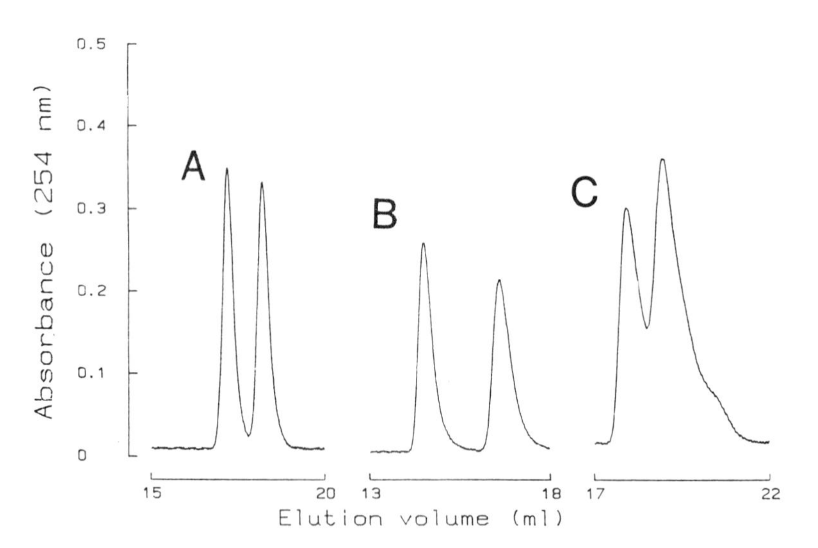

Fig. 3. 'Pirkle test compound', (c) in text, tested for separation of stereoisomers.
Columns: **A**, company-supplied, 'S5 Chiral 1'; **B**, phenyl; **C**, naphthyl. Eluent: 1.5% (v/v) isopropanol in hexane.

#ncC-9

A Note on

HPLC ANALYSIS OF 3-HYDROXYPYRIDIN-4-ONES: NOVEL ORALLY ACTIVE IRON CHELATORS

R.O. Epemolu, R.C. Hider and ⊗L.A. Damani

Chelsea Department of Pharmacy,
King's College London, Manresa Road, London SW3 6LX, U.K.

The requirement for an orally active, non-toxic alternative to desferrioxamine has stimulated considerable research effort to design specific iron chelators [1]. The 3-hydroxypyridin-4-ones offer promise in these respects, effectively decreasing iron overload in animal models and hepatocytes [2].

General structure of the 3-hydroxypyridin-4-ones

Compound	Code	R_1	R_2
1	CP028	-H	$-CH_3$
2	CP020	$-CH_3$	$-CH_3$
3	CP051	$-(CH_2)_2OCH_3$	$-CH_3$
4	CP094	$-C_2H_5$	$-C_2H_5$
5	CP040	$-(CH_2)_2OH$	$-CH_3$
6	CP052	$-(CH_2)_3OC_2H_5$	$-C_2H_5$
7	CP099	-H	$-C_2H_5$

In studying the *in vivo* behaviour of some candidate compounds, we had to develop assay methods, to include potential metabolites, as described here for an HPLC approach that allows selective and sensitive assay of the free ligands in biological matrices, e.g. blood, urine and tissue homogenates.

On C-18 silica columns the compounds ran poorly (Fig. 1A), possibly due to interaction with iron or free silanols: broad asymmetrical and sometimes multiple peaks were often encountered*. Approaches investigated to improve peak symmetry included (a) change of pH and ionic strength of buffers, (b) addition of competing organic compounds such as triethylamine, and (c) addition of substitute chelators like picolinic acid [3], or of an analyte analogue (5 mg/L). Only 'analyte' addition was even partially successful, but there was no selectivity between the different compounds in this series and there was often the complication of 'negative' peaks.

⊗addressee for any correspondence. *Adsorption/desorption phenomena are considered in a 1989 Supelco leaflet (T.L. Ascah *et al.*).

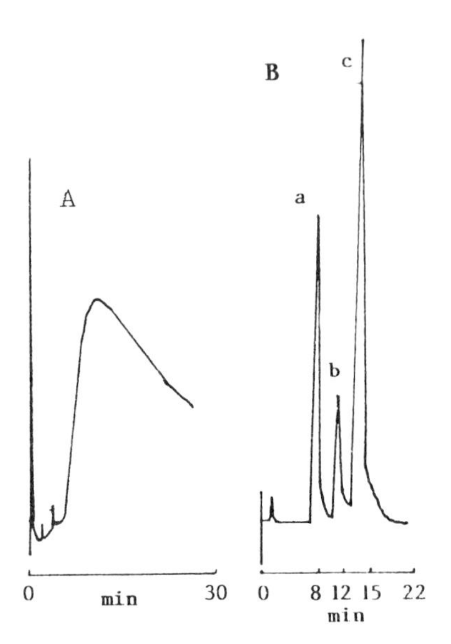

Fig. 1. HPLC patterns (detection at 286 nm).
A: CP020 on 5 µm C-18 silica; mobile phase: Na_2HPO_4 (10 mM, pH 6)-ethanol, 95:5 by vol.
B: CP020 (**c**), CP028 (**a**) and CP099 (**b**), on Hypercarb after extraction from simulated microsomal incubates; mobile phase: Na_2HPO_4 (10 mM, adjusted to pH 3 with H_3PO_4)-acetonitrile, 91:9; 1 ml/min. The Hypercarb column (very stable, and easily regenerated) was 10 × 0.47 cm.

The loads are typically in the range 0.1-15 µg on-column. Pre-treatment of biological samples (1 ml, pH 7.4) is by extraction with 2 × 5 ml dichloromethane (extraction efficiency 94 ±5%).

Peak symmetry was greatly improved by using silica columns with a high carbon loading (Ultracarb®) or non-silica columns, e.g. the Polymer RP (PRP) column⊗ or (shown in Fig. 1B) porous graphitized carbon columns (Hypercarb*). It is inferred that silanol binding sites were the cause of the asymmetric peak seen with C-18 silica.

The peak-shape problem is not confined to HPLC: it was also observed in TLC analysis of these compounds, the spots always being streaked even when C-18 plates were employed. The streaking was eliminated using the strategy of adding 'analyte' to the solvent, as explored for HPLC. It entailed pre-conditioning the TLC plates in the solvent system with CP094/50 present (5 mg/50 ml), mimicking the actual analytes.

In conclusion, TLC and HPLC systems have been developed for quantitation of the 3-hydroxypyridin-4-ones. Resolution of three of these is exemplified in Fig. 1B for HPLC.

Acknowledgement.- The authors thank the British Technology Group (BTG) for financial support.

References

1. Hider, R.C. (1984) *Structure and Bonding 58*, 25-87.
2. Pippard, M.J., Johnson, D.K. & Finch, C.A. (1981) *Blood 58*, 685-692.
3. Roberts, D.W., Ruane, R.J. & Wilson, I.D. (1989) *J. Chromatog. 471*, 433-441 (& see art. #ncC-9, this vol.).

⊗Polymer Labs., Church Stretton *Shandon Scientific, Runcorn

#ncC-10

A Note on

PROBLEMS IN THE HPLC OF SOME METAL-CHELATING COMPOUNDS

R.J. Ruane, D.W. Roberts† & I.D. Wilson

ICI Pharmaceuticals, Mereside, Alderley Park,
Macclesfield, Cheshire SK10 4TG, U.K.

Recently while developing methods for the analysis of certain metal-chelating compounds, serious problems were encountered due to poor chromatographic peak shape [1]: peak tailing and peaks with peculiar unresolved trailing edges were found with RP columns. These problems, which we have now observed with three distinct chemical series (Fig. 1), have been observed with tetracyclines by other workers and overcome by including related compounds or additives such as EDTA in the mobile phase [2-4]. These additives probably stop complexation with metal ions in the chromatographic system. We now outline how the chromatographic problems associated with our compounds were overcome.

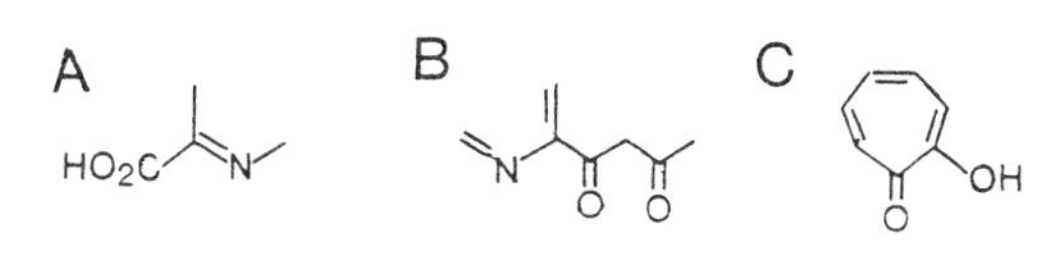

Fig. 1. Part-structures of series of metal-chelating compounds: **A**, heterocyclic acids; **B**, β-diketones; **C**, tropolones.

METHODS

The HPLC system comprised a LC-XPD pump (Pye Unicam), an injector fitted with a 50 μl loop (Rheodyne 7125), a variable wavelength UV detector (Milton Roy/LDC, Spectromonitor 3000) operating at 300 nm, and a chart recorder (Bryans Southern Insts., Mitcham; BS273). All solvents were of HPLC grade. Chemicals were of Analar grade or equivalent.

Compounds from series **A** or **C** were chromatographed using Zorbax C8 silica (Jones Chromatography) packed into a stainless steel column (150 × 4.5 mm i.d.). Compounds from series **B** were chromatographed using a polymer-based resin, 5 μm PLRP-5 (Polymer Labs.) with a similar column. The mobile phases were as stated in the Fig. legends.

RESULTS AND DISCUSSION

Fig. 2A illustrates the type of peak tailing observed when chromatographing a series **A** compound using acetonitrile admixed with butylamine-phosphoric acid buffer. Exhaustive attempts were made to eradicate this problem using, for example, extremes of pH and addition of competing complexing agents such as EDTA, ferric ions or ion-pair reagents. All proved to be unsuccessful.

†*now at* Sterling-Winthrop Res., Alnwick

Fig. 2. Peak shapes for a compound of the heterocyclic acid series on a Zorbax C8 column. Mobile phase: a 1:1 (by vol.) mixture of acetonitrile and pH 2.5 phosphoric acid-butylamine buffer - **A**, unsupplemented; **B**, containing 1 mM picolinic acid. Flow-rate 1 ml/min.

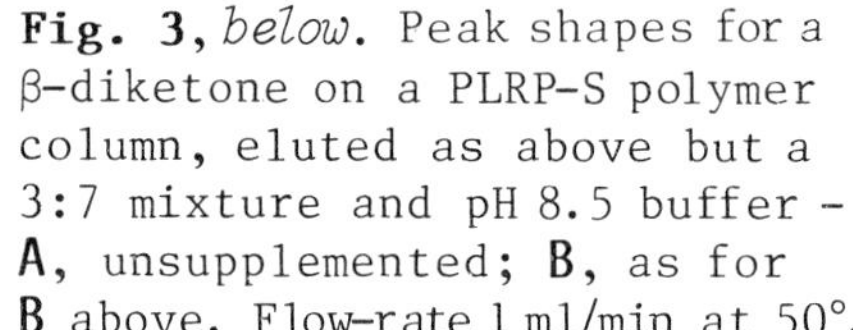

Fig. 3, *below.* Peak shapes for a β-diketone on a PLRP-S polymer column, eluted as above but a 3:7 mixture and pH 8.5 buffer - **A**, unsupplemented; **B**, as for **B** above. Flow-rate 1 ml/min at 50°.

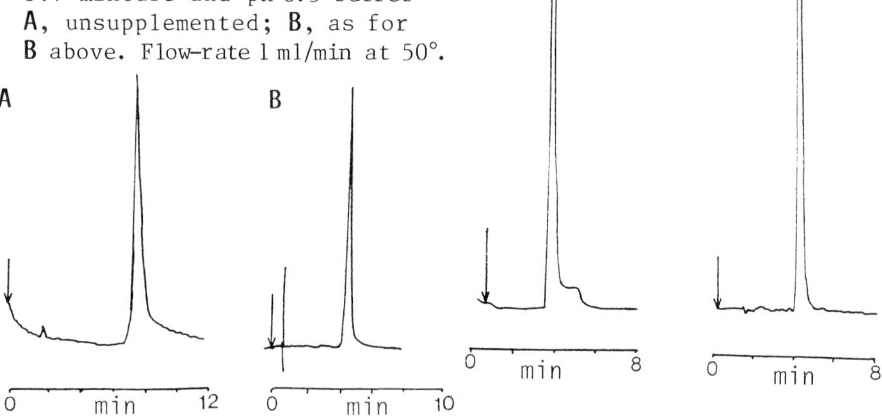

It seemed likely that these problems were associated with interaction between the analytes and a component of the chromatographic system, not residual silanols because changing the pH had no effect on peak shape. We therefore added the compounds themselves to the mobile phase (~1 mM) to try to saturate these interactions. This dramatically improved peak shape but also markedly increased the UV background of the mobile phase resulting in loss of sensitivity. In order to alleviate this problem, picolinic acid (pyridine-2-carboxylic acid) was added to the mobile phase instead of the compound of interest. This resulted in the symmetrical peak shape shown in Fig. 2B. To confirm that the peak-shape improvement was specific to picolinic acid, other structurally related analogues were added instead [1], namely pyridine-3-carboxylic acid and 2-methylpyridine. They produced peak shapes similar to that in Fig. 2A, suggesting that the observed interaction was relatively specific for carboxyl groups adjacent to heterocyclic N atoms.

The β-diketones were not amenable to RP-HPLC, but behaved somewhat better on the polymer column (Fig. 3A) although the peak

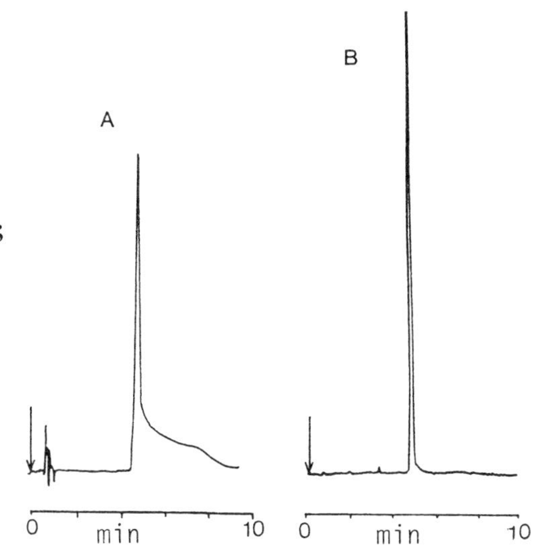

Fig. 4. Peak shapes for a tropolone on a Zorbax C8 column using a 45:55 (by vol.) mixture of acetonitrile and pH 5.0 phosphoric acid-butylamine buffer - **A**, unsupplemented; **B**, containing 5 mM tropolone itself (not the analogue). Flow-rate 1.5 ml/min.

shape did not allow the development of a sensitive assay. Addition of the compound itself to the mobile phase (~1 mM) gave no improvement in peak shape. As a last resort we added picolinic acid to the mobile phase (~1 mM) which resulted in the symmetrical peak shape shown in Fig. 3B.

Compounds from the 'tropolone' series also gave poor peak shapes when chromatographed using an acetonitrile-buffer eluent and C-8 bonded silica (Fig. 4A). Picolinic acid added to the eluent did not give a dramatic improvement in peak shape. The compounds from this series, as the name suggests, are analogues of tropolone. Therefore tropolone itself was added to the mobile phase (~5 mM). This resulted in the symmetrical peak shape shown in Fig. 4B.

The results obtained for all three series of compounds suggest that the interaction which is taking place is highly specific and can only be saturated by addition of a suitable metal-chelating molecule to the mobile phase. Based on these results we have been able to develop sensitive and specific assays for a wide range of compounds of these types in plasma.

References

1. Roberts, D.W. (1989) *J. Chromatog. 471*, 437-441.
2. Böcker, R. (1980) *J. Chromatog. 187*, 439-441.
3. Ersborg, S. (1981) *J. Chromatog. 208*, 78-82.
4. Knox, J.H. & Jurand, J. (1975) *J. Chromatog. 110*, 103-115.

#ncC-11

A Note on

RADIO-TLC OF ^{14}C-PARACETAMOL AND ITS METABOLITES WITH OFF-LINE IDENTIFICATION USING FAB-MS WITHOUT ANALYTE ELUTION

Tracey Spurway, Paul J. Phillips, Ian D. Wilson and Alan Warrander

ICI Pharmaceuticals, Mereside, Alderley Park, Macclesfield, Cheshire SK10 4TG, U.K.

Paracetamol (acetaminophen) is an important analgesic, widely used therapeutically and also [e.g. 1] (cf. #ncC-6, this vol.) as a model compound for the study of drug metabolism, especially conjugation reactions involving glucuronidation and sulphation. Besides, it may be metabolized to GSH* adducts and various GSH-derived metabolites, such as 3-cysteinyl, 3-mercapturic acid, 3-thiomethyl- and 3-methyl-sulphonylparacetamol. In the course of continuing studies on the metabolism of [^{14}C]-paracetamol in the rat [2], we have routinely used TLC for the separation of the drug and its metabolites. However, in common with others [3-5] we have had to employ for these separations a number of different TLC systems, including 2-D methods which made quantification rather impractical. We have therefore investigated the use of multiple-development and RP TLC to ascertain whether either might offer any benefits over our normal TLC systems. In addition we have investigated the use of TLC-FAB-MS for the identification of paracetamol metabolites on TLC plates without the need for prior elution from the phase.

MATERIALS AND METHODS

Compounds.- [^{14}C]Paracetamol (ring label, sp. act. 7.5 mCi/mol; radiochemical purity 98%) was purchased from Sigma Chemical Co., and paracetamol from Aldrich Chemicals. Dr J.K. Nicholson (Chemistry Dept., Birkbeck College) provided reference samples of paracetamol metabolites. All other chemicals were at least of reagent grade and came from BDH Chemicals.

Animals and treatments.- Each of 4 male Wistar rats (200-300 g) was given i.p. a single 300 mg (~8 µCi) dose of [^{14}C]paracetamol, dissolved along with unlabelled drug in 30% ethanol in physiological saline to give 150 mg/ml with sp. act. ~0.10 µCi/mg. The rats were individually housed in metabowls and a complete 0-24 h urine collection made after dosing.

**Abbreviations.-* GSH, glutathione; FAB-MS, fast atom bombardment mass spectrometry; NP, normal phase; RP, reversed-phase.

Radioactivity measurement was by liquid scintillation counting (Beckman 'Ready Value' cocktail, Intertechnique SL30 counter). A 50 μl aliquot of each sample was diluted with 1 ml water and mixed with 10 ml of scintillant. Counting efficiency was determined using the external standards channel-ratio method.

Conjugate hydrolysis.- Urine samples (0.25 ml), adjusted to pH 5, were incubated overnight at 37° with 1.0 μl of *Helix pomatia* digestive juice (Sigma Chemical Co.) as a source of β-glucuronidase and aryl sulphatase.

NP-TLC was performed on 20 x 20 cm silica gel TLC plates containing a fluorescent indicator (E. Merck, #5629; from BDH Chemicals). Authentic paracetamol metabolites were dissolved in ethanol (~1-2 mg/ml) and applied to the origin as 1 cm bands using disposable micropipettes. A Camag IV Linomat was used to apply 10 μl aliquots of the urine samples to the plates as 1 cm bands. Ascending chromatography was performed in glass TLC tanks using multiple development with two solvent systems: system **A**, chloroform-methanol-acetic acid (95:15:1 by vol.); system **B**, butan-1-ol-water-acetic acid (80:10:10). For double development the plates were first run from the origin up to 150 mm in solvent **A**, air-dried and re-chromatographed in the same solvent. After two developments in solvent **A** the plates were transferred to solvent **B** and the solvent front allowed to migrate 65 mm up the plate.

Ion-pair RP-TLC was performed on 10 x 20 cm bonded TLC plates containing a fluorescent indicator (E. Merck, #15423), pre-coated with the ion-pair reagent tetrabutylammonium bromide (0.2 M) dissolved in methanol. Authentic paracetamol metabolites and samples were applied as for NP-TLC. Ascending chromatography was performed in glass TLC tanks using solvent system **C**, methanol-water (35:65). The plates were allowed to develop 150 mm.

Analyte location and quantification of paracetamol and its metabolites.- Developed plates were visualized under UV light at 254 nm. Radioactive metabolites present in the urine samples were located by autoradiography using X-ray film. These metabolites were also located and quantified using a Berthold LB 2842 linear analyzer. Preliminary identification of urinary metabolites was achieved by reference to the R_F values of authentic specimens.

Mass spectrometry.- Authentic paracetamol metabolites were analyzed by FAB-MS in both positive- and negative-ion modes using a Kratos MS80RF mass spectrometer equipped with a DS55 data system. Radioactive bands were located on

Table 1. TLC R_F's for authentic compounds (**P1** denotes paracetamol): **a**, NP-TLC (solvent **A** front travelled 2 × 150 mm, **B** front 68 mm); **b**, ion-pair RP-TLC. For **a**, 'overall' signifies **A** + **B**.

Compound	**a**: solv. **A**	**a**: solv. **B**	**a**: overall	**b**
P1	0.52	-	0.52	0.33
3-thiomethyl-**P1**	0.73	-	0.73	0.15
3-methylsulphonyl-**P1**	0.50	-	0.50	0.30
3-cysteinyl-**P1**	0.0	0.41	0.18	0.63
P1 mercapturate	0.02	0.62	0.28	0.28
P1 glucuronide	-	0.35	0.16	0.60
P1 sulphate	0.02	0.59	0.26	0.11

the TLC plate (see above), and these areas of silica were removed from the plate, mixed with matrix (glycerol) and analyzed by FAB-MS. The metabolites were identified by comparison of their spectra with those of authentic specimens.

RESULTS AND DISCUSSION

TLC of authentic metabolites.- Due to the wide range of polarities exhibited by paracetamol and its metabolites we could not design a solvent system capable of separating all of them in a single development. We therefore attempted to devise methods based on multiple development with solvents of different eluotropic strengths. Double development in system **A** was used first to chromatograph and separate paracetamol and its non-polar metabolites (e.g. the thiomethyl metabolite). The polar metabolites (e.g. sulphate and glucuronide conjugates) remained at the origin, and were separated by a further development in system **B**. The R_F's for the authentic metabolites are given in the '**a**' columns of Table 1. As shown, the sulphate and the mercapturate had the same R_F, but the separation of other metabolites was good.

Since none of the NP systems tested gave adequate resolution of the sulphate and mercapturate, we investigated RP-TLC. However, it soon became apparent that the use of an ion-pair system would be required in order to obtain satisfactory chromatography. An ion-pair RP-TLC system was therefore developed with C-18 bonded silica gel to separate the paracetamol sulphate from the mercapturate. With the finally adopted solvent system (**C**), authentic compounds gave the R_F's shown in the last column ('**b**') of Table 1. Separation of sulphate and mercapturate was achieved, but the latter remained prone to streaking.

Although neither NP nor RP systems on their own were satisfactory for paracetamol and its metabolites, the two

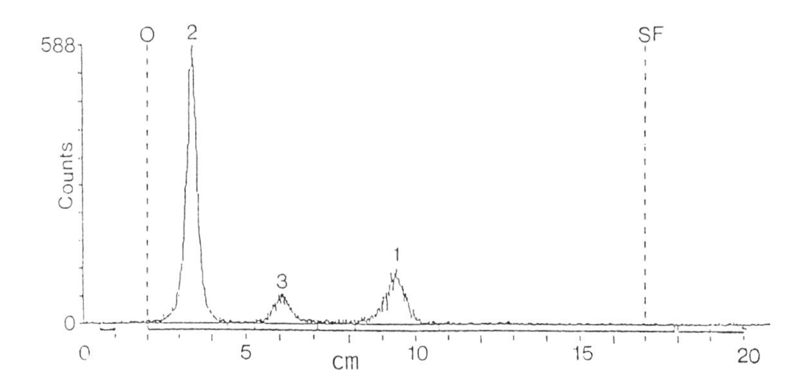

Fig. 1. Typical NP-TLC linear-analyzer trace (double development in solvent **A**, single development in **B**) of rat-urine paracetamol (3) and metabolites: 1, glucuronide; 2, sulphate. O = origin, SF = solvent front.

systems in conjunction enabled all the metabolites to be separated and tentatively identified by their R_F's.

TLC of rat urine samples was performed with the systems described above. NP-TLC separated the radioactivity into the 3 peaks shown in Fig. 1; peaks 1 and 3 co-chromatographed with authentic compounds, and peak 2 corresponded in R_F to the sulphate or mercapturate. Incubation of the urine with the enzymes resulted in loss of both peaks 1 and 2 and increased peak 3, providing further evidence that the metabolites present in 2 and 1 were sulphate and glucuronide conjugates of paracetamol. **RP-TLC** was also performed on the urines, to confirm identities. The radioactivity again appeared in three peaks (Fig. 2) whose identities were as in the NP-TLC study as likewise confirmed by co-chromatography and enzymic effects. There was no radiolabel in the paracetamol mercapturate position.

TLC FOLLOWED BY MS

The identity of the urinary radioactive peaks following NP-TLC was verified using TLC-FAB-MS by comparison of spectra with those obtained from standards. All the latter did give spectra, but with rather poor sensitivity which led to difficulty when urine samples were analyzed by TLC-FAB-MS. However, the area corresponding to paracetamol glucuronide (peak 1, Fig. 1) gave a spectrum by negative-ion FAB-MS which corresponded to that similarly produced by the authentic glucuronide (Figs. 3A & 3B). The putative paracetamol sulphate (peak 2, Fig. 1) similarly gave a spectrum (Fig. 4B) which corresponded to that of the authentic compound (Fig. 4A).

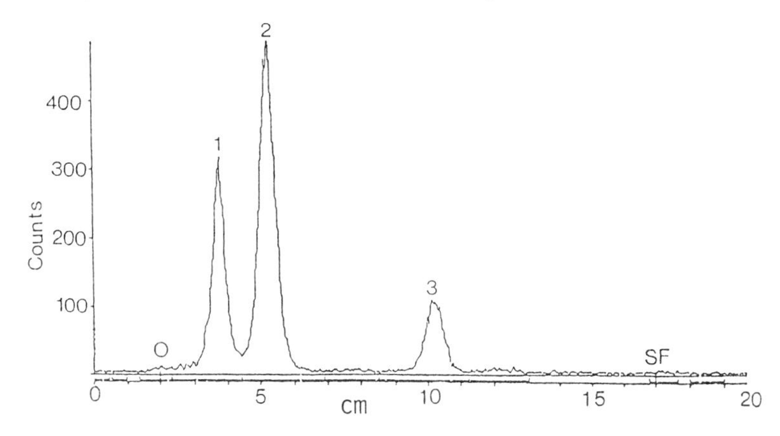

Fig. 2. As for Fig. 1, but ion-pair RP-TLC (solvent system **C**); same numbering for peaks.

CONCLUSIONS

Amongst the paracetamol metabolites available to us, all were separable by NP-TLC on silica gel, using multiple development with two solvent systems, with the exceptions of the sulphate and the mercapturate. These were separable by ion-pair RP-TLC albeit with poor spot shape for the mercapturate. The identity of the sulphate and glucuronide conjugates could be confirmed by FAB-TLC-MS, but with a lack of sensitivity which, however, was a property of the compounds in the chosen FAB matrix rather than an effect of the silica gel onto which they were adsorbed. Selection of the matrix for the FAB-MS of a given compound can profoundly affect sensitivity (cf. G.C. Bolton's art., #ncC-12). Glycerol, the most commonly used matrix, may not have yielded the optimum response in this case but sufficed for our purposes.

References

1. Nicholson, J.K., Sadler, P.J., Tulip, K. & Timbrell, J.A. (1986) in *Bioactive Analytes, including CNS Drugs, Peptides and Enantiomers*⊗ (Reid, E., Scales, B. & Wilson, I.D., eds.), 321-335.
2. Warrander, A., Allen, J.M. & Andrews, R.S. (1985) *Xenobiotica 15*, 891-897.
3. Davis, M., Simmons, C.J., Harrison, N.G. & Williams, R. (1976) *Quart. J. Med. 45*, 181-191.
4. Jollow, D.J., Thorgeirsson, S.S., Potter, W.Z., Hashimoto, M. & Mitchell, J.R. (1974) *Pharmacology 12*, 251-271.
5. Pang, K.S., Yuen, V., Fayz, S., Koppele, J.M. & Mulder, G.J. (1986) *Drug. Metab. Disp. 14*, 102-111.

⊗Vol. 16, this series; Plenum, New York

[Figs. 3 & 4: OVERLEAF

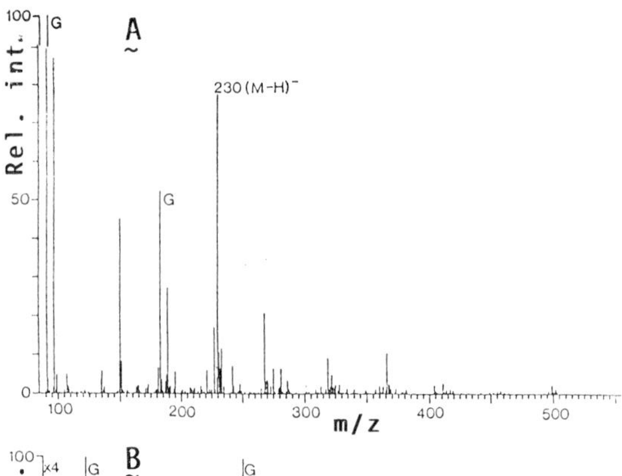

Fig. 3. Negative-FAB-MS of paracetamol glucuronide standard (A) and peak 1 (B), obtained directly from the TLC spot. G = glycerol.

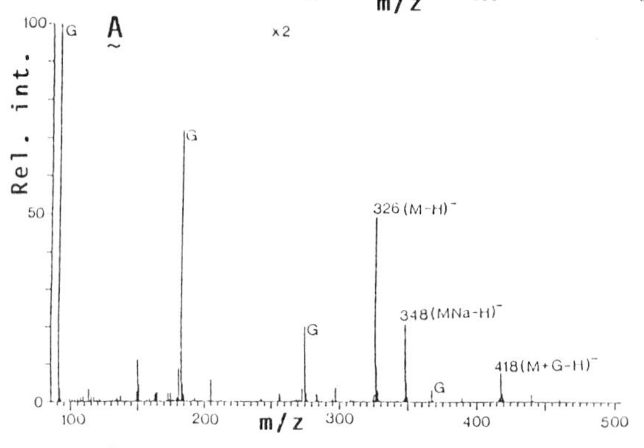

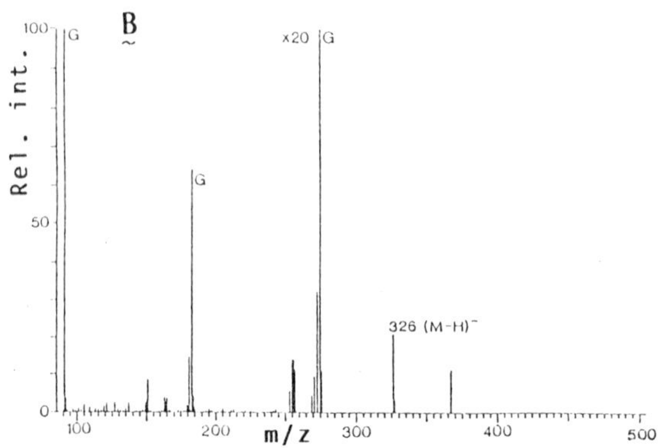

Fig. 4. As for Fig. 3, but paracetamol sulphate standard (A) and peak 2 (B).

#ncC-12

A Note on

THE POTENTIAL USE OF A TLC-FAB-MS INTERFACE IN DRUG DEVELOPMENT

G.C. Bolton, G.D. Allen, M. Nash and H.E. Proud

Department of Drug Metabolism and Pharmacokinetics,
Beecham Pharmaceuticals Research Division,
Medicinal Research Centre, Coldharbour Road,
The Pinnacles, Harlow, Essex CM19 5AD, U.K.

Experiments have been carried out to evaluate the potential usefulness of a recently introduced Jeol TLC-MS interface in drug development with particular emphasis on drug metabolism studies. A number of different drugs have been investigated following TLC as neat solutions and when spiked into dog urine. Conditions for obtaining spectra have been explored, particularly the application of the fast atom bombardment (FAB) matrix to the TLC plates and the scanning speed of the magnetic focussing mass spectrometer. Two FAB matrices were tested in these experiments: glycerol and thiodiethanol.

Drugs investigated were applied as spots, ~4 mm diameter, to Keiselgel $60F_{254}$ TLC plates (layer thickness 0.2 mm) either as pure solutions in methanol or as solutions in dog urine. The latter was used to determine the possible interference from endogenous materials when obtaining spectra for drugs in a biological fluid. The plates were examined by TLC-MS after suitable application of the FAB matrix using one of the following methods:
- matrix spread onto the plate;
- spraying with matrix/methanol;
- dipping in matrix/methanol (various proportions);
- matrix dissolved in the mobile phase.

Appropriate regions of the plates were cut into strips (0.8 × 5 cm) and inserted into the TLC-MS interface for direct analysis. Rates of plate movement and scan speed were also investigated, with the aim of optimizing reproducibility and sensitivity.

Instrument conditions were as follows: Jeol DX303 mass spectrometer (magnetic); ionization by FAB (xenon gas); resolution, nominal 1000; gun volts, 2 kV; optimum step distance, 0.5 mm/scan; optimal scan time, 10 sec.

Results.- FAB spectra were successfully obtained from all the tested compounds, which differed in detection limits. For solutions in dog urine the following limit values (μg applied to plate) were obtained: atenolol, 2.5; methylatropine, 0.5; chlorpromazine, 2.5; cimetidine, 5.0; furosemide, 5.0; mupirocin, 2.0; metoclopramide, 2.5; naloxone, 2.5; renzapride, 2.5; scopolamine, 0.5. Fig. 1 shows MS results for renzapride.

The results from the experiment using four different methods of applying the matrix to the TLC plate clearly demonstrated that the best method was dipping in a 1:1 matrix-methanol mixture. Application of the matrix by spraying, using different proportions of methanol and matrix, did not readily give reproducible results. Application of neat matrix generally gave a high background and the plate had to pass through the FAB beam several times before a signal for the drug could be detected. Incorporation of the matrix into the TLC solvent did not allow sufficient matrix to be deposited on the plate for a successful result without deterioration of the chromatography.

Use of thiodiethanol as the FAB matrix sometimes gave more sensitive results but with less reproducibility than glycerol. Glycerol performed better overall, and good spectra were easily obtainable for the compounds tested. Interference from endogenous compounds in neat urine was not a problem under the conditions of these experiments. We conclude that this TLC-FAB-MS interface will be useful in drug development, notably in drug metabolism studies where identification of metabolites is required, and will be a valuable complement to HPLC/GC-MS techniques.

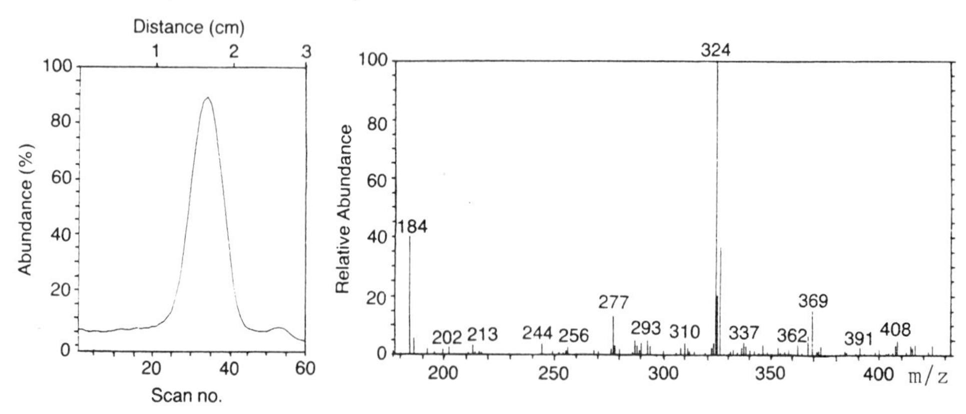

Fig. 1. Illustrative results, with the FAB mass spectrum, following TLC-MS of renzapride in dog urine. TLC developing solvent: chloroform-methanol-aq. ammonia, 10:4:1 by vol. The peak located by selected-ion monitoring *(left)* gave a spectrum as illustrated *(right)*.

#ncC-13

A Note on

LASER-INDUCED FLUORESCENCE AS A DETECTION MODE IN COLUMN LIQUID CHROMATOGRAPHY

H. Lingeman, R.J. van de Nesse, U.A.Th. Brinkman, C. Gooijer and N.H. Velthorst

Department of General and Analytical Chemistry,
Free University, De Boelelaan 1083,
1081 HV Amsterdam, The Netherlands

Multi-sample analysis of trace-level drugs and drug-derived or endogenous compounds still calls for more powerful as well as speedier bioanalytical procedures [1]. Where the cardinal need is a gain in sensitivity rather than in selectivity, LIF* in conjunction with LC† is a promising technique [2]. However, until recently use has been made only of mainly continuous wave line lasers providing a limited number of lasing wavelengths in the UV and visible regions. Hence this technique is inapplicable to analytes for which the intrinsic fluorescence is nil, low or at unfavourable wavelengths [3]. Possible remedies are the use of variable wavelength dye lasers [2], of derivatization techniques (pre-, on- or post-column) [1, 4], or of shorter **ex.** wavelengths [5].

This article follows up the LIF survey that appeared earlier in this series [6]. Whilst it opens with an outline of principles and of LIF-LC instrumentation, the focus is on approaches that extend applicability, such as frequency doubling, variable wavelength dye lasers, and derivatization techniques. Brief consideration is also given to some techniques that improve selectivity, such as 2-photon excitation and time-resolved fluorescence.

LASER-INDUCED FLUORESCENCE DETECTION: PRINCIPLES AND EQUIPMENT

Lasers give photons that are identical in phase, direction and amplitude, and the output beam is highly uni-directional, intense, monochromatic and coherent. LIF not only gives a significantly higher photon flux than CIF but also minimizes stray-light effects. Furthermore, due to the high monochromacity, Rayleigh and Raman bands are relatively small, resulting in a favourable optical window. However, in comparison

†*Editor's abbreviations*.- Column liquid chromatography, LC (CLC was the authors' abbreviation); *prefix* cap.- = capillary; **ex.**, excitation; **em.**, emission; nm *connotes* wavelength. **Other abbreviations*.- CIF/LIF, conventional-/laser-induced fluorescence (detection *usually implied*); DPIH, 2-diphenylacetyl-1,3-indandione-1-hydrazone; NBD, nitrobenzoxadiazole; NDA, naphthalenedialdehyde; OPA, *o*-phthaldialdehyde; RP, reversed-phase. **Ar** = argon, **H** = hydrogen, *etc. (Ed.'s policy).*

with CIF these bands are much more intense, resulting in a higher background signal especially when the laser output is increased. Another benefit of LIF is that the beam can be accurately focused onto the flow-cell, minimizing excitation energy loss and allowing the analysis of small sample volumes [4, 7].

In the now popular **Ar**-ion gas laser [2], there is a resonance tube housing a plasma tube (Fig. 1) within which **Ar** atoms are excited by means of electrons. The principle of stimulated emission is that the excited **Ar** ion will spontaneously emit a photon while returning to its electronic ground state. When this photon is emitted parallel to the resonance tube it will hit another excited **Ar** ion, after a certain period of time. This ion also returns to its electronic ground state, resulting in two identical photons which can hit two other excited molecules resulting in 4 identical photons. Thereby the laser's photon flux can be increased enormously. Because one of the two mirrors at the end of the resonance tube is only partially reflective, some of the generated light can escape and be used for detection purposes.

Equipment.- Because in a LC-LIF detection device the laser serves only as a light source, there are requisite components besides those for LC:- some lenses or optical fibres, a flow-cell, a monochromator or interference filter, stray-light filter(s), a photon-counting unit, and some recording facilities [4, 5, 7]. No commercial equipment is yet available, but we (with colleagues) have found it easy to construct an optical train [8]. Between the flow-cell and the photon-counting unit, which normally is a blue/green-sensitive cooled photomultiplier tube, are placed a condenser lens (to focus the light onto the latter tube), an interference filter or monochromator (transparent only for the emitted fluorescence light), and an optical glass or liquid dye filter to eliminate the interfering Rayleigh scattering. So as to achieve a high and repeatable signal, all the optical components should be linearly positioned. To help with positioning and repeatability, an optical fibre can be installed between the flow-cell and the photomultiplier tube and/or between the laser and the flow-cell.

For conventional LC the flow-cell design is relatively simple: thus a purchased 25-μl quartz cell can be used, allowing fluorescence measurements in the in-plane 90° geometry, with an optical pathlength of 1.5 mm and an aperture of 11 × 1.5 mm [7] (Fig. 2A). Another possibility is to place the flow-cell in a metal box with two holes in a straight line for the incident and the transmitted laser beam [5].

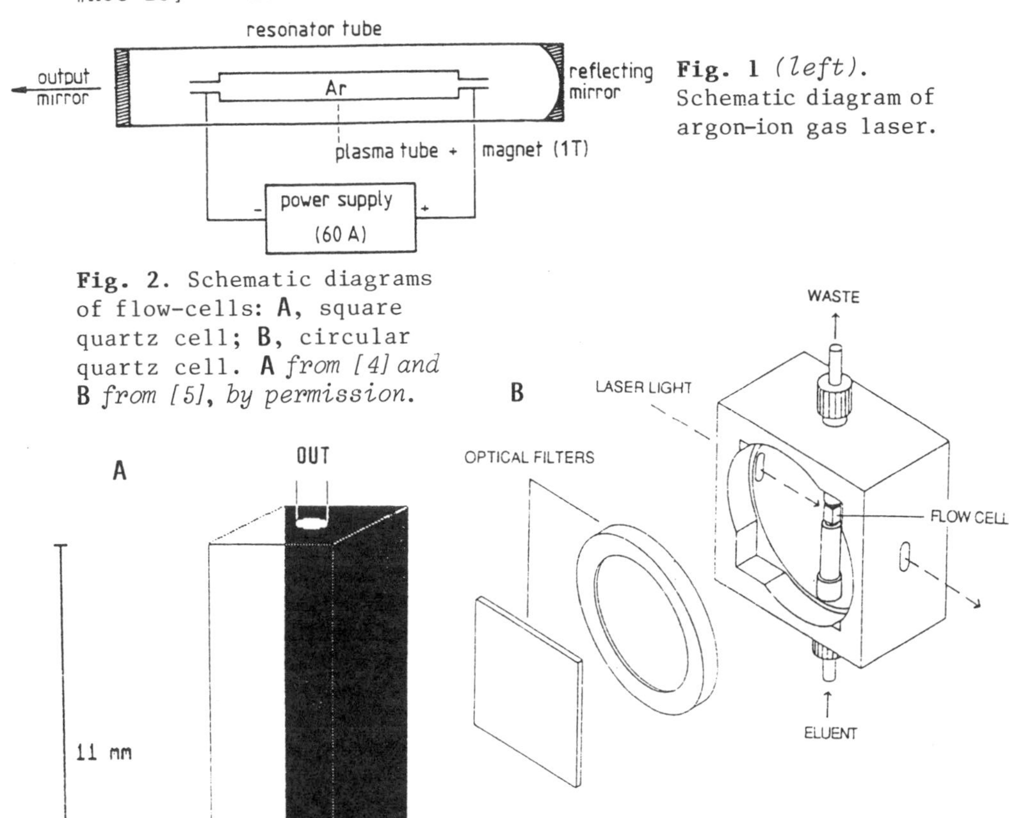

Fig. 1 *(left).* Schematic diagram of argon-ion gas laser.

Fig. 2. Schematic diagrams of flow-cells: **A**, square quartz cell; **B**, circular quartz cell. **A** *from [4] and* **B** *from [5], by permission.*

So as to avoid reflections the inside of the box is coated with a suspension of graphite. The quartz cell, home-made or purchased, consists of a square block of quartz with a 1.1 mm internal circular bore. The flow-cell housing is mounted directly in front of the photomultiplier tube (Fig. 2B).

CHOICE OF LASING SYSTEM

In considering what type to adopt, and whether to choose a continuous-wave, a pulsed or a dye laser, it should be realized that all systems other than the variable wavelength dye lasers emit one or a limited number of characteristic wavelengths, and that only the 250-500 nm range applies for

LIF of biologically important analytes. In the following list (partly from [2]) of lasing wavelengths (nm), the gas lasers are numbered thus: **1**, **Ar**-ion; **2**, **He**-**Cd**; **3**, **N** (nitrogen); **4**, **Kr**-ion; **5**, **Kr** fluoride; **6**, **Xe** fluoride.-

248: **5**, 0.15 J	363.8: **1**, 1.0 W	465.8: **1**, 0.75 W
308: **6**, 0.06 J	406.7: **4**, 0.9 W	468.0: **4**, 0.3 W
325: **2**, 5 W	413.1: **4**, 0.5 W	472.7: **1**, 1.2 W
334: **1**, 0.3 W	415.4: **4**, 0.3 W	476.2: **4**, 0.4 W
337.1: **3**, 0.01 J	441.6: **2**, 0.02 W	476.5: **1**, 2.7 W
351: **6**, 0.05 J	454.5: **1**, 1.1 W	482.5: **4**, 0.4 W
351.1: **1**, 1.1 W	457.9: **1**, 1.35 W	488.0: **1**, 6.5 W
		496.5: **1**, 2.5 W

W *refers to* maximum available power, *and* J *to* pulse energy

Because of costliness, **4** and the excimer lasers (e.g. **Ar** fluoride and **5** and **6**) are hardly used for detection purposes. Furthermore, continuous-wave lasers are, in principle, preferable to pulsed lasers, because of system instability, and to variable wavelength dye lasers because these are relatively costly and inefficient [2]. It is obvious why **1-3** are favoured.

As the number of available lasing lines in the UV and visible regions is limited, many analytes - even if possessing native fluorescence - cannot be detected by LIF because they cannot be excited efficiently. For the same reason derivatization procedures developed for CIF detection in LC are often precluded. Moreover, Raman scatter due to the eluent may interfere with the analyte's fluorescence [2, 5, 7-10] as is particularly important with analytes having a relatively small Stokes shift [2]. Possible remedies are the application of variable wavelength dye lasers, derivatization techniques or the use of shorter **ex.** wavelengths.

APPLICATION OF RELATIVELY SHORT EXCITATION WAVELENGTHS

Advantageously, molar absorptivities are normally higher in the UV region and diverse analytes have UV absorbance and thus can be excited. As a rule of thumb, moreover, analytes which fluoresce at >350 nm also fluoresce at ~250 nm. Hence frequency-doubling of the 514 nm **em.** line of the **Ar**-ion laser, resulting in the 257 nm lasing line, will be an attractive possibility, with benefit not only to the number of eligible solutes but also to the background signal. In LIF this signal is governed by Rayleigh scatter, refractions and reflections, by Raman scatter from the eluent, and by luminescence from cell walls and impurities. For example, with **ex.** at 458 nm, the Raman spectrum extends to ~560 nm, so that sensitive fluorescence detection can be performed only in one of the windows or at >560 nm (Fig. 3). The 257 nm line obtained by frequency doubling enables all wavelengths over 290 nm to be used for detection [5]. In other

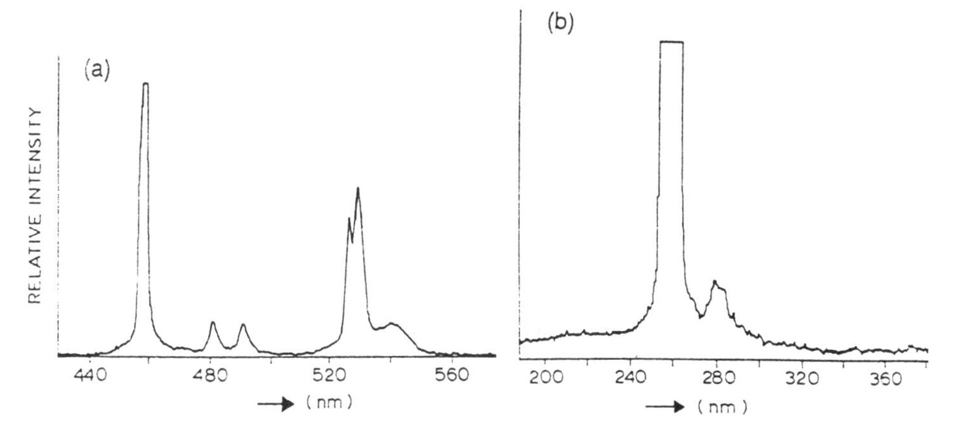

Fig. 3. Raman spectra of methanol-water (90:10 by vol.) at (**a**) 458 nm and (**b**) 257 nm laser **ex**. *From [5], by permission.*

words, the advantage of shorter wavelengths is that Raman scatter of the eluent no longer interferes with the analyte's fluorescence.

However, frequency doubling of an **Ar**-ion laser is rather inefficient. An initial laser power of ~500 mW of the 514 nm line results in ~5 mW of 257 nm line, providing detection limits for polyaromatic hydrocarbons at the 50 ppt level [5]. That the gain in selectivity, compared to CIF selection, is limited can be readily explained because in LIF the bandwidth of the **ex**. beam is only a few picometres, which means that it is extremely difficult to excite exactly at the **ex**. maximum, while in conventional fluorescence the applied bandwidths are between 40 and 80 nm which makes it easier to achieve the analyte's **ex**. maximum.

APPLICATION OF VARIABLE WAVELENGTH DYE LASERS

The second possible way to extend the LIF application range is use of a variable wavelength dye laser. These systems consist of two coupled lasers [2]: the first, an **N**d-YAG, **N** or **Ar**-ion laser, serves as a pump laser for the second, in which an organic dye or dye mixture is present. The disadvantages of such lasers are that they are rather costly and not very efficient, especially since normally frequency-doubling is needed to obtain the required wavelength for the pump laser. Furthermore the efficiency of the second (dye) laser is relatively low; but the advantage is that these lasers can be tuned over a certain wavelength range because of the incorporated dye. Because the energy difference between the electronic and vibrational states of these organic molecules is relatively small, the separate peaks will overlap and so a more-or-less continuous spectrum is created. Using the **N** laser as the pump source, for example, any wavelengths between 537 and 618 nm can be isolated when rhodamine 6G is the chosen dye, and if butyl PBD is applied the tuning range is 356-390 nm [booklet (1986) by U. Brackmann; Lambda Physik GmbH, Göttingen].

A nice application of a **N**-pumped dye laser is the determination of NBD derivatives of primary amines [11], exemplified by octylamine (g in Fig. 4). The features of 2-photon excitation, which significantly increases sensitivity, will be explained later. However, frequency-doubling of the pump laser, or derivatization, is still usually unavoidable, as comparatively few analytes have sufficient native fluorescence at >300 nm **ex.**

APPLICATION OF DERIVATIZATION PROCEDURES

For the derivatization approach to extending the LIF application range, there are now available a number of fluorescence probes with high reactivity towards one or more functional groups and with **ex.** wavelengths which match one of the available lasing lines [1, 2]. Thus, dopamine can be derivatized with OPA, and 16 pg detected with the 350-360 nm **Ar**-ion laser multi-line [2, 12]. The following data represent a comparison between NDA [10] - a structural analogue of OPA - and *(in italics) OPA*, the **Ar**-ion laser **ex.** of the derivatives being at 458 nm (900 mW) *or 334-363 nm (350 mW)*.-

ex. 420 *(340)*, **em.** 490 *(455)*; detection limits (fmol): 10 *(200)* by CIF, 0.2 *(5)* by LIF; fluorescence quantum yield 0.54 *(0.11)*; relative intensity of fluorescence 2.8 *(1)*.

Evidently NDA possesses more favourable fluorescence characteristics: a higher quantum yield, and longer **ex.** and **em.** wavelengths; hence it gives ~20-fold better detection limits with CIF, ~25-fold better with LIF, and 50-fold better sensitivity when CIF and LIF are compared. Both probes serve for derivatizing primary amines and can be excited with the **Ar**-ion laser. The latter was also the case for a second example - fluorescein and fluorescein derivatives - where ~490 nm **ex.** is the fluorphore's maximum, matching the 488 nm line which is the most intense in the **Ar**-ion laser's spectrum [2, 7]. Moreover, fluorescein derivatives are usable for all kinds of functional groups, such as fluorescein isothiocyanate which can derivatize primary amino groups [8] and gives detection limits of ~1 fg after RP-LC and LIF [4, 7].

The use of the **He-Cd** laser along with derivatization (coumarin) is exemplified by the determination (down to 10 fg) of solvolyzed plasma steroids [13], by cap.-RP-LC with a cap.-flow-cell (100 nl volume). Appropriate flow-cells were surveyed earlier in this series [6]. The most suitable type is made of fused silica tubing of 100 µm i.d.; a video camera helps position the PM tube to optimize signal-to-noise [14].

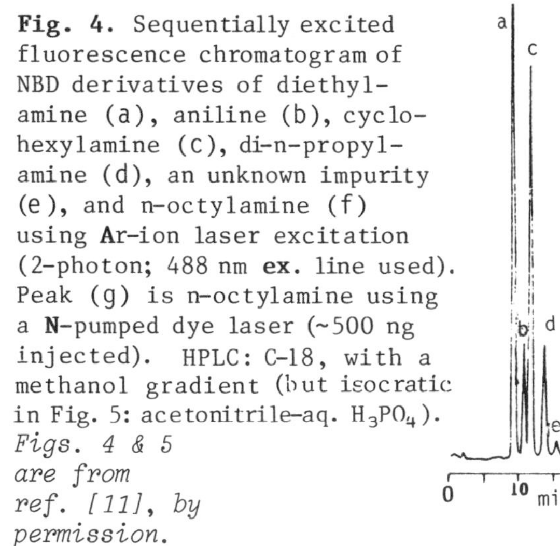

Fig. 4. Sequentially excited fluorescence chromatogram of NBD derivatives of diethylamine (a), aniline (b), cyclohexylamine (c), di-n-propylamine (d), an unknown impurity (e), and n-octylamine (f) using **Ar**-ion laser excitation (2-photon; 488 nm **ex.** line used). Peak (g) is n-octylamine using a **N**-pumped dye laser (~500 ng injected). HPLC: C-18, with a methanol gradient (but isocratic in Fig. 5: acetonitrile–aq. H_3PO_4). *Figs. 4 & 5 are from ref. [11], by permission.*

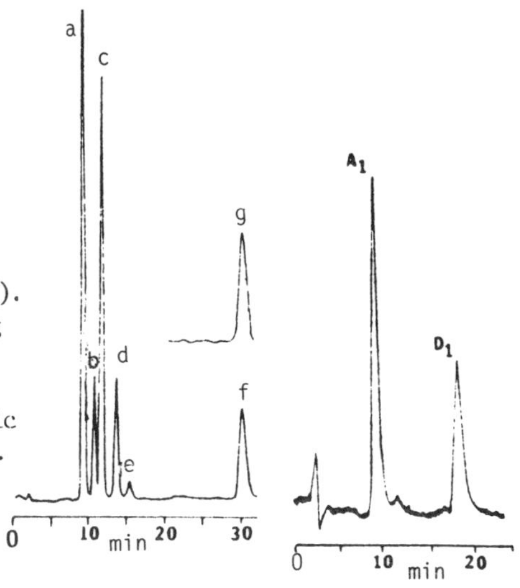

Fig. 5 *(right)*. Chromatogram of adriamycin (A_1) and daunorubicin (D_1) with detection as in Fig. 4: **Ar**-ion, **ex.** 488 nm (2.0 W); 41 ng of each drug was injected.

Simple carboxylic acids such as fatty acids can be derivatized using 4-bromomethyl-7-methoxycoumarin [14]; caproic acid was detectable down to 1.3 fmol with cap.-LC. A further example is the derivatization of ketosteroids with DPIH, which reacts selectively with aldehydes and ketones [15]; since these hydrazone derivatives have maximal **ex.** at ~315 nm, the **He-Cd** laser which gives 325 nm suits well.

APPLICATION OF SELECTIVITY-ENHANCING TECHNIQUES

Other detection systems besides those surveyed above can be applied to improve selectivity, notably 2-photon excitation, time-resolved fluorescence and optical activity detection. The improved selectivity of sequentially excited 2-photon excitation arises from the fact that, contrary to normal fluorescence, the emission wavelength is shorter than the excitation wavelength, thereby eliminating interferences from scattered light [2]. In the example shown in Fig. 5 for two anti-tumour agents, the sensitivity gain compared with conventional fluorescence is ~5-fold, but far more important is the gain in selectivity [11].

Another way to increase selectivity in LIF is the use of time-resolved fluorescence. For instance, a number of drugs, such as tetracyclines, form complexes with the europium ion which give a long-lived fluorescence, unlike the matrix and background fluorescence which is short-lived. Hence by using a delay time between the excitation pulse and measuring the fluorescence signal, most of the interfering

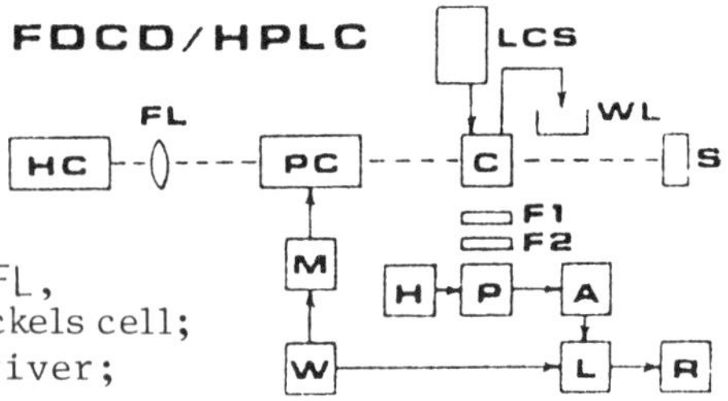

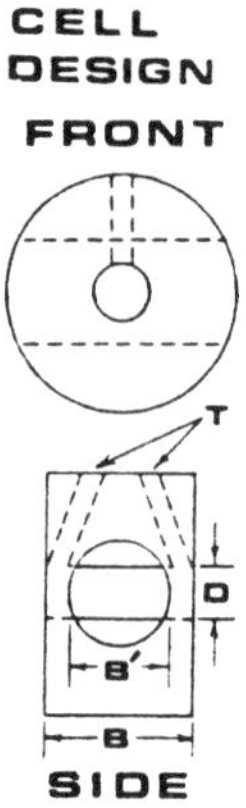

Fig. 6. Fluorescence-detected circular dichroism LC system. HC, He-Cd laser; FL, focal lens; PC, Pockels cell; M, modulation driver; W, wave-form generator; LCS, LC system; WL, waste; C, detection cell; S, beam stop; F, filter; P, PMT; H, power supply; A & L, amplifiers; R, recorder. Detection cell *(right)*. B, total pathlength (1.8cm); B', observed pathlength (1.2 cm); D, quartz tubing (1.0 mm i.d.); T, chromatography tubing.

From ref. [16], by permission.

emissions can be eliminated. This approach is, of course, applicable only with pulsed lasers (e.g. Nd-YAG, excimer).

Fluorescence-detected circular dichroism [16] is a new development in detection modes, combining the selectivity of fluorescence and that of optical activity. (Circular dichroism is defined as the difference in absorbance of left- and right-circularly polarized light.) A He-Cd laser operating at 325 nm with ~8 mW power is used with electro-optic modulation at 150 kHz (Fig. 6). The detection cell has an observed light path of 1.2 cm. Riboflavin was the test compound in this study [16] since it exhibits strong fluorescence and optical activity. With CIF the detection limit is in the pg range, but in complex matrices only circular dichroism provides the required selectivity; the detection limit is then ~170 pg using conventional RP-LC.

CONCLUSIONS

Nowadays there are available a number of fluorescence probes that possess high reactivity for one or more functional groups, and excitation wavelengths which match a type of continuous-wave system that is often used - the argon- (Ar-)ion gas laser: one of its most important lasing wavelengths (514 nm) yields, with frequency-doubling, the 257 nm line. The main advantages of using shorter wavelengths are the higher excitation coefficients and the ability to eliminate the interferences from Raman scatter of the eluent.

The salient techniques that extend applicability, in priority order, are frequency-doubling, derivatization, and the use of dye lasers.

Although not many applications have been described so far, the use of a frequency-doubled or -tripled neodymium-YAG (Nd-YAG) pulse laser or an excimer pulse laser [2] seems to be especially promising. Particular reasons are that the majority of analytes that fluoresce at higher wavelengths will also fluoresce at 265 nm, and that it is possible to apply selectivity-enhancing techniques such as 2-photon excitation or time-resolved techniques. Furthermore, these lasers can be used for the analysis of very small sample volumes.

References

1. Lingeman, H. & Underberg, W.J.M., eds. (1990) *Detection-Oriented Derivatization Techniques in Liquid Chromatography*, Marcel Dekker, New York.
2. Van den Beld, C.M.B. & Lingeman, H. (1990) in *Comprehensive Handbook on Luminescence Spectroscopy* (Baeyens, W.R.G., ed.), Marcel Dekker, New York.
3. Van der Greef, J., Lingeman, H., Niessen, W.M.A. & Tjaden, U.R. (1987) in *Topics in Pharmaceutical Sciences 1987* (Breimer, D.D. & Speiser, P., eds.), Elsevier, Amsterdam, pp. 137-149.
4. Lingeman, H., Tjaden, U.R., Van den Beld, C.M.B. & Van der Greef, J. (1988) *J. Pharm. Biomed. Anal. 6*, 687-695.
5. Van de Nesse. R.J., Hoornweg, G.Ph., Gooijer, C., Brinkman, U.A.Th. & Velthorst, N.H. (1989) *Anal. Chim. Acta 227*, 173-179.
6. Brinkman, U.A.Th., De Jong, G.J. & Gooijer, C. (1988) in *Bioanalysis of Drugs and Metabolites* [Vol. 18, this series], (Reid, E., Robinson, J.D. & Wilson, I.D., eds.), Plenum, N.York, 321-338.
7. Van den Beld, C.M.B., Lingeman, H., Van Ringen, G.J., Tjaden, U.R. & Van der Greef, J. (1988) *Anal. Chim. Acta 205*, 15-27.
8. Lingeman, H. (1987) *Chem. Mag.*, 829-830.
9. Hulshoff, A. & Lingeman, H. (1985) in *Molecular Luminescence Spectroscopy: Methods and Applications, Part I* Schulman, S.G., ed.), Wiley-Interscience, New York, pp. 621-716.
10. Roach, M.C. & Harmony, M.D. (1987) *Anal. Chem. 59*, 411-415.
11. Huff, P.B., Tromberg, B.J. & Sepaniak, M.J. (1982) *Anal. Chem. 54*, 946-950.
12. Brinkman, U.A.Th., Frei, R.W. & Lingeman, H. (1989) *J. Chromatog. 492*, 251-298.
13. Gluckman, J., Shelly, D. & Novotny, M. (1984) *J. Chromatog. 317*, 443-453.
14. Tsuda, T. & Noda, H. (1989) *J. Chromatog. 471*, 311-319.
15. Heindorf, M.A. & McGuffin, V.L. (1989) *J. Chromatog. 464*, 186-194.
16. Synovec, R.E. & Yeung, E.S. (1986) *J. Chromatog. 368*, 85-93.

#ncC-14

A Note on

THE USE OF IMMOBILIZED ANTIBODIES FOR AUTOMATED ON-LINE SAMPLE PRE-TREATMENT IN HPLC

A. Farjam, H. Lingeman[†], P. Timmerman, A. Soldaat, A. Brugman, N. van de Merbel, G.J. de Jong, R.W. Frei and U.A.Th. Brinkman

Department of Analytical Chemistry, Free University, De Boelelaan 1083, 1081 HV Amsterdam, The Netherlands

An Ab[*] can be 'tailored' for almost any organic solute of interest, exhibiting high selectivity and affinity for this analyte (antigen). Pre-columns containing these Ab's in an immobilized form are a powerful tool in sample pre-treatment. Samples containing antigenic analytes can be cleaned and enriched with high efficiency, due to the strong and selective Ab-antigen interaction. In the present study a system has been developed which comprises an immuno pre-column coupled on-line with an HPLC system. The system allows the *direct* injection of large volumes of biological samples (e.g. 20 ml urine, 6 ml plasma, 10 ml milk) combined with fully automated sample pre-treatment and HPLC analysis. The performance of the system is now demonstrated with three different immuno pre-columns specific for anabolic steroids, oestrogens and aflatoxins [1, 2].

The critical step in the on-line coupling of an immuno pre-column and the separation column is the effective transfer of analytes from one system to the other: the trapped analytes must be desorbed in such a way that they can be reconstituted in a small band on the separation column. Two different techniques have been evaluated.

(1) THE IMMUNOSELECTIVE DESORPTION TECHNIQUE

With this technique the trapped analytes are desorbed from the immuno pre-column and reconcentrated on a short RP pre-column (C-18; Fig. 1) which is switched in series with it. The desorbing solution has a high content of a compound which, similarly to the analytes, has a high affinity for the immobilized Ab, and so displaces them from

---

[†]addressee for any correspondence

[*]*Abbreviations*.- Ab, antibody; RP, reversed-phase; s.s., stainless steel. The term HPLC is used rather than the authors' term CLC (column liquid chromatography); where column dimensions are given, the second is the i.d. For pre-columns the prefix immuno signifies immunoaffinity.

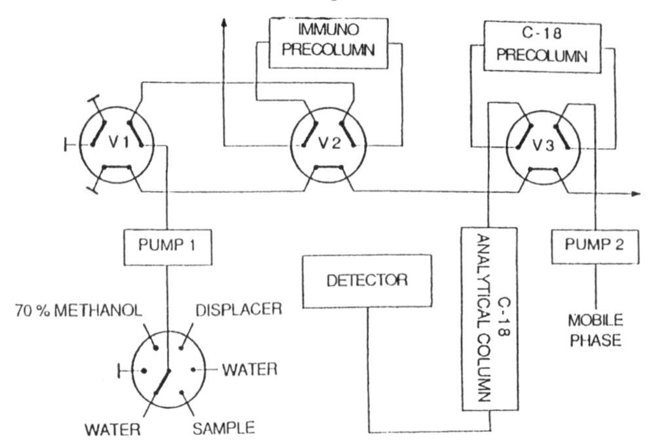

From ref. [2], courtesy of Elsevier.

Fig. 1. Set-up for the immunoselective desorption technique. V1-V3: High-pressure switching valves, computer-controlled. HPLC pumps - **1**, for sample preparation, computer-controlled, and **2**, for the analytical column. Immuno pre-column: s.s. (10 × 10 or 10 x 4 mm), containing agarose-immobilized Ab's. C-18 Pre-column: s.s. (10 × 2 or 20 × 2 mm), containing 40 µm C-18 silica material. Analytical column: glass (10 × 3 mm), containing 5 µm C-18 silica material, packed in-house.

the active sites on the Ab's. The analytes of interest and the excess of displacer, which have been refocused on the C-18 pre-column, are finally transferred to the analytical column. Fig. 1 shows the set-up and the analytical schedule, which is as follows (with letters denoting the positions of the 3 valves: V1, V2, V3; depicted in position A):

(**a**) flushing of the immuno pre-column with water in order to precondition it by displacing the 70% (v/v) methanol which is still present from the previous run (A, A, A);

(**b**) flushing of the capillaries with sample (B, A, A);

(**c**) flushing of the immuno pre-column with sample so that the sample is pumped through it and the antigenic analytes are selectively trapped (A, A, A);

(**d**) flushing of the capillaries with water (B, A, A);

(**e**) flushing of the immuno pre-column with water so that the residual sample is flushed away (A, A, A);

(**f**) as for (**d**);

(**g**) flushing of the C-18 pre-column with water in order to precondition it (B, A, B);

(**h**) flushing of the capillaries with desorbing solution, which contains the cross-reacting antigen (B, B, B);

(**i**) flushing of the immuno pre-column and the C-18 pre-column with desorbing solution, whereby the analytes are desorbed, transferred and re-trapped - the desorption being achieved by displacement which, through use of a high excess of the antigen (200 µg/L), is complete (A, B, A);

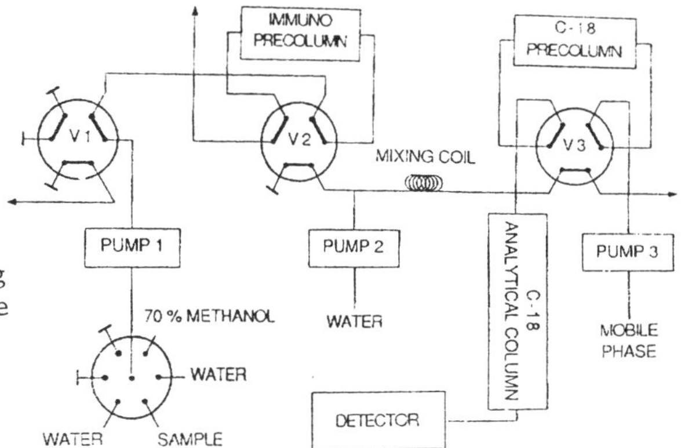

Fig. 2. Set-up for the non-selective desorption technique. Components as for Fig. 1, except that pump **3** corresponds to pump **2** of Fig. 1; pump **2** is for diluting the effluent from the immuno pre-column.

(**j**) switching of the C-18 pre-column and analytical column so that the compounds trapped on the former are transferred to the analytical column and separated (A, B, A);

(**k**) flushing of the immuno pre-column with methanol-water so that it is regenerated with methanol, having become completely saturated with displacer during the desorption step (A, A, A).

(2) THE NON-SELECTIVE DESORPTION TECHNIQUE

Fig. 2 shows the set-up for this approach. The desorption is performed with an aqueous solution containing an organic modifier, viz. 70% (v/v) methanol. After desorption the effluent is diluted with water *via* a T-piece and subsequently reconcentrated on a C-18 pre-column. The dilution decreases the % methanol in such a way that the analytes can be reconcentrated on the C-18 pre-column, whereafter they are transferred to the analytical column. The schedule for automated analysis by this approach is as follows:

(**a**)-(**d**): as for (**a**)-(**d**) opposite;

(**e**) flushing of the immuno pre-column with water so that the residual sample is flushed away by pump 1 (A, A, B), and flushing of the C-18 pre-column with water so as to precondition it by pump 2;

(**f**) flushing of the capillaries with methanol-water (B, A, B);

(**g**) flushing of the immuno pre-column with methanol-water, dilution with water and reconcentrating on the C-18 pre-column, whereby the analytes are desorbed from the immuno pre-column with 2 ml of methanol-water, this solution then being diluted on-line by adding water *via* the T-piece, so reducing the methanol to ~4% and allowing the reconcentration to occur (A, B, B);

(**h**) flushing of the immuno pre-column with methanol-water, so that strongly bound analytes which could still be present in this pre-column are flushed to waste by pump 1 (A, A, B),

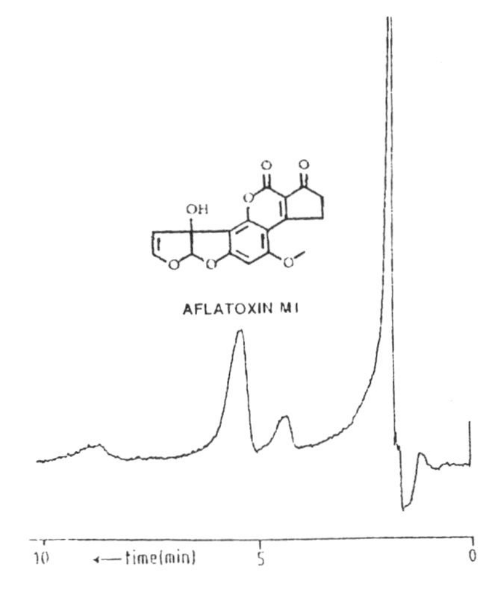

Fig. 3. Aflatoxin M1 in crude milk (applicable too to B1, G1, G2 and related compounds). Monoclonal anti-aflatoxin Ab on pre-column (10 × 10 mm). No off-line pre-treatment of the sample; 10 ml injected. Desorption non-specific, by 1.6 ml methanol-water (70:30). Mobile phase: water-methanol-acetonitrile (60:24:16 by vol.). Detection by fluorescence (ex 365, em 440 nm); 10 ppt detectable with 10 ml.

and flushing of the C-18 pre-column with water so that any residual analytes still in the mixing coil are flushed to the C-18 pre-column by pump 2;

(**i**) switching of the C-18 pre-column and analytical column, allowing the compounds trapped on the C-18 pre-column to be transferred to the analytical column and separated (A, A, A).

Contact with an organic modifier such as methanol will denature the immobilized Ab, resulting in the release of trapped antigen. After desorption the activity of the Ab can be restored by flushing the immuno pre-column with water (renaturation).

APPLICATIONS AND PERFORMANCE OF THE SYSTEM

Different immuno pre-columns were used for the 3 examples now given. In the aflatoxin example (Fig. 3) the analysis time was 45 min. The immuno pre-column served for only one analysis: in a first analysis when new, the recovery was 80%, but gradually decreased in runs that followed. Method optimization to allow repeated use of the immuno pre-column without loss of recovery is under investigation.

In contrast, surprising stability was found with anti-nortestosterone and anti-oestrogen pre-columns, which remained serviceable after at least 500 and 50 runs respectively. In the oestrogen examples (Figs. 4 & 5) the two desorption techniques were compared. Clearly the non-selective technique (Fig. 5) had a shorter analysis time (23 instead of 45 min) and provided a cleaner chromatogram. Moreover, the method development with this technique is faster, because the time-consuming search for a suitable displacer can be omitted. If, for a new application, a different immuno pre-column

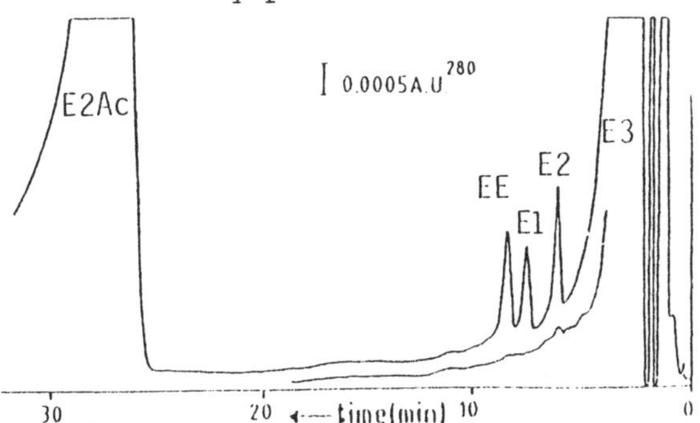

Fig. 4. Oestrogens in urine, with immunoselective desorption: displacement with 60 ml of a solution containing 260 µg each of oestriol (**E3**) and oestradiol 17-acetate (**E2Ac**) as displacers. *Applicable to Fig. 5(a) also.*- Immuno pre-column (10 × 10 mm): polyclonal anti-oestrogen Ab immobilized on agarose. Sample injected: 15 ml of male urine spiked with 5 µg/L each of oestradiol (**E2**), oestrone (**E1**) and ethynyl oestradiol (**EE**) [blank also run for the same urine]; off-line dilution with water (1:1), then filtration. Mobile phase: a 90:10 (by vol.) mixture of methanol-water (35:65) and tetrahydrofuran. Detection at 280 nm; limit 0.1–0.5 ppb (15 ml sample); linear range 1–5 µg/L. Other possible analytes: oestriol; other 16- and 17-substituted oestrogens. *For Fig. 4* the analysis time was 45 min; the recoveries for the 3 analytes were 60, 60 and 67% in comparison with a loop injection.

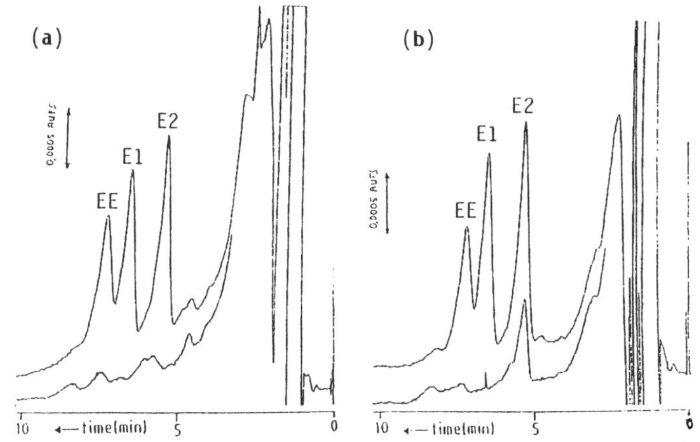

Fig. 5. Oestrogens in (**a**) urine, (**b**) plasma, with non-selective desorption by 3.7 ml methanol-water (95:5). *See Fig. 4 legend for* (**a**); analysis time 23 min; recoveries 76, 53 and 56%. For (**b**), 6 ml of plasma was spiked with 12.5 µg/L of each oestrogen, and the dilution was 1:5; recoveries were 71, 64 and 50%. Otherwise the (**a**) description applies to (**b**) also.

has been introduced into the system, the only parameters to be optimized are the HPLC separation and the dilution-concentration step. This simple procedure is feasible because any immuno pre-column can in principle be desorbed with an organic modifier.

It was unfortunate that the anti-aflatoxin immuno pre-column lost its efficacy so quickly that only one analysis was feasible. The use of automated pre-column exchange and the falling price of monoclonal Ab's will certainly help overcome this problem.

CONCLUDING COMMENTS

The use of immuno pre-columns for pre-treating biological samples, with large-volume (6-26 ml) direct 'injections', opens new possibilities for HPLC automation. The flexible system presented allows quick changing from one application to another, by merely exchanging the immuno pre-column. The whole analysis, including sample pre-treatment, can be run unattended. Such systems have been successful with different types of biological samples - urine, plasma and milk. Selectivity, reliability and speed are the characteristic features of the approach. A future option is to combine different types of immuno pre-column in one cartridge so as to create tailor-made analytical screening systems.

References

1. Haasnoot, W., Schilt, R., Hamers, A.R.M., Huf, F.A., Farjam, A., Frei, R.W. & Brinkman, U.A.Th.(1989) *J. Chromatog. 489*, 157-171.
2. Farjam, A., Frei, R.W., Brinkman, U.A.Th., Haasnoot, W., Hamers, A.R.M., Schilt, R. & Huf, F.A. (1988) *J. Chromatog. 452*, 419-433.

....

ADDENDUM: *excerpts from a 'companion' contribution to the Forum by* **U.A.Th. Brinkman**

RECENT DEVELOPMENTS IN PRE-COLUMN SAMPLE HANDLING FOR HPLC

Pre-columns (pC's) as routinely used for on-line (OL) trace-enrichment and clean-up of biological samples lack selectivity. For nucleosides, 5-FU and AZT in urine and plasma, selectivity was achieved with a Ag-loaded pC phase; a very steep pH jump effected desorption. Another such approach, with an immobilized Ab, appears in the foregoing article.

A completely different approach is the use of OL pC derivatization techniques. Much progress has been made in the area of phase-transfer catalysis and micelle-mediated techniques, e.g. for the rapid labelling of carboxylic acids and for the 2-phase dansylation of phenolic steroids in untreated plasma and urine. Another alternative, the OL combination of dialysis and pC trace-enrichment, has been applied for the determination of various types of veterinary drugs in milk and meat samples. Already this approach is being employed for routine analysis.

Comments on **#C-1:** R. Woestenborghs - RIA SELECTIVITY
#C-7: G.L. Evans - APPROACHES FOR LACIDIPINE

Comment to R.W. by J.D. Robinson.- Mostly your Ab's are of low titre (1:500 to 1:1000). Increased specificity might be achieved with higher titres such as might be obtained by re-thinking your immunization protocols and your immunogens. **Woestenborghs, replying to R.J.N. Tanner.-** We immunize 3 times, but often find that specificity is greater after the first than after the last. **Reply to G. Muirhead**, who asked whether immunoradiometric assays (IRMA's) might increase specificity and sensitivity:- time has not allowed this; generally our approach is pragmatic.

Remarks to Evans by R.P. Quinn.- The RIA might have worked without extraction, if the affinity constant were high enough to displace the drug from plasma protein. Higher sensitivity might have been achieved with higher ^{3}H specific activity in the antigen, or with labelled iodine.

Comments on **#ncC-5:** F.Y.K. Ghauri *et al.*: ^{19}F and ^{1}H NMR STUDIES
#ncC-6 K.E. Wade *et al.*: ^{15}N NMR STUDIES
and on a survey not in this book: I.D. Wilson - SCOPE OF NMR

Wilson, answering R. Whelpton.- The sensitivity of NMR for urine samples depends on the compound: e.g. 'isolated' methyls give large peaks; detectability is achievable with compounds dosed at ~50 mg/day. **Replies to L.A. Damani.-** Indeed sensitivity could be improved many-fold by first subjecting urine to a sample preparation stage, but if this were complicated, NMR would no longer have the benefit of speed, and other techniques such as HPLC or MS might as well be used. **Wade, answering U.A.Th. Brinkman.-** With ^{15}N NMR the time needed to get a useful outcome may range from 1 h to a weekend, depending on the problem and on the information required. **Wilson, answering S.S. Good:** discovering unknown metabolites in a ^{1}H NMR spectrum comes from proton-shift assignments and from comparison with pre-dose samples. **Damani asked** whether metabolic reactions in intact cells, e.g. hepatocytes, can be monitored directly in an NMR tube. **Reply.-** NMR will detect only free 'tumbling' molecular species, whereas in cells the compounds are likely to be bound to proteins; but J.K. Nicholson has investigated hepatocyte reactions after rupturing the cells.

Comments on **#ncC-7:** C.J. Olliff - BIOSENSORS
#ncC-9: L.A. Damani, R.O. Epemolu - IRON CHELATORS
#ncC-10: R.J. Ruane, I.D. Wilson - METAL CHELATORS

Olliff, replying to G.S. Land: as we are investigating, electrode construction and condition does affect selectivity; Pt must be present.

Remark by Wilson to Epemolu.- Your chromatographic problems are likely to be attributable not to silanol groups but to interactions with metal ions present as impurities in the silica itself. **Reply.**- We feel that besides an iron effect, hard to discern, there is a silica effect*. **Answer to H. Eggers/M.J. Ruane.**- Various additives, including amines and EDTA, have been tried besides the compounds themselves, but without success. **Replies by Ruane**: after we solved the non-linear baseline problem by adding picolinic acid, we did not try adding back metal ions to find whether they were the culprits; (**answering K. Borner**) iron is suspected to be the metal ion that causes the tailing; (**answering C. Town**) whilst believing that metal ions are involved, we have no idea why picolinic acid but not nicotinic acid is effective with our β-diketone system.

Relevant refs. noted by Senior Editor

In a thorough study [1] of HPLC assay conditions for nitroxoline (8-hydroxy-5-nitroquinoline), severe tailing was encountered with several modes - RP, ion-exchange (IEC), and 'dynamic IEC' with cetrimide in the mobile phase as already investigated [2] for naldixic acid and its hydroxy metabolite. A suspected cause of the tailing, besides free silanols, was the known presence of trace elements in silica gel. With the RP (μBondapak C-18) mode as adopted, the tailing was abolished by adding a structural analogue, 8-hydroxyquinoline, to the mobile phase, possibly through complexation since some decrease in tailing was produced by a metal-chelator (Titriplex IV) using the dynamic IEC mode ('soap chromatography'). With the RP mode (mobile phase: methanol-pH 7.4 phosphate buffer), plasma loaded direct after deproteinization with acetonitrile caused a chromatographic disturbance due seemingly to temporary stripping of the 8-hydroxyquinoline from the column by the acetonitrile. This was obviated by an altered plasma treatment: chloroform extraction, drying down, and re-constitution in the mobile phase.

In a recent report, 3-hydroxy-pyridin-4-one iron chelators in body fluids were assayed by RP-HPLC after PCA deproteinization [3]. Tailing problems are discussed. The mobile phase comprised methanol-pH 2 phosphate with heptane- or octane-sulphonic acid. After analyte elution, late-running material was cleared from the column by a brief gradient step.

1. Sorel, R.H.A., Snelleman, C. & Hulshoff, A. (1981) *Chromatog. 222*, 241-248.
2. Sorel, R.H.A., Hulshoff, A. & Snelleman, C. (1980) *J. Chromatog. 221*, 129-137.
3. Goddard, J.G. & Kontoghioghas, G.J. (1990) *Clin. Chem. 36*, 5-8.

**Ref. noted by Ed.:* Pfleiderer, B. & Bayer, E. (1989) *J. Chromatog. 468*, 67-71.

Editor's précis of a Forum Abstract (no text furnished) on **Immunoaffinity chromatography** for extraction prior to MS
- C.P. Goddard & L. Statham (Glaxo Group Res., Ware)

For sample preparation, SPE is especially advantageous if highly specific binding characteristics are conferred, as by an immunoaffinity chromatography approach. This has been investigated in connection with the GC-MS analysis of salbutamol, a bronchiodilator. Anti-salbutamol Ab's have been attached to a number of different supports using various cross-linking reactions of general applicability [cf. O. Cromwell's art. in Vol. 2, this series. - *Ed.*]. The orientation of the Ab's in relation to the support was a major consideration, as steric hindrance by the binding site can adversely affect the interaction of antigen with Ab. Hence methods allowing control of Ab orientation were used in preference to methods for random attachment. Various conditions were also investigated for the binding of salbutamol to the immobilized Ab's and its subsequent elution from the solid phase. The main objective is to get a good recovery of salbutamol from plasma in a medium compatible with the GC-MS analytical mode.

Comments on #**ncC-14** - Ab PRE-COLUMNS, & #**ncC-13** - USE OF LASERS *and on a contribution summarized on p. 370*
- *Amsterdam contributors:* H. Lingeman, U.A.Th. Brinkman

Concerning pre-columns with immobilized Ab's, **Brinkman agreed with H. de Bree** that there is a dilemma: the more selective are the Ab's, the more difficult it is to find a suitable displacer. This may help explain why trial of immobilized receptors has so far been unsuccessful.

Brinkman, replying to D. Dell: the 'ASTED' system normally works even if there is high protein binding, since there is always a C-18 pre-column after the dialysis unit and hence any protein-bound drug is likely to be released. **R. Woestenborghs asked Lingeman** whether laser-induced fluorescence detectors might soon be on the market. **Reply.**- Not for at least 5 years, in view of present lack of consensus on what type of laser source to use. **Reply to K. Borner** concerning bioanalytical applications of time-resolved fluorescence for HPLC detection.- There is a reported method for europium complexes of hexocycline.

'State-of-the-art' refs. noted by Senior Editor

For RP-HPLC-UV of phenobarbital in urine, and diazepam and metabolites in plasma, a **micellar mode** (SDS; cf. J.G Dorsey's art. in Vol. 18) was used for pre-column loading, followed by back-flushing [Koenigbauer, M.J. & Curtis, M.A. (1988) *J. Chromatog.* *427*, 277-285]. Sample and plate-matrix effects on **TLC quantitation** have been reviewed [Mack, M. & Hauck, H-E. (1989) *J. Planar Chromatog.* *2*, 180-193].

Supercritical Fluid Chromatography [book (1988): Smith, R.M., ed.; Roy. Soc. Chem., Cambridge] contains useful guidance. The coupling of SFC to MS has been surveyed, with powerful advocacy, by D.E. Games *et al.* [(1987) *Anal. Proc. 24*, 371-372] and, along with CZE-MS, by R.D. Smith & H.R. Udseth [(1988) *Chem. Br. 24*, 350-352].

***Chiral** refs. noted by Senior Ed. [cf. Vols. 16 & 18 Index entries]*

With extracted biological fluids, pre-chromatographic **diastereoisomer formation** was performed as a way to separate enantiomers: **tiaprofenic acid** by RP-HPLC and **etodolac** by NP-HPLC [Mehvar, R., & co-authors (1987) *J. Pharm. Sci*, *76*, S10], and **methoxyphenamine** and metabolites by GC [Srinavas, N.R., *et al.* (1989) *J. Chromatog. 487*, 61-72]. HPLC with **chiral additives** in the mobile phase has been surveyed, with **naproxen** as a drug example [Pettersson, C. (1988) *Trends Anal. Chem. 7*, 209-217], and applied to **propranolol** with solvent recycling [Gupta, M.B., *et al.* (1988) *J. Chromatog. 424*, 189-194]. The scope of HPLC **chiral stationary phases** for drug enantiomers has been reviewed [Mehta, A.C. (1988) *J. Chromatog. 426*, 1-13], and separations achieved using β-cyclodextrin [Edholm, L-E., *et al.* (1988) *J. Chromatog. 424*, 61-72] with SPE-separated **terbutaline.**

*Various **drug-assay** refs. (usually on plasma) noted by Senior Ed. - especially for therapeutic classes featured in previous Vols. (certain **anti-inflammatories** appear in previous para.)*

***Anti-cancer drugs** (Vol. 14)- see earlier, p. 235. Another example:* **adriblastin** monitoring in plasma, urine, saliva and liver biopsies was by RP-HPLC (UV, fluorimetric and EC detection investigated) [Czeka, M.J. & Georgopoulos, A. (1988) *J. Chromatog. 424*, 182-188]: in effect there were two pre-columns, with switching.

***CNS drugs** (Vol. 16).*- HPLC methology for **antidepressants** has been reviewed [Wong, S.H.Y. (1988) *Clin. Chem. 34*, 848-855]. Capillary GC-NPD was used for **flunixin** in horse urine after methylation of the carboxyl group [Johansson, M. & Anler, E-L. (1988) *J. Chromatog. 427*, 55-66], and for **trazodone** after SPE with 2 elution 'cuts' [Rifai, N., *et al.* (1988) *J. Anal. Toxicol. 12*, 150-152]. For **pemoline** [Aoyama, T. .... (1988) *J. Chromatog. 430*, 351-360], solvent extraction and RP-HPLC were used (plasma, urine, tissues).

***'CV' drugs** (Vol. 18).*- GC-ECD after solvent extraction and derivatization was used for **amlodipine** [Beresford, A.P., *et al.* (1987) *J. Chromatog. 420*, 178-183], and GC-MS after SPE and derivatization for **benazeprilat** and its pro-drug [Sioufi, A., *et al.* (1988) *J. Chromatog. 434*, 239-246], for which there is an enzymic method too [Graf, P., *et al.* (1988) *J. Chromatog. 425*, 353-361]. GC-ECD after derivatization in a 2-phase system was used for **guanadrel** [Kaiser, D.G., *et al.* (1988) *J. Chromatog. 434*, 135-143]. RP-HPLC with fluorimetry was used for **quinidine** + hydroxyquinidine [Debruyne, D., *et al.* (1989) *Int. J. Clin. Pharm. Res. 9*, 319-325].

=========

Analyte Index

Key overleaf to the 10-category **chemical classification** (collation based on some analytically relevant features).

Use of a compound as an internal standard is **not** indexed. Hyphen '-' as in '17-' connotes *et seq.*, i.e. treatment in depth.

Prefixes to some page entries, *besides* ch = *chiral distinction*:-

Superscript, e.g. [1], signifies that the study included **metabolite(s)** of the listed compound (**see over**).
Subscript $_r$ signifies that pharmacologically relevant **'real' samples** were assayed, usually including plasma and/or urine. Entries lacking this prefix mostly concern pure compounds, or (e.g.) residues in food.
Prefix *p* denotes a study comprehending a **precursor** or **prodrug**.

If **Index-searching in earlier vols.** (listed on p. 316; same 10 categories), note that **cumulative** indexing featured in Vol. 12 [and is anticipated in Vol. 22 (1992)].

CATEGORY I (no amino group, nor cyclic N except maybe imide)

#**Ia**: acid *other than conjugate*, or ester *(criterion: SEE OVER)*

#**Iy**: *not* acid *(unless conjugate)* or ester, & no halo, N or P

ASSIGNMENT 'CATECHISM' *(See previous p. for* ***other guidance****)*

\# **Metabolites** are not separately listed; the parent molecule's entry is preceded by a superscript: *Phase I* metabolite(s) denoted [1] or, if including *N*-desalkyl or -desacyl, **[1]** (**bold**); *Phase II (conjugates)*, [2].

Parent molecules as indexed generally bear generic names as listed (with formulae) in the *Merck Index*.

\# **Assignment as 'acidic'** (to Ia, IIa or IIIa) applies where the pKa is <6; this excludes phenols, and **conjugates are excluded** since only the parent molecule is listed (prefixed [2]; see above). Also 'acidic' (notional) are **esters** yielding an acidic group in the *main* moiety if hydrolysed (as may happen *in vivo*).

Cyclic *N* (conducive to UV absorbance?) is never treated as 'amino'; it is 'imide' (possible category: Ic) if -CO-**N**-CO-, but otherwise *may* be basic.

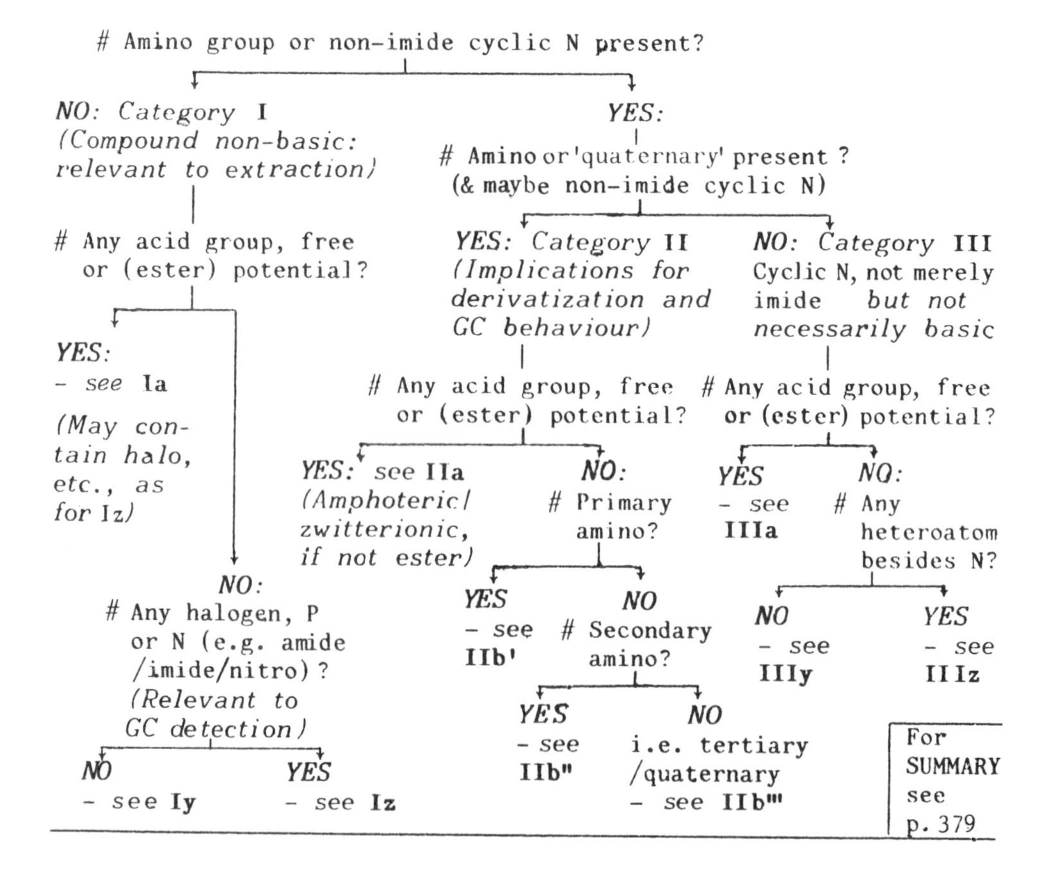

#IIIy: N-hetero (not merely imide) but no other hetero atoms; *no* amino; main moiety *not* an acid

#IIIz: heteroatom besides N; otherwise as for #IIIy (*no* amino)

SUMMARY OF CATEGORIES

	I	II	III
Amino?	no	✓	no
Non-imide hetero-N?	no	maybe	✓
Acid or potential acid (not conjugate)?	✓ = Ia	✓ = IIa	✓ = IIIa
– no! (and not an ester)	Halo, P or N? – no: Iy – ✓ = Iz	Primary amino? ✓ = IIb' If no: 2^y = IIb" 3^y or 4^y = IIb"'	Hetero atom besides N? – no = IIIy, ✓ = IIIz

Only **parent compound** listed; prefix [1] if Phase I metabolites studied [or **[1]** (**bold**) if dealkylated amino], & [2] if Phase II (conjugate); prefix $_r$ = 'real' (e.g. plasma) sample. ***Full Key:*** *p.* 376. *Prefixes* ch, *p* – *see p. 375.*

Chemical classification, including key functional groups, features likewise in a useful **bibliography of metabolites:** BIOTRANSFORMATIONS, ed. D.R. Hawkins, Vols. 1 (1989) & 2 (1990), publ. Royal Society of Chemistry, Cambridge.

General Index

This Index deals mainly with features studied and with approaches and points of technique, indexed similarly to previous 'A' vols. (listed on p. 316) so as to facilitate back-searching. The preceding Analyte Index deals with compounds investigated; exceptionally, a few types are also listed below according to their nature, e.g. 'Antivirals'.

In a page entry such as '17-', the '-' means '*et seq.*', i.e. coverage in depth.

[*See next p.*

[continued

[continued